Integrated Wastewater Management for Health and Valorization: A Design Manual for Resource Challenged Cities

Integrated Wastewater Management for Health and Valorization: A Design Manual for Resource Challenged Cities

Stewart M. Oakley

Published by   IWA Publishing
Unit 104–105, Export Building
1 Clove Crescent
London E14 2BA, UK
Telephone: +44 (0)20 7654 5500
Fax: +44 (0)20 7654 5555
Email: publications@iwap.co.uk
Web: www.iwapublishing.com

First published 2022
© 2022 IWA Publishing

Apart from any fair dealing for the purposes of research or private study, or criticism or review, as permitted under the UK Copyright, Designs and Patents Act (1998), no part of this publication may be reproduced, stored or transmitted in any form or by any means, without the prior permission in writing of the publisher, or, in the case of photographic reproduction, in accordance with the terms of licenses issued by the Copyright Licensing Agency in the UK, or in accordance with the terms of licenses issued by the appropriate reproduction rights organization outside the UK. Enquiries concerning reproduction outside the terms stated here should be sent to IWA Publishing at the address printed above.

The publisher makes no representation, express or implied, with regard to the accuracy of the information contained in this book and cannot accept any legal responsibility or liability for errors or omissions that may be made.

Disclaimer
The information provided and the opinions given in this publication are not necessarily those of IWA and should not be acted upon without independent consideration and professional advice. IWA and the Editors and Authors will not accept responsibility for any loss or damage suffered by any person acting or refraining from acting upon any material contained in this publication.

British Library Cataloguing in Publication Data
A CIP catalogue record for this book is available from the British Library

ISBN: 9781789061529 (Paperback)
ISBN: 9781789061536 (eBook)
ISBN: 9781789061543 (ePub)

This eBook was made Open Access in September 2022

© 2022 The Author; All photos in this book were taken by the author unless otherwise noted.

This is an Open Access eBook distributed under the terms of the Creative Commons Attribution Licence (CC BY-NC-ND 4.0), which permits copying and redistribution for non-commercial purposes with no derivatives, provided the original work is properly cited (https://creativecommons.org/licenses/by-nc-nd/4.0/). This does not affect the rights licensed or assigned from any third party in this book.

Contents

Preface xiii

Chapter 1
Integrated wastewater management for reuse in agriculture *1*
1.1 Introduction 1
    1.1.1 Wastewater and agriculture 1
        1.1.1.1 Increasing water scarcity and stress 1
        1.1.1.2 Population growth 2
        1.1.1.3 Wastewater as a resource 2
    1.1.2 The end-of-pipe paradigm for wastewater discharge 3
        1.1.2.1 Global wastewater production, treatment, reuse, and discharge 3
        1.1.2.2 Water resources and wastewater discharges 4
        1.1.2.3 Global discharge of nitrogen and phosphorus 4
        1.1.2.4 Energy use in mechanized wastewater treatment 7
    1.1.3 The integrated wastewater management paradigm 9
        1.1.3.1 Wastewater as a water resource 10
        1.1.3.2 Semi-arid climates: irrigation water requirement 1500 mm/yr 10
        1.1.3.3 Valorization of nutrients (N and P) in wastewater 11
        1.1.3.4 Value as fertilizer, 2021 prices 12
        1.1.3.5 Energy saved from fertilizer production 12
        1.1.3.6 $CO_{2,equiv}$ emissions saved from not using synthetic fertilizers 13
        1.1.3.7 Valorization of energy from anaerobic processes 13
1.2 Wastewater Reuse in Agriculture and Development of End-of-Pipe Paradigm 14
    1.2.1 Historical use of wastewater in agriculture: 3000 BCE–1915 CE 14
    1.2.2 Decline of wastewater reuse with end-of-pipe paradigm: 1915–1990 14
    1.2.3 End-of-pipe paradigm with resource recovery in EU and North America: 2000–2020 17
        1.2.3.1 Secondary treatment with tertiary processes and resource recovery 17
        1.2.3.2 Wastewater reuse in agriculture in the EU and the US 18
    1.2.4 Wastewater treatment and resource recovery in China: 1980–2020 18

		1.2.4.1	Wastewater treatment and discharge of excess nitrogen to surface waters 18
		1.2.4.2	Resource recovery in a Chinese 'concept wastewater treatment plant'....20
	1.2.5	End-of-pipe paradigm in resource-limited cities/peri-urban areas: 2000–2020......23	
		1.2.5.1	Indirect reuse of wastewater in agriculture............................23
		1.2.5.2	Direct reuse of inadequately treated wastewater in agriculture 24
		1.2.5.3	Direct reuse in agriculture with effluent wastewater meeting WHO guidelines...25
1.3	Wastewater Treatment for Agricultural Reuse in Resource-Limited Regions 29		
	1.3.1	Urban population growth 29	
	1.3.2	Coverage of wastewater treatment in the EU and North America 30	
	1.3.3	Coverage of wastewater treatment in resource-limited SDG regions 30	
	1.3.4	Effectiveness of wastewater treatment in resource-challenged urban areas..........33	
		1.3.4.1	Bolivia: waste stabilization ponds and wastewater reuse 33
		1.3.4.2	Honduras: pathogen reduction in waste stabilization ponds.............35
		1.3.4.3	Ouagadougou, Burkina Faso: protozoan cyst and helminth egg removal in the WSP system 36
		1.3.4.4	Lima, Peru: *Vibrio cholera* reduction in the San Juan de Miraflores WSP-reuse system..38
		1.3.4.5	Mendoza, Argentina: Campo Espejo waste stabilization ponds with reuse in agriculture 38
1.4	The Sustainable Development Goals and Integrated Wastewater Management 40		
	1.4.1	The 2030 Agenda for Sustainable Development............................40	
	1.4.2	Sustainable development goals relevant for integrated wastewater management 40	
		1.4.2.1	Goal 2: end hunger, achieve food security, improve nutrition, promote sustainable agriculture 42
		1.4.2.2	Goal: 3 ensure healthy lives and promote well-being for all ages 42
		1.4.2.3	Goal 6: ensure availability and sustainable management of water and sanitation for all 42

Chapter 2
Selection of natural systems for wastewater treatment with reuse in agriculture 43

2.1	Introduction 43		
2.2	Wastewater Characteristics and Traditional Levels of Treatment.......................44		
	2.2.1	Characteristics of domestic wastewater 44	
		2.2.1.1	Screenings and grit..45
		2.2.1.2	Pathogens...45
		2.2.1.3	Total suspended solids 47
		2.2.1.4	Biodegradable organics 47
		2.2.1.5	Nutrients 48
	2.2.2	Levels of wastewater treatment 48	
2.3	Pathogen Reduction in Wastewater Treatment Processes 49		
	2.3.1	High-rate treatment processes 49	
	2.3.2	Pathogen reduction data from operating high-rate treatment systems 51	
		2.3.2.1	Activated sludge treatment plants without disinfection in Tunisia.........51
		2.3.2.2	Activated sludge treatment plant with chlorine disinfection in the US 51
		2.3.2.3	Activated sludge treatment plants with microfiltration and disinfection in Spain...52
	2.3.3	Natural system treatment processes 52	

Contents

2.4	Natural System Treatment Processes for Integrated Wastewater Management		54
	2.4.1	Facultative→maturation pond systems	55
		2.4.1.1 Simplicity	55
		2.4.1.2 Land requirements	55
		2.4.1.3 Low cost	55
		2.4.1.4 Minimal sludge handling	56
		2.4.1.5 Process complexity and operation and maintenance requirements	56
		2.4.1.6 Energy consumption	57
		2.4.1.7 Process stability and resilience	57
	2.4.2	Anaerobic→secondary facultative→maturation pond systems	57
	2.4.3	UASB→secondary facultative→maturation pond systems	60
	2.4.4	UASB→trickling filter→batch stabilization reservoir	61

Chapter 3
Wastewater flows, design flowrate, and flow measurement ... *65*

3.1	Sources of Wastewater		65
3.2	Wastewater Flows		65
	3.2.1	Domestic wastewater flow and urban water consumption	65
	3.2.2	Infiltration and inflow	67
	3.2.3	Industrial wastewater flows	67
3.3	Design Flowrate		69
	3.3.1	Design flowrate from wastewater flow data: the ideal case	69
	3.3.2	Design flowrate by equation: the non-ideal case (but most common)	73
3.4	Design Example: Design Flowrates for the City of Trinidad, Honduras		75
3.5	Case Study: Design Flowrate for Saylla, Peru		77

Chapter 4
Preliminary treatment ... *81*

4.1	Introduction		81
4.2	Removal of Coarse Solids: Bar Screens		81
	4.2.1	Design of bar screens	82
	4.2.2	Design equations for bar screens and approach canal	84
	4.2.3	Final disposal of screenings	87
4.3	Grit Removal: Design of Grit Chambers		89
	4.3.1	Free-flow Parshall flume equations for the design of grit chambers	89
	4.3.2	Design of rectangular grit chambers	93
4.4	Bypass Channel Design		97
4.5	Procedure for Preliminary Treatment Design with the Parshall Flume		98
	4.5.1	Case study design: preliminary treatment, WSP system, Catacamas, Honduras	99
4.6	Final Disposal of Screenings and Grit		102

Chapter 5
Theory and design of facultative ponds ... *103*

5.1	Natural Processes as the Driving Force in Facultative Ponds		103
	5.1.1	Algal and bacterial processes in the aerobic zone	104
	5.1.2	Bacterial processes in the anaerobic zone	109
	5.1.3	Process analysis: methane emissions from facultative pond, Catacamas, Honduras	110

5.2	Theory of Design of Facultative Ponds			116
	5.2.1	Maximum organic surface loading		116
		5.2.1.1	Sources of solar radiation data	117
		5.2.1.2	Water temperature and algal growth	118
		5.2.1.3	Case study: surface loading and facultative pond performance, Nagpur, India	123
		5.2.1.4	Case study: organic overloading of facultative ponds in Honduras	125
	5.2.2	Wind effects in facultative ponds		128
	5.2.3	Hydraulic considerations		128
		5.2.3.1	Longitudinal dispersion	128
		5.2.3.2	Thermal stratification and hydraulic short circuiting	129
		5.2.3.3	Sludge accumulation effect on hydraulic short circuiting	130
	5.2.4	Pathogen reduction		130
		5.2.4.1	Helminth egg reduction	131
		5.2.4.2	*E. coli* or fecal coliform reduction	133
	5.2.5	BOD_5 removal		134
	5.2.6	TSS removal		135
	5.2.7	Sludge accumulation		135
		5.2.7.1	Sludge accumulation reported in the literature	135
		5.2.7.2	Projection of sludge accumulation with flowrates and solids loadings	135
		5.2.7.3	Design example part 1: projection of sludge accumulation for $TSS = 200$ mg/L	137
		5.2.7.4	Design example part 2: projection of sludge accumulation for $TSS = 350$ mg/L	138
		5.2.7.5	Discussion of design example results	139
5.3	Facultative Pond Design Procedure			141
5.4	Design Example: Facultative Pond Redesign for Agricultural Reuse, Cochabamba, Bolivia			144

Chapter 6
Theory and design of maturation ponds *153*

6.1	Maturation Ponds and Pathogen Reduction			153
	6.1.1	Factors affecting pathogen reduction		153
		6.1.1.1	Sunlight	153
		6.1.1.2	Temperature	154
		6.1.1.3	Hydraulic retention time	154
		6.1.1.4	Sedimentation	154
		6.1.1.5	Predation	156
	6.1.2	Design strategies for pathogen reduction		156
		6.1.2.1	Sunlight exposure	157
		6.1.2.2	Depth	158
		6.1.2.3	Maximize theoretical hydraulic retention time and minimize dispersion	158
		6.1.2.4	Longitudinal dispersion and mean hydraulic retention time	158
		6.1.2.5	Residence time distribution analysis to assess longitudinal dispersion	160
		6.1.2.6	Limitations of residence time distribution studies	161
		6.1.2.7	Case study: residence time distribution analysis to assess fecal coliform reduction in a maturation pond, Corinne, Utah, USA	162
		6.1.2.8	Determination of residence time distribution parameters	163

		6.1.2.9	Estimation of fecal coliform reduction using the Wehner and Wilhem equation... 165
		6.1.2.10	Comment on Corinne maturation pond case study 166
		6.1.2.11	Wind abatement 167
		6.1.2.12	Overflow rate... 167
		6.1.2.13	Rock filters.. 167
6.2	Design of Maturation Ponds... 167
	6.2.1	Unbaffled ponds 168
		6.2.1.1	Hydraulic retention time 168
		6.2.1.2	Depths 169
		6.2.1.3	Length to width ratios 169
		6.2.1.4	Inlet/outlet structures... 171
		6.2.1.5	Case study: unbaffled maturation ponds in series, Belo Horizonte, Brazil.. 171
	6.2.2	Baffled ponds 175
		6.2.2.1	Depths 175
		6.2.2.2	Length to width ratios 175
		6.2.2.3	Transverse baffle equations: baffles parallel to width......... 175
		6.2.2.4	Longitudinal baffle equations: baffles parallel to length 175
		6.2.2.5	Design example: comparison of transverse and longitudinal baffled ponds 177
		6.2.2.6	Design strategies for baffled and unbaffled ponds............. 177
		6.2.2.7	Case study: *E. coli* reduction in transverse baffled maturation pond, Helidon, Australia 178
	6.2.3	Rock filters for pathogen reduction 182
		6.2.3.1	Design of rock filters.. 182
		6.2.3.2	Case study: rock filter design, City of Biggs, California waste stabilization pond system.. 184
6.3	Maturation Pond Design Procedure 188
6.4	Maturation Pond Design Example: Redesign for Effluent Reuse, Cochabamba, Bolivia 190

Chapter 7
Wastewater reuse in agriculture: guidelines for pathogen reduction and physicochemical water quality 197

7.1	The Safe Use of Wastewater for Reuse in Agriculture... 197
7.2	Pathogen Reduction... 197
	7.2.1 WHO guidelines for wastewater use in agriculture............................. 197
	7.2.1.1 Development of the WHO guidelines................................ 197
	7.2.1.2 Health risks and pathogen reductions for wastewater reuse 200
	7.2.1.3 Verification monitoring of wastewater treatment................ 205
	7.2.1.4 Verification monitoring of health protection measures 206
	7.2.1.5 Pathogen reduction in sludges produced in wastewater treatment....... 206
	7.2.2 Case studies: pathogen reduction for wastewater and sludge use in agriculture 209
	7.2.2.1 Night soil use in China and East Asia: the original valorization of human excreta... 210
	7.2.2.2 Use of fecal sludge from ecosan toilets in Africa: Burkina Faso and Malawi 211
	7.2.2.3 Wastewater reuse in Ouagadougou, Burkina Faso 218
	7.2.2.4 Fecal coliform mortality on wastewater-irrigated cattle fodder, Aurora II Sanitary Engineering Research Center, Guatemala City 222

		7.2.2.5	Wastewater reuse in the Municipality of Sololá, Lake Atitlán Basin, Guatemala 228

		7.2.2.6	Pathogen reduction for wastewater reuse in agriculture at the Campo Espejo waste stabilization pond system, Mendoza, Argentina. . . . 233

7.3 Physical-Chemical Water Quality for Irrigation with Wastewater 241
    7.3.1 Physical-chemical guidelines for interpretation of water quality for irrigation 241
        7.3.1.1 Salinity. 241
        7.3.1.2 Infiltration: sodium adsorption ratio 243
        7.3.1.3 Specific ion toxicity. 244
        7.3.1.4 Miscellaneous effects 245
    7.3.2 Case studies of physicochemical characteristics of wastewater used in agriculture 246
        7.3.2.1 Physicochemical parameter concentrations in wastewater effluent in the US and Germany 246
        7.3.2.2 Salinity, infiltration, and sodium ion toxicity potential problems, Cochabamba, Bolivia 246
        7.3.2.3 Potential salinity and pH irrigation problems, Ouagadougou WSP system effluent 248

Chapter 8
Design of wastewater irrigation systems with valorization of nutrients. 251

8.1 Introduction 251
    8.1.1 Types of irrigation systems 251
    8.1.2 Design equations and parameters for irrigation requirements. 251
        8.1.2.1 Crop evapotranspiration and reference evapotranspiration 253
        8.1.2.2 Effective precipitation. 254
        8.1.2.3 Leaching factor and irrigation efficiency 254
        8.1.2.4 Irrigation area 254
        8.1.2.5 Nutrient loading rates. 255
        8.1.2.6 Carbon emissions saved by using wastewater in lieu of synthetic fertilizers. 256
        8.1.2.7 Valorization of nitrogen and phosphorus. 257
8.2 Wastewater Irrigation without Reservoirs in Dry Climates 257
    8.2.1 Design procedure for irrigation without reservoirs 258
    8.2.2 Design example: pond redesign for wastewater irrigation, Cochabamba, Bolivia. 259
    8.2.3 Case study design example: wastewater irrigation at the Campo Espejo ACRE 265
8.3 Wastewater Irrigation with Stabilization Reservoirs in Wet Climates 270
    8.3.1 Wastewater stabilization reservoirs. 270
        8.3.1.1 Single reservoir 270
        8.3.1.2 Batch stabilization reservoir with secondary reservoir 270
        8.3.1.3 Batch stabilization reservoirs in parallel 271
    8.3.2 Design procedure for single reservoirs 271
    8.3.3 Case study design example: irrigation with a single reservoir, Sololá, Guatemala 273
    8.3.4 Design procedure for three batch stabilization reservoirs in parallel 276
    8.3.5 Case study design example: batch stabilization reservoirs, Sololá, Guatemala 278
8.4 Wastewater Irrigation without Reservoirs in Wet Climates using Land Application 279
    8.4.1 Design equations for land application during the wet season 280

		8.4.1.1	Hydraulic loading rate based on soil permeability 280
		8.4.1.2	Land area requirements without reservoir varying irrigated area each month 281
		8.4.1.3	Hydraulic loading based on nitrogen limits. 281
	8.4.2	Design procedure for irrigation without reservoirs in wet climates 281
	8.4.3	Case study design example: crop irrigation and land application, Sololá, Guatemala 282

Chapter 9
Physical design and aspects of construction 287

- 9.1 Introduction 287
- 9.2 Site Selection 287
- 9.3 Geotechnical Investigations 291
- 9.4 Water Balance 292
- 9.5 Preliminary Treatment and Flow Measurement 296
- 9.6 Hydraulic Flow Regime 298
- 9.7 Hydraulic Structures 302
  - 9.7.1 Flow distribution devices 303
  - 9.7.2 Inlets and outlets 304
  - 9.7.3 Discharge structures for final effluent 305
  - 9.7.4 Drainage structures for facultative ponds 305
  - 9.7.5 Overflow weirs and bypass channels 306
- 9.8 Embankments and Slopes 310
  - 9.8.1 Interior slopes 310
  - 9.8.2 Exterior slopes 310
  - 9.8.3 Embankment and access ramps 310
  - 9.8.4 Fences 310
  - 9.8.5 Operation building 316

Chapter 10
Operation and maintenance 317

- 10.1 Introduction 317
- 10.2 Operation and Maintenance Manual 317
- 10.3 Basic Operation 318
  - 10.3.1 Initial start-up 318
  - 10.3.2 Flow measurement 318
  - 10.3.3 Water level control 319
  - 10.3.4 Bypass channel operation 319
  - 10.3.5 Adjusting the discharge level with bottom sluice gate 319
  - 10.3.6 Sensory detections: odors and colors 319
  - 10.3.7 Sludge depth measurement 319
- 10.4 Routine Maintenance 320
  - 10.4.1 Bar screens 320
  - 10.4.2 Grit chamber 320
  - 10.4.3 Removal of scum and floating solids 320
  - 10.4.4 Grass, vegetation and weeds, and aquatic plants 320
  - 10.4.5 Mosquitoes, flies, rodents, and other animals 322
  - 10.4.6 Embankment slopes 323

		10.4.7 Fences and roads 323
		10.4.8 Equipment and maintenance tools 330
10.5	Field Records for Basic Operation and Maintenance 330
10.6	Operation for Performance Control: Analytical Monitoring 332
	10.6.1 Laboratory sampling and testing program 333
	10.6.2 Presentation and interpretation of the results of monitoring programs 333
10.7	Sludge Removal in Facultative Ponds 338
10.8	Required Personnel 338
10.9	Hygienic Measures for Operators 339
10.10	Operational Problems and their Solution 340
	10.10.1 Signs of well-functioning facultative and maturation ponds 340
	10.10.2 Problems of operation in facultative and maturation ponds 340
	10.10.3 Accumulation of scum and floating solids 340
	10.10.4 Odors 340
	10.10.5 Abnormal colorations 340
	10.10.6 Weed growth 341
	10.10.7 Mosquitoes and other insects 341

References 343

Preface

Appropriate, sustainable wastewater treatment that protects public health and the environment has often been a failure in resource-challenged cities worldwide. These failures continue to occur in spite of decades of funding from national and international organizations. A typical scenario in a small municipality follows the historical end-of-pipe paradigm of wastewater collection, treatment, and discharge:

(1) The national government obtains funds for water and sanitation projects to distribute to various municipalities.
(2) Municipalities receiving funds are obligated to build wastewater treatment plants as part of the project to comply with national wastewater discharge regulations, which typically are modeled after the European Union (EU) or the US.
(3) Local government officials, engineers, and the public rank wastewater treatment lowest in priority, below roads, buildings, drinking water, even solid waste disposal. This is a result of the high costs for construction, and operation and maintenance, with an unproductive effluent discharge that has no benefit to the local population, or the current municipal government.
(4) Wastewater treatment plant design is performed by external consultants who typically select inappropriate technologies.
(5) Plant design and construction is often a slow process. It is not uncommon for the construction process to take years, sometimes passing from one government to the next, at which time the design may be changed, or the project abandoned all together. The engineers involved in the project may have never seen a successful wastewater treatment plant in operation.
(6) Wastewater reuse in agriculture is rarely considered. This is a result of the dominance of the end-of-pipe paradigm taught in engineering schools and accepted by governments and funding agencies, and enforced by regulatory agencies, where the purpose of wastewater treatment is to produce an effluent that can be safely discharged to surface waters. Only the wastewater typically discharged is poorly treated at best, and surface waters used for irrigation, which are often drinking water sources, are heavily contaminated with excreted pathogens and nutrients (nitrogen and phosphorus).

© 2022 The Author. This is an Open Access book chapter distributed under the terms of the Creative Commons Attribution Licence (CC BY-NC-ND 4.0), which permits copying and redistribution for noncommercial purposes with no derivatives, provided the original work is properly cited (https://creativecommons.org/licenses/by-nc-nd/4.0/). This does not affect the rights licensed or assigned from any third party in this book. The chapter is from the book *Integrated Wastewater Management for Health and Valorization: A Design Manual for Resource Challenged Cities*, Stewart M. Oakley (Author).

(7) The prevalence of excreta-related infections remains high as a result of poor wastewater management.
(8) Local farmers are often encouraged to use synthetic fertilizers, often sold at low prices with subsidies from national governments, while the nutrients in wastewater discharges flow by their fields to the ocean, or to inland lakes where they contribute to eutrophication.

Lake Atitlán, Guatemala, where I have worked for 10 years, is a typical example of this paradigm failure. At least 20 wastewater treatment plants have been built in small cities throughout the lake basin during the last 15 years. Some plants were designed by local engineers, others by engineers from international organizations, and all were financed by national or international sources. Each plant has a distinct process design based on organic matter removal, all discharge directly or indirectly to the lake, and none were designed to remove pathogens or nutrients, the principal problems of lake water quality. None of the plants has adequate operation and maintenance, or adequate sludge management. As a result, the lake water quality, which is the drinking water source for 125,000 persons, continues to deteriorate in terms of pathogen and nutrient concentrations. While there are no epidemiological studies of excreta-related diseases, *Cryptosporidium* prevalence has been reported in children in two cities, and excess nutrients have caused several large cyanobacteria blooms. No lessons have been learned, and cities continue with plans to build more wastewater treatment plants, including some that have abandoned their existing plant as they receive funds from the new municipal governments. At the same time, small farmers in the basin use synthetic fertilizers, which also contribute to nutrient loads to the lake. This story is repeated in resource-challenged cities worldwide where wastewater treatment plants have been built with the end-of-pipe paradigm.

This book incorporates the paradigm of integrated wastewater management for health and valorization without surface water discharge. In this paradigm, wastewater is considered a valuable resource that should be (i) treated for pathogen removal and (ii) applied to the land for its water and fertilizer value. The natural biogeochemical cycles of water, carbon, nitrogen, and phosphorus are utilized, eliminating the linear, one-way discharges of the end-of-pipe paradigm. This is especially relevant to the problems of climate change with increasingly less water availability for agriculture, and global warming caused by greenhouse gases, in which synthetic fertilizer production plays a role.

In the integrated wastewater management paradigm, the purpose of treatment is (i) to protect public health by reducing pathogens using natural treatment systems and (ii) to produce an effluent that is valorized for its water and fertilizer value in agriculture. Methane production as a sustainable energy source could be considered for those applications where it is appropriate, typically very large installations that have the resources to operate and maintain anaerobic processes.

International organizations such as the United Nations Environment Program and the World Health Organization have long recognized the problems of the end-of-pipe paradigm and have published policy and design manuals on integrated wastewater management. My experience in consulting and teaching professional courses and workshops, however, has shown that many design engineers, engineering students, and other stakeholders working in municipal development are either unaware or indifferent to the alternatives of integrated wastewater management for health and valorization. And national and international funding agencies continue to fund end-of-pipe wastewater treatment projects regardless of the continued evidence of failure.

The purpose of this book is to introduce professionals and engineering students to the integrated wastewater management for health and valorization paradigm. (The word treatment is intentionally avoided because of the persistent overemphasis of this term with the implication of discharge.)

This paradigm for resource-challenged cities worldwide uses one of the simplest and most environmentally sustainable technologies, the waste stabilization pond. Properly designed and operated waste stabilization ponds can reduce the concentrations of the principal pathogens to low risk levels, producing an effluent acceptable for restricted or unrestricted irrigation. If anaerobic ponds are used under the appropriate conditions, the methane produced can be used for its heat value in small systems and for the production of electricity in large systems. (Anaerobic systems must be used with caution as many fail because of increased operation and maintenance requirements.)

This book will also be relevant for small cities in the EU and the US that have waste stabilization pond systems that often cannot meet stricter discharge requirements, but that also cannot afford the capital and operational costs of a new mechanized plant. These cities could benefit with an upgrade of the pond system and a change from effluent discharge to agricultural reuse. As an example, many existing waste stabilization pond systems in small cities in California cannot meet the new discharge requirement for ammonia, and engineering consulting firms recommend abandoning the systems for a mechanized treatment process. An alternative solution, however, for those systems close to agricultural land is to change from surface water discharge to land application where the nitrogen, which is seen as a problem for effluent discharge, can be a valorized resource.

I wish to thank the following colleagues who I have collaborated with over the last 30 years and whose work contributed to the case studies and design examples presented in this book:

Argentina: Tec. Fernanda González, Ecol. Fernando Santos
Australia: Dr. Charles Lemckert
Bolivia: Ing. Olver Coronado, Ing. Álvaro Rodolfo Mercado
Brazil: Dra. Luciana Coêlho Mendonça, Me. Eng. Sérgio Rolim Mendonça, Dr. Marcos von Sperling
Burkina Faso: Dr. Ynoussa Maiga
China: Professor Jingxia Yang
Colombia: Dra. Elizabeth Carvajal Florez, MSc. Ing. Rubén Pinzón, Ing. Ítalo Gandini
El Salvador: Arq. Julián Monge
Guatemala: MSc. Ing. Joram Gil, Dr. Adán Pocasangre, MSc. Ing. Pedro Saravia
Honduras: MSc. Ing. Luis Eveline, Ing. Carlos Flores
India: Dr. Narayanan Jothikumar
Mexico: Dr. Ernesto Espino, Dr. Juan Manuel Morgan
Nicaragua: Ing. Arturo Coco, Ing. Francisco Saavedra
Peru: MSc, Ing. Osmara Agramonte Ochoa, Ing. Julio Moscoso Cavallini
United Kingdom: Dr. Barry Lloyd
USA: Dr. Menahem Libhaber, Dr. James Mihelcic, Dr. Kara Nelson, Dr. Matthew Verbyla

doi: 10.2166/9781789061536_0001

Chapter 1
Integrated wastewater management for reuse in agriculture

'In the English language, we normally speak of sewage as wastewater, thus defining it as useless. We raise the question whether this framing of sewage as a sub-section of water management further reinforces linear, end-of-pipe solutions. A shift toward more iterative and adaptive solutions might be facilitated if sewage were redefined outside the water management umbrella (for example, energy and nutrients), with organizations responsible for delivering sewage services organized accordingly'. (Öberg *et al.*, 2014)

1.1 INTRODUCTION
1.1.1 Wastewater and agriculture
Wastewater is becoming more widely recognized as an important resource for global agriculture. This is especially important in resource-limited urban and peri-urban areas worldwide. The principal driving forces behind wastewater use in agriculture have been defined by WHO (2006):

- Increasing water scarcity and stress;
- Population growth with increased demand for food and fiber; and
- Recognition of the resource value of wastewater for its water, nutrient, and energy value.

1.1.1.1 Increasing water scarcity and stress
Water scarcity is defined as the lack of volumetric availability, while water stress is the inability to satisfy demand as a result of additional factors such as water quality and accessibility (see Figure 1.1). Key problems of water scarcity and stress include the following (UNEP, 2017):

- Two-thirds of the global population lives in
- areas of water scarcity for at least 1 month per year.
- 500 million persons live in areas where water consumption exceeds locally renewable water resources by a factor of two.

© 2022 The Author. This is an Open Access book chapter distributed under the terms of the Creative Commons Attribution Licence (CC BY-NC-ND 4.0), which permits copying and redistribution for noncommercial purposes with no derivatives, provided the original work is properly cited (https://creativecommons.org/licenses/by-nc-nd/4.0/). This does not affect the rights licensed or assigned from any third party in this book. The chapter is from the book *Integrated Wastewater Management for Health and Valorization: A Design Manual for Resource Challenged Cities*, Stewart M. Oakley (Author)

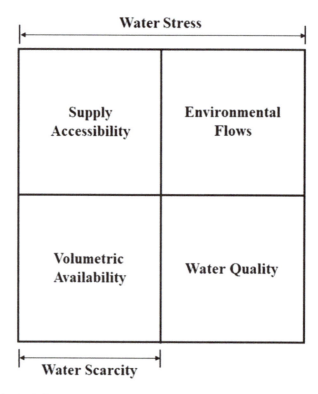

Figure 1.1 Water scarcity is defined as the lack of volumetric availability of water; water stress is the inability to satisfy water demand as a result of various factors, such as contamination (water quality) and lack of access. *Source*: Diagram developed from Schulte (2014).

- Growing competition between agriculture and urban areas for high-quality freshwater continues to increase pressure on freshwater resources.
- Increased discharges of untreated or poorly treated wastewater further degrades surface water quality, and particularly threatens resource-poor populations in arid climates.

1.1.1.2 Population growth

By 2050, it is estimated that greater than 40% of the global population will live in countries facing water scarcity and stress (WHO, 2006). Most of this population growth is projected to occur in resource-limited urban and peri-urban areas. The growth of this urban population will influence the production, treatment, and use of wastewater in the following ways (UNEP, 2017):

- Higher population densities will generate more wastewater with little or no treatment.
- Wastewater flows will increase as a result of increased per capita demand in urban areas.
- Sewerage will increase as a result of the population increase.
- Urban agriculture using wastewater will play an important role in supplying food.
- Domestic wastewater will become the sole water source for farmers in water-stressed areas.

1.1.1.3 Wastewater as a resource

Agriculture is the largest user of freshwater in the world, consuming 70% of all freshwater extractions worldwide (UNEP, 2017). As freshwater becomes scarcer due to population growth, urbanization and climate change, the use of wastewater in agriculture will continue to increase as farmers learn the water and nutrient value of wastewater as a resource. The wastewater resource is lost, however, if treated wastewater is discharged to surface waters, as is the dominant global practice.

The use of wastewater in agriculture can help replace synthetic fertilizer use, restore the natural N and P cycles, supply organic matter to the soil, and provide a year-round, steady supply of water that does not change significantly with the seasons. Under the right conditions of operation and maintenance in larger wastewater treatment/reuse systems, anaerobic processes can also be used to produce methane as a sustainable energy source for cooking, and the generation of electricity. (Anaerobic processes should only be recommended where resources and technical personnel are available for long-term operation and maintenance. Otherwise, failure will likely occur, which is a common outcome.)

1.1.2 The end-of-pipe paradigm for wastewater discharge

Historically, in Europe and the US, wastewater generated from urban areas was used for irrigation of agricultural land, from the 1500s until the early 1900s. As urban areas grew larger land area for irrigation disappeared, and by necessity raw wastewater was progressively discharged to nearby surface waters, a strategy in resource-limited cities worldwide that continues to this day. Surface waters soon became seriously polluted, however, causing noxious conditions and posing serious public health risks, necessitating the gradual development of wastewater treatment before discharging to surface waters (Metcalf & Eddy/AECOM, 2006). As new problems with effluents were discovered, treatment plants were continuously redesigned and upgraded with add-on processes to maintain receiving water quality. The end-of-pipe paradigm predominates to this day, but with the caveat that wastewater must be treated to the appropriate level before discharge in high and upper-income urban areas.

For resource-challenged urban areas, the end-of-pipe paradigm with downstream discharge also predominates, but with untreated or inadequately treated wastewaters, with high concentrations of excreted pathogens and nutrients.

1.1.2.1 Global wastewater production, treatment, reuse, and discharge

Table 1.1 presents estimates of global domestic wastewater production, collection, treatment, reuse, and surface water discharge for 2015 (Jones *et al.*, 2021). Of the 359.4×10^9 m^3 of wastewater produced, it is estimated that 88.7% (318.7×10^9 m^3) was discharged to surface waters, with 11.3% (40.7×10^9 m^3) of treated wastewater reused in agriculture. When the data are classified according to economic classification as shown in Table 1.2, the high and upper middle-income regions are found to produce 80.3% of global wastewater flows, largely due to increased per capita consumption in higher income regions. (Treatment means a plant exists, not that it working properly, or designed to solve the necessary public health and water quality problems. Globally, outside of the EU and the US, wastewater treatment plants are still designed mostly to remove organic matter; some may have disinfection with chlorine, but chlorination is ineffective for bacteria and viruses unless the final effluent is if high quality, which is not likely. Chlorine also does not inactivate protozoan cysts or helminth eggs. Removal of N and P is not generally practiced except in local regions where eutrophication is a threat.)

In Table 1.2, wastewater reuse as a percentage of treated wastewater ranges from 19.2 to 25.4% for the high, upper-middle, and lower-middle economic regions, which together comprise 91.4% of the global population. Reuse as a percentage of production, however, is lower, and the largest percent of the global population, the lower-middle economic classification, only reuses 6.6% of produced wastewater. This is the population most in need of sustainable wastewater treatment with valorization of the effluent for reuse in agriculture.

Global wastewater production is predicted to grow over present levels by 24% to the year 2030, and 51% by 2050 (Qadir *et al.*, 2020). If present trends continue, increasingly more wastewater, the majority untreated and inadequately treated, will be discharged to surface waters rather than valorized for its water and fertilizer value in agriculture.

Table 1.1 Global domestic wastewater production, collection, treatment, and reuse, 2015.

Domestic Wastewater	Annual Flowrate (10⁹ m³/yr)	Percent of Production (%)
Production	359.4	
Collection	225.6	62.8
Treatment	188.1	52.3
Reuse	40.7	11.3
Discharge to surface waters	318.7	88.7
(1) Discharge with treatment	147.4	41.0
(2) Discharge w/o treatment	171.3	47.7

Source: Data from Jones *et al.* (2021).

Table 1.2 Global domestic wastewater flows by economic classification, 2015.

Economic Classification	Percent of Global Population	10⁹ m³/yr Production	Collection	Treatment	Reuse	Reuse as Percent of Treatment (%)	Reuse as Percent of Production (%)
High	16.1	149.1	121.7	110.4	21.2	19.2	14.2
Upper middle	34.8	139.5	74.8	60.2	15.1	25.1	10.8
Lower middle	40.5	66.8	28.8	17.3	4.4	25.4	6.6
Low	8.6	4.0	0.4	0.2	0	0	0
Total	100	359.4	225.7	188.1	40.7		

Source: Data from Jones *et al.* (2021).

1.1.2.2 Water resources and wastewater discharges

The estimated global volume of wastewater discharged to surface waters, 318.7 billion m³/yr, is a significant water resource that could be used for agricultural irrigation after appropriate treatment for pathogen reduction. This volume of wastewater is near the volume of agricultural water withdrawals in China in 2015, and much greater than withdrawals in the US, as shown below:

Agricultural Water Withdrawals and Global Wastewater Discharge and Reuse, 2015	10⁹ m³/yr
India	688
China	385
Global Wastewater Discharges	**319**
US	175
Brazil	45
Global Wastewater Reuse	**41**

Source: https://ourworldindata.org/water-use-stress#agricultural-water-withdrawals.

As water scarcity and stress increase worldwide, wastewater will be one of the few reliable sources of water for agriculture.

1.1.2.3 Global discharge of nitrogen and phosphorus

When wastewaters are discharged to surface waters, the fertilizer value of nitrogen and phosphorus for agriculture is lost, while at the same time, surface waters become threatened by eutrophication and hypoxia. The global mean concentration of total nitrogen (TN) and total phosphorus (TP) in raw wastewater has been estimated at 44 and 8 mg/L, respectively (Qadir *et al.*, 2020). Using the mean concentrations and the annual discharge of raw wastewater, the following masses of TN and TP are estimated to be discharged to global surface waters:

$$\text{Total Nitrogen (TN)} = 0.001 \left(\frac{\text{kg/m}^3}{\text{mg/L}} \right) (44 \, \text{mg/L}) \left(318.7 \times 10^9 \, \text{m}^3/\text{yr} \right)$$

$$= 14.0 \times 10^9 \, \text{kg/yr} \left(\frac{1.0 \, \text{Mt}}{10^9 \, \text{kg}} \right) = 14.0 \, \text{Mt/yr}$$

$$\text{Total Phosphorus (TP)} = 0.001 \left(\frac{\text{kg/m}^3}{\text{mg/L}} \right) (8 \, \text{mg/L}) \left(318.7 \times 10^9 \, \text{m}^3/\text{yr} \right)$$

$$= 2.55 \times 10^9 \, \text{kg/yr} \left(\frac{1.0 \, \text{Mt}}{10^9 \, \text{kg}} \right) = 2.55 \, \text{Mt/yr}$$

Nitrogen. The global demand for N in fertilizer is estimated to be 115.5 Mt/yr (Qadir *et al.*, 2020). The mass of TN from wastewater discharge is thus 12.1% of global demand:

$$\% \, \text{TN}_{\text{wastewater}} = \frac{14.0 \, \text{Mt/yr}}{115.5 \, \text{Mt/yr}} (100\%) = 12.1\%$$

The nitrogen in wastewater exists in the forms of NH_3/NH_4^+ and organic N, and is part of a group of nitrogen compounds known as reactive nitrogen, N_R, which react biochemically in aquatic environments. The mass of N_R in the biosphere, which was in equilibrium up to the early 1900s, has increased significantly in the last 80 years as a result of the Haber–Bosch process for nitrogen fertilizer production, shown in Figure 1.2. Nitrogen fixation from the Haber–Bosch process now greatly exceeds natural denitrification, causing an increase in N_R. The excess N_R in aquatic environments, of

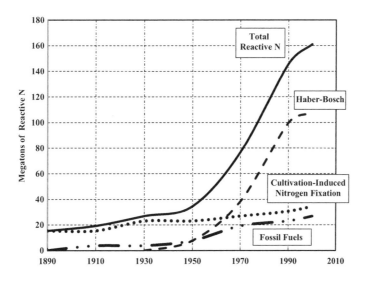

Figure 1.2 The increase in mass of global reactive nitrogen, N_R, as a result of the Haber–Bosch process for fertilizer production. The excess N_R, which has increased fourfold since the 1950s, contributes to the nitrogen cascade, where excess N_R in wastewater discharges and agricultural runoff fosters the growth of organisms in aquatic ecosystems, leading to eutrophication and hypoxia. *Source*: Figure redrawn from Galloway *et al.* (2003).

which wastewater discharge is a significant contributor, along with cropland runoff, is responsible for eutrophication, hypoxia, loss of biodiversity, and habitat degradation in coastal aquatic environments, in what is called the nitrogen cascade (Galloway *et al.*, 2003). The wastewater nitrogen cascade flows with the end-of-pipe paradigm to final discharge:

N_2
↓
Synthetic Fertilizer Haber-Bosch Process → **Food-N** → **Wastewater-N** → **Rivers, lakes, coastal waters**

The reuse of N in wastewater in agriculture models the natural N-cycle, avoiding N water quality problems, and creating an important resource for sustainable agriculture.

Phosphorus. The global demand for P in fertilizer is estimated to be 43.8 Mt/yr (Qadir *et al.*, 2020). The mass of TP from wastewater discharge is thus 5.8% of global demand:

$$\% \text{ TP}_{\text{wastewater}} = \frac{2.55 \,\text{Mt/yr}}{43.8 \,\text{Mt/yr}} (100\%) = 5.8\%$$

As with nitrogen, the phosphorus cycle is no longer a cycle: In the 1950s, the mining of phosphate rock for fertilizer surpassed the natural sources of P in animal and human manure, and guano, and the mass of P from fertilizer on the earth's surface has increased fourfold since 1960 (Figure 1.3). Excess P, as with N, is discharged with the linear end-of-pipe paradigm:

Phosphate Rock → **P Fertilizer** → **Food-P** → **Wastewater-P** → **Rivers, lakes, coastal waters**

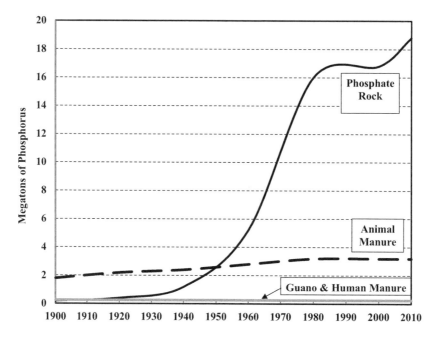

Figure 1.3 Historical sources of phosphorus fertilizers used globally in agriculture. Animal and human manure, and guano, were the predominant natural sources until the 1950s, when the mining of phosphate rock grew dramatically up to the present day. The mass of P used in fertilizer from phosphate rock has increased nearly fourfold since 1960. Much of this excess P is removed in crop harvesting, and after human consumption is discharged in domestic wastewaters. *Source*: Redrawn from Ashley *et al.* (2011).

The 2.55 Mt P/yr discharged in global wastewater is close to what is still being used in animal manure, ≈3.0 Mt/yr (Figure 1.3), which is a sustainable nutrient resource as wastewater could be if it were not discharged. It is estimated that accessible phosphate rock reserves will be depleted in 60–130 years. It is also estimated that approximately 25% of mined phosphorus ends up in aquatic environments, also contributing to eutrophication and hypoxia along with wastewater discharges (WHO, 2006). As with N, the reuse of P in wastewater closes the P-cycle, eliminating environmental problems while creating an important resource for sustainable agriculture.

1.1.2.4 Energy use in mechanized wastewater treatment

Wastewater treatment with the activated sludge process is well known to be a consumer of electricity, and it is important to compare this electricity consumption to the other water sector processes. Table 1.3 lists the principal water sector processes as defined by the International Energy Agency.

The electricity consumption in billions of kWh/yr for each water process sector is plotted in Figure 1.4 for the US, EU, and India, for the year 2014. The results are summarized as follows:

- Wastewater treatment is by far the highest consumer of electricity in the US and the EU, ranging from 1.6 to more than 2 times greater than any other water sector process. This consumption is due to the technologies selected for organic matter removal, principally activated sludge, and tertiary processes designed to meet strict effluent quality requirements to protect surface waters.
- Wastewater treatment is the lowest consumer of electricity of all the water sector processes in India, where there are at least 108 wastewater treatment plants that use natural processes that do not require energy input. The natural process systems include (Kumar & Asolekar, 2016):
  - 74 waste stabilization pond systems, with 22 systems reusing effluents in agriculture;
  - 15 polishing pond systems, with 5 systems reusing effluents in agriculture;
  - 10 constructed wetlands;
  - 5 Karnal technology installations (wastewater is applied with furrow irrigation with trees planted on furrow ridges); and
  - 4 duckweed pond systems, with all reusing effluents in agriculture.

Water supply and treatment processes are the highest consumers of electricity as a result of extensive groundwater pumping: India accounts for 40% of global groundwater use.

The high energy consumption in wastewater treatment in the EU and the US derives from the end-of-pipe paradigm, where the aerobic activated sludge process was developed to rapidly remove the organic oxygen demand in wastewater so effluent discharges would not deplete dissolved oxygen in surface waters. The activated sludge process historically has consumed 50% of the electricity demand in wastewater treatment plants (Metcalf & Eddy/AECOM, 2014).

Table 1.3 Water sector processes as defined by the international energy agency (IEA).

Water Sector Process	Description
Supply and treatment	Extraction from groundwater and surface water; drinking water treatment
Desalinization	Potable water made by reverse osmosis or thermal processes; necessary as a result of water stress (primarily Middle East and North Africa)
Distribution	Pumping from the water treatment plant to end-users through a pressurized distribution network
Wastewater treatment	Transport and treatment to meet regional effluent standards
Transfer	Large-scale inter-basin transfer projects

Source: Adapted from IEA (2016).

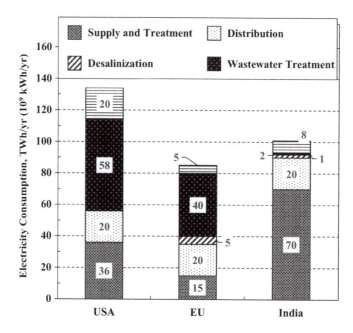

Figure 1.4 Electricity consumption in the five water sector processes for the US, EU, and India in 2014. Wastewater treatment is the highest consumer of electricity in the US and the EU, from 1.6 to 2 times greater than other processes. Electricity consumption for wastewater treatment is the lowest of all in India: most treatment plants use natural processes, principally waste stabilization ponds, that do not require electricity; water supply and treatment processes are the highest as a result of extensive groundwater pumping – India accounts for 40% of global groundwater use. *Source*: Adapted from OECD (2016).

When the first activated sludge treatment systems were developed not much was known about the energy characteristics of domestic wastewater. An alternative approach to the high-energy consuming activated sludge process becomes apparent when the energy characteristics of organic matter and nutrients in domestic wastewater are examined. Table 1.4 presents the energy characteristics of raw domestic wastewater in terms of organic matter (suspended and dissolved BOD), and nitrogen and

Table 1.4 Energy characteristics of raw domestic wastewater.

Parameter	Typical Concentration (mg/L)		Energy (kWh/m³)		
			Consumption in Activated Sludge Treatment Plants (US Average)	Maximum Energy Production from Organic Oxidation	Required to Produce Synthetic Fertilizers[1]
Ultimate BOD$_L$	320				
Suspended	175			0.67	
Dissolved	145			0.56	
Total nitrogen	40				0.77
Total phosphorus	8				0.02
		Total	0.60	1.23	0.79

Source: From McCarty *et al.* (2011).
[1] From the Haber–Bosch process for nitrogen, and the mining of phosphate rock for phosphorus.

phosphorus in terms of energy consumption, energy production from embedded organic matter, and the energy required to produce synthetic nitrogen and phosphorus fertilizers at the same concentrations found in domestic wastewater.

Energy consumption in activated sludge treatment plants. As shown in Table 1.4, the US average energy consumption was reported as 0.6 kWh/m³ by McCarty *et al.* (2011). This value is close to the reported median value of 0.77 kWh/m³ for 1377 wastewater treatment plants, with a median flow of 11,356 m³/d, by Energy Star (2015).

The energy consumption in an activated sludge treatment plant is in the form of electricity to power the blowers for activated sludge aeration ($\approx$55% of total energy demand), with the remainder of energy used for pumping, heating, lighting, and sludge dewatering (Metcalf & Eddy/AECOM, 2014).

Maximum energy production from organic matter oxidation. If the biodegradable organic matter exerting an ultimate BOD_L of 320 mg/L could be 100% oxidized to CO_2 and H_2O, it would have a maximum energy production of 1.23 kWh/m³, twice the amount of energy consumed in an average activated sludge plant. This energy yield cannot be reached in practice, but if anaerobic processes were used, perhaps 50–70% of the embedded energy in organic matter can be converted to methane as a sustainable energy source. In this case, wastewater treatment could, in theory, produce energy rather than consume it for the removal of organic matter.

Energy required to produce synthetic N and P fertilizer. The energy required to produce synthetic N and P fertilizers, which are globally the most commonly used fertilizers (Figures 1.2 and 1.3), is estimated to be 19.3 kWh/kg N for the Haber–Bosch process, and 2.11 kWh/kg P for the mining and processing of phosphate rock (McCarty *et al.*, 2011). For the typical raw wastewater concentrations of 40 mg/L TN and 8 mg/L TP reported by McCarty *et al.* (2011), the energy required to produce these concentrations with Haber–Bosch and phosphate rock processes would be 0.79 kWh/m³. To obtain this equivalent energy in practice as fertilizer without an energy input, it is only necessary to irrigate with the wastewater effluent after pathogen reduction.

For resource-challenged urban areas and small cities, an appropriate approach to wastewater treatment and valorization of embedded energy would be to treat wastewater for pathogen reduction, followed by reuse of the effluent in agriculture for its water and nutrient value. With domestic wastewater, the energy value of the embedded N and P will be higher, and the treatment processes simpler, than that likely to be obtained with anaerobic processes.

1.1.3 The integrated wastewater management paradigm

The integrated wastewater management (IWWM) paradigm incorporates wastewater into the water resources framework, where its value can be realized as outlined in Table 1.5.

Integrated wastewater management paradigm:

- Management of domestic wastewater should focus on public health as a first priority, with the removal of pathogens as the principal objective of wastewater treatment.
- The best available technology for accomplishing this goal is the wastewater stabilization pond system, which can most easily meet the World Health Organization guidelines for wastewater reuse in agriculture than any other technology.
- In order to resolve the problem of agricultural demand for water and sustainability of wastewater treatment, wastewater treatment should be integrated with the productive reuse of the treated wastewater in agriculture.
- Natural system concepts and the natural cycles of water, carbon, nitrogen, and phosphorous are incorporated into the design.
- N and P concentrations in wastewater are valorized for agricultural reuse, lowering or eliminating the need for synthetic fertilizers, and helping to restore the natural N and P cycles.
- Reuse in agriculture fosters food and water security, and livelihood in agriculture.
- Wastewater is a water resource, with continuous flow independent of local climate conditions, that should be used in agriculture.

Table 1.5 Integrated wastewater management: wastewater as part of water resources.

Resources in Wastewater	Resource Management Options	Treatment Option[1]	Potential Benefits
Water Nutrients (N, P, K) Organic Matter Energy Content	• Reuse in agriculture • Water for irrigation • Nutrients replace synthetic fertilizers • Organic matter as soil conditioner • Anaerobic processes for methane production[2]	Waste stabilization Ponds stabilization Reservoirs	Pathogen reduction Health protection Livelihood in agriculture Water security Food security Improvement in surface water quality Adaptation to climate change and water scarcity Natural cycles of carbon, nitrogen, and phosphorus are incorporated into the treatment/reuse processes

Source: Developed from UNEP (2017).
[1]Waste stabilization ponds are the only viable option for pathogen reduction in resource-limited areas where wastewater is reused in agriculture; if other treatment technologies must be used because of space limitations, stabilization reservoirs should be used to ensure required pathogen reduction (Chapter 8).
[2]Anaerobic processes are only recommended in larger cities where resources for adequate operation and maintenance are ensured.

- Under the right conditions, the energy characteristics of wastewater can be valorized for the production and use of methane. (Successful production and use of methane in small treatment systems are not common and most attempts fail. This option should only be considered under circumstances where resources exist for successful design, construction, and permanent operation and maintenance with skilled operators. The energy value of the embedded N and P in wastewater used in agriculture is likely greater than energy from methane production as shown in Table 1.4, and efforts should first be focused on reuse in agriculture.)

1.1.3.1 Wastewater as a water resource

Wastewater is a valuable water resource for agriculture. In arid and semi-arid environments, wastewater is a reliable, continuous source of water that has a relative constant flowrate throughout the year. The following calculations show a rough estimate of potential irrigated land area in semi-arid climates with the global wastewater currently discharged to surface waters.

1.1.3.2 Semi-arid climates: irrigation water requirement 1500 mm/yr

Much of the global irrigated area is in dry climates. Mendoza, Argentina, is an example that has a semi-arid climate requiring 1.0 m/yr of irrigation water (see Chapter 8). The global area that could be irrigated with the 318.7×10^9 m^3/yr of wastewater applied at 1.5 m/yr is calculated as

(1) Irrigation water requirement = 1.0 m^3/m^2 yr in semi-arid climates.
(2) Potential global irrigation area with the annual wastewater volume of 318.7×10^9 m^3/yr is currently discharged to surface waters:

$$\text{Irrigated area} = \frac{\left(318.7 \times 10^9 \text{ m}^3/\text{yr}\right)}{(1.0 \text{ m}^3/\text{m}^2 \text{ yr})} = 318.7 \times 10^9 \text{ m}^2 = 31.9 \times 10^6 \text{ ha}$$

This irrigated area is close to the reported global area of 29.3 million ha of cropland irrigated with surface waters from catchments with low levels of wastewater treatment (Thebo *et al.*, 2017). It is assumed in this example that (i) the wastewater would be adequately treated for pathogen reduction by sustainable treatment technologies such as waste stabilization ponds (WSPs) and (ii) the wastewater is not mixed with surface waters but applied directly to the fields.

1.1.3.3 Valorization of nutrients (N and P) in wastewater

In valorization, the nutrients flow within their natural cycles (Figure 1.5):

(1) Wastewater is treated to remove pathogens.
(2) Nutrients are applied with wastewater and are taken up by crops.
(3) Crops are harvested with embedded nutrients.
(4) Nutrients return to wastewater in excreta.
(5) Wastewater is treated to complete the cycle.

Nutrient application rates:

Using the 31.9×10^6 ha of irrigated land calculated previously, the N and P areal loadings as fertilizer are calculated below:

$$TN_{Applied} = \frac{14.0 \times 10^9 \text{ kg/yr}}{31.9 \times 10^6 \text{ ha}} = 440 \text{ kg TN/ha yr}$$

$$TP_{Applied} = \frac{2.55 \times 10^9 \text{ kg/yr}}{31.9 \times 10^6 \text{ ha}} = 80 \text{ kg TP/ha yr}$$

These values fall within the ranges of forage crops shown in Table 1.6.

Yields have consistently been higher when crops are irrigated with wastewater rather than fresh water, and even with freshwater with added fertilizer (Mara, 2003; Moscoso, 2016). Figure 1.6 compares crop yields for eight important crops grown in the coastal areas of Peru for (i) irrigation with waste stabilization pond effluent and (ii) irrigation with well water with added fertilizer. Yields with wastewater irrigation were higher for every crop, ranging from 50 to 250% greater than yields from irrigation with well water with added fertilizer (Moscoso, 2016).

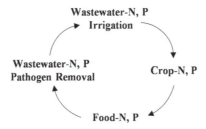

Figure 1.5 The natural N and P cycles in wastewater-irrigated agriculture.

Table 1.6 Nitrogen and phosphorus uptake rates for forage and field crops.

Crop Type	Nutrient Uptake Rate (kg/ha yr)	
	Nitrogen	**Phosphorus**
Forage crops	130–675	20–85
Field crops	75–250	15–30

Source: From USEPA (1981).

12   Integrated Wastewater Management for Health and Valorization

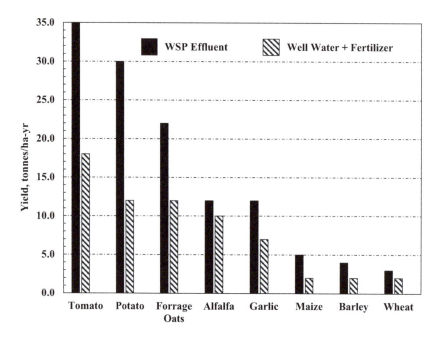

Figure 1.6 Crop yields in tonnes/ha yr (10³ kg/ha yr) from irrigation with (i) waste stabilization pond effluent and (ii) well water with added fertilizer nutrients (N, P, and K), in Tacna, Peru. The WSP effluent produced higher yields than the well water with added nutrients for all eight crops, ranging from 40 to 250% greater yields than the well water-fertilizer mix. *Source*: Figure developed from data by Moscoso (2016).

1.1.3.4 Value as fertilizer, 2021 prices

The annual mean value of N and P in diammonium phosphate (DAP) fertilizer in 2021 was calculated from IndexMundi, as shown in Section 8.2.2 (www.indexmundi.com/commodities/?-commodity=dap-fertilizer&months=12) (indexmundi.com):

Value of N and P per kg (2021):

N : US$3.74/kg N

P : US$3.11/kg P

The annual value of N and P can then be calculated with the masses of N and P discharged to surface waters:

$$\text{Value of N} = (14.00 \times 10^9 \text{ kg N/yr})(\text{US\$3.47/kg N}) = \text{US\$48,659,116,000}$$
$$\text{Value of P} = (2.55 \times 10^9 \text{ kg P/yr})(\text{US\$3.11/kg P}) = \underline{\text{US\$7,929,256,000}}$$
$$\text{Total: US\$56,588,372,000}$$

The annual value of N and P in discharged wastewater is thus estimated to be US$56.6 billion/yr if they could be used as a fertilizer rather than discharged to surface waters.

1.1.3.5 Energy saved from fertilizer production

Using the energy required per kg of N and P fertilizer production, 19.3 kWh/kg N for the Haber–Bosch process, and 2.11 kWh/kg P for the mining and processing of phosphate rock (McCarty *et al.*, 2011), the energy saved by using the nutrients in wastewater is estimated as:

$$\text{Energy Saved for N Production} = (14.0 \times 10^9 \text{ kg N/yr})(19.3 \text{ kWh/kg N}) = 271 \times 10^9 \text{ kWh/yr}$$
$$\text{Energy Saved for P Production} = (2.55 \times 10^9 \text{ kg P/yr})(2.11 \text{ kWh/kg P}) = \underline{5.38 \times 10^9 \text{ kWh/yr}}$$
$$\text{Total: } 276 \times 10^9 \text{ kWh/yr}$$

The total energy saved by using N and P natural fertilizers rather than buying synthetic ones is 276 billion kWh/yr. This consumption is equivalent to the annual electricity consumption of 284 million persons consuming electricity at a rate of 972 kWh/person yr, which is the mean per capita consumption in India for 2020 (Ritchie & Roser, 2020).

1.1.3.6 $CO_{2,equiv}$ emissions saved from not using synthetic fertilizers

Carbon dioxide equivalent emissions from the manufacture of synthetic fertilizers are saved when wastewater is reused in agriculture. The emission factors for synthetic fertilizer manufacture are shown below (see Section 8.1.2):

$$\text{Nitrogen Emission Factor} = \left(\frac{2.522 \text{ kg } CO_{2,equiv,N}}{\text{kg TN}}\right)$$

$$\text{Phosphorus Emission Factor} = \left(\frac{0.472 \text{ kg } CO_{2,equiv,P}}{\text{kg TP}}\right)$$

The metric tons (t) per year of carbon dioxide equivalent emissions (t $CO_{2,equiv}$) are calculated as

$$t\, CO_{2,equiv,N} = (14.0 \times 10^9 \text{ (kg N/yr)}) \left(\frac{2.522 \text{ kg } CO_{2,equiv,N}}{\text{kg TN}}\right)\left(\frac{1.0 \text{ t}}{1000 \text{ kg}}\right) = 35{,}365{,}502\, t\, CO_{2,equiv,N}/\text{yr}$$

$$t\, CO_{2,equiv,P} = (2.55 \times 10^9 \text{ (kg P/yr)}) \left(\frac{0.472 \text{ kg } CO_{2,equiv,P}}{\text{kg TP}}\right)\left(\frac{1.0 \text{ t}}{1000 \text{ kg}}\right) = 1{,}203{,}411\, t\, CO_{2,equiv,P}/\text{yr}$$

$$\text{Total Emissions Saved} = 36{,}568{,}913\, t\, CO_{2,equiv,T}/\text{yr}$$

The total emissions saved by using the nutrients in wastewater are thus estimated at 36.6 million metric tons of carbon dioxide equivalent.

Table 1.7 summarizes the valorization results for globally discharged N and P if they were used in agriculture. As wastewater production and discharges increase by more than 50% of present levels by 2050, the valorization of water and nutrients could play a key role in global agriculture in resource-challenged areas.

1.1.3.7 Valorization of energy from anaerobic processes

Anaerobic processes produce methane that can be captured and used as a sustainable energy source. Small-scale anaerobic ponds and reactors, however, often have many operational problems such

Table 1.7 Valorization of nutrients in globally discharged wastewater for reuse in agriculture.[1]

Domestic Wastewater	Discharged to Surface Waters (10^9 kg/yr)	Value as Fertilizer (US$/yr)	Energy Equivalent Fertilizer Production (10^9 kWh/yr)	$CO_{2,equiv}$ Saved (t $CO_{2,equiv}$/yr)
Total N	14.0	$48,659,116,000	271	35,365,502
Total P	2.55	$7,929,256,000	5.38	1,203,411
Total		$56,588,372,000	276	36,568,913

[1]Flowrate is equal to 318.7×10^9 m³/yr.

as corrosion, sludge build-up requiring frequent desludging, and poor gas yields. Many small-scale installations fail after they fill with sludge. Thus, anaerobic processes must be selected with caution and are more appropriate in large-scale applications where resources exist for proper operation and maintenance.

1.2 WASTEWATER REUSE IN AGRICULTURE AND DEVELOPMENT OF END-OF-PIPE PARADIGM

1.2.1 Historical use of wastewater in agriculture: 3000 BCE–1915 CE

Table 1.8 presents a brief history of the importance of excreta and wastewater reuse in agriculture, which has existed for at least 5000 years.

3000 BCE–330 CE. One of the first documented examples, beginning in 3000 BCE, was the use of night soil in China for both agriculture, and later aquaculture, which was still used in China until the 1990s (Figure 1.7). For 5000 years, up to the 1900s, night soil and wastewater were commonly used in agriculture in various cultures in Eurasia and the Americas.

Urban centers in Crete, the Indus Valley, Greece, and Rome, developed sewerage systems, some very sophisticated, to collect wastewaters, and sometimes stormwaters, which were then drained by gravity to agricultural fields and surface waters. Collection basins and cisterns were also used to store wastewaters and rainwater, which were later conveyed to agricultural fields. While the large Cloaca Maxima in Rome was used only to collect surface waters and groundwaters from urban areas, a sewerage system connected to individual houses also existed, and drained to agricultural fields.

In addition to China, other Asian countries, including Japan and Korea, used night soil as fertilizer for agriculture and aquaculture.

500–1500 CE. The Middle Ages have been called the Sanitary Dark Ages, where high death rates from excreta-related and water-related diseases. It is estimated that 25% of the population of Europe died during this period due to cholera and other excreta- and water-related diseases, and, the plague (Angelakis *et al.*, 2018). In some larger cities, such as Paris, sewers were constructed, but caused further problems in downstream neighborhoods.

During the same period in the Americas, the Aztecs' agriculture developed using chinampas, or floating gardens, that used animal and human wastes as fertilizer along with sediments and vegetation. This agricultural system supported 250,000 people in the Valley of Mexico when the Spanish arrived, when both London and Paris had populations less than 20,000 (Angelakis *et al.*, 2018).

1530–1900. Reuse of wastewater in agriculture, or on what was also called sewage farms, was common throughout Europe and North America from the end of the middle ages until the early 1900s. Europe had wastewater irrigation areas up 4400 ha, while in the US the largest reported was 1600 ha.

In the 1890s, Mexico City exported raw wastewater to agricultural lands in the Mezquital Valley by gravity, irrigating 90,000 ha; this system is still in operation today, but uses treated wastewater from a large wastewater treatment plant in Mexico City with a design flowrate of 3,000,000 m^3/d.

1.2.2 Decline of wastewater reuse with end-of-pipe paradigm: 1915–1990

As urban areas grew larger in Europe and North America, less agricultural land was available for wastewater reuse and soil infiltration, and as a result wastewater reuse began to decline. By necessity wastewaters were increasingly discharged to downstream surface waters, creating the end-of-pipe paradigm that exists to this day.

The end-of-pipe paradigm had unforeseen consequences, however. As surface waters became overloaded with untreated domestic wastewaters, water quality degraded in terms of high pathogen concentrations, and loss of dissolved oxygen from biochemical oxygen demand. From the 1900s to the

Table 1.8 Brief history of excreta and wastewater reuse in agriculture and aquaculture.

Year	Location	Type of Reuse
3000 BCE	China	Night soil was used as fertilizer in agriculture, continuing until the 1990s.
3000 BCE	Crete	Developed sewerage systems used for agricultural irrigation with wastewaters and discharge to surface waters; cisterns used to store wastewater for irrigation.
2600 BCE	Indus Valley	Developed sophisticated sewerage systems discharging to agricultural fields.
1100 BCE	China	Use of human wastes and wastewaters in aquaculture.
500–400 BCE	Greece	Wastewater and stormwater are collected in the sewerage system, stored in a collection basin, and conveyed to agricultural fields.
100 BCE–330 CE	Rome	A sewerage system connected to individual houses drained to agricultural fields used for crop irrigation.
500–1500 CE	Europe	Sanitary dark ages. High death rate from cholera and other excreta- and water-related diseases due to poor sanitation.
1400–1500s	Mexico City	Reuse of animal and human wastes, sediments, and vegetation in chinampas built over wetlands by the Aztecs, are considered to be one of the most productive and sustainable forms of agriculture. Smaller areas of chinampas are still in use to the present day.
1531–1896	France, UK, Germany, Italy, Poland	Development of sewage farms in Europe where urban wastewater was drained to agricultural lands for disposal and reuse, with areas from 160 to 4400 ha.
1700–1800s	New York City	Night soil companies haul excreta for disposal and land application.
1890–Present	Mexico City	Drainage canals were built to export wastewater from Mexico City to agricultural lands in the Mezquital Valley; still in operation to this day irrigating approximately 90,000 ha with treated wastewater.
1876–1915	USA	Development of sewage farms where urban wastewater was drained to agricultural lands for disposal and reuse, with areas ranging from 5 to 1600 ha.
1915–1990s	EU and USA	Significant decline of wastewater reuse in agriculture with the development of the end-of-pipe paradigm for wastewater discharge to surface waters due to: (1) Disappearance of agricultural lands near urban areas with the growth of large cities; (2) Development of gradually intensified add-on treatment processes to protect surface waters rather than reuse of wastewater in agriculture; (3) N and P considered pollutants to remove rather than resources.
2000–2020	EU and USA	Gradual emphasis on onsite resource recovery during treatment, and producing a high-quality effluent for discharge; some reuse with strict water quality standards. Agricultural reuse is still not common.
2000–2020	Resource-poor cities and urban areas worldwide	The end-of-pipe paradigm dominates as urban areas grow at 3.5% per year, but the majority of cities have inadequate wastewater treatment, or none at all. Many downstream surface waters are seriously contaminated. Direct or indirect use of wastewater in agriculture is common, with an estimated 29.3 million ha irrigated in 2017, but with high risks of excreta-related infections. Farmers are more interested in the reuse of wastewater than design engineers, and local and national government officials. Excellent examples exist of waste stabilization ponds with agricultural reuse that should be modeled.

Source: Developed from Angelakis *et al.* (2018), Ashley *et al.* (2011), Lofrano and Brown (2010), and Thebo *et al.* (2017).

Figure 1.7 The night soil system in China used human excreta for fertilizer for thousands of years, in perhaps the first well-documented example of valorization of human wastes for their nutrient value. Excreta collected in carts from latrines (top left) was taken to storage pits at the edge of fields (top right), where it was later collected in buckets and applied in the fields (bottom photos). These photos were taken in a peri-urban area of Chengdu, China, in 1991. The night soil system no longer exists in Chengdu, nor in most of China except possibly in remote rural areas.

1970s, wastewater treatment in Europe and North America gradually developed in a linear manner to solve these problems as outlined by Feachem *et al.* (1983):

- Increasing awareness of public health risks in large cities led to the construction of large sewers that discharged raw wastewaters into rivers.
- Raw wastewater discharges depleted dissolved oxygen in rivers, which often became open sewers.
- Various treatment technologies were developed to reduce suspended solids loads and the oxygen demand of the discharged wastes. (For example, The Imhoff tank is an example of one of the first technologies to address these issues.) In 1900, the UK Royal Commission on Sewage Disposal proposed effluent standards of <30 mg/L TSS and <20 mg/L BOD, values not much different than those promulgated today in countries worldwide.
- In the 1950s to 1970s, a growing awareness of increasing public health and environmental problems, coupled with growing urban populations, led to the development of tertiary treatment technologies to address issues such as eutrophication caused by nutrients in wastewater.
- At the same time, it became clear that none of the treatment technologies were efficient at inactivating pathogens in wastewaters. As a result, effluent disinfection with chlorine was borrowed from the water treatment industry as a way to inactivate bacteria in effluents, a practice that continues worldwide to this day. Protozoa (oo)cysts and helminth eggs, however, are highly resistant to chlorine.
- Conventional wastewater treatment as developed historically up to the 1980s, with a focus on TSS and BOD removal, was never designed to remove pathogens, and the full range of excreted pathogens in the influent can be present in the effluent in all types of wastewater treatment processes (Oakley & Mihelcic, 2019).

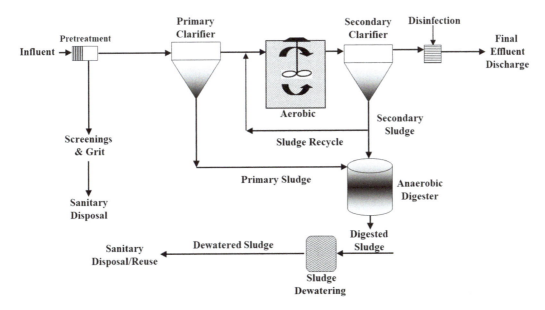

Figure 1.8 Conventional secondary treatment flow diagram as developed historically (1960s–1990s). The aerobic unit process originally consisted of trickling filters or activated sludge, but today is almost exclusively activated sludge in large cities worldwide. Disinfection was an ad-on process in the linear treatment train after problems with pathogens in effluents became apparent in the 1960s–1970s (Feachem et al., 1983). Removal of N and P was also implemented at select locations where eutrophication was a problem (e.g., Great Lakes, Chesapeake Bay, Florida).

A flow diagram of the unit processes in a conventional activated sludge treatment plant developed up to the 1990s is shown in Figure 1.8. This conventional design is common in large cities worldwide, but many still have problems with operation and maintenance, and overloading from increasing flows from population growth.

1.2.3 End-of-pipe paradigm with resource recovery in EU and North America: 2000–2020

During the last 20 years, additional wastewater treatment processes have been developed that go beyond the original objectives of BOD, TSS, and nutrient removal, and disinfection, prior to discharge to surface waters. Increased emphasis is also focused on onsite resource recovery during treatment, and producing a high-quality effluent for discharge, or high-end reuse with strict water quality standards (Metcalf & Eddy/AECOM, 2007, 2014).

1.2.3.1 Secondary treatment with tertiary processes and resource recovery

Figure 1.9 shows a flow diagram of secondary treatment, tertiary processes, and resource recovery. Add-on processes in the treatment train include:

- nitrogen removal with nitrification–denitrification,
- chemical precipitation of phosphorus,
- microfiltration for enhanced removal of suspended solids and microorganisms, and
- disinfection with UV or ozone.

Resource recovery includes:

- nitrogen and phosphorus recovery from sludges and sidestreams, and
- the capture of methane for heating and electricity generation.

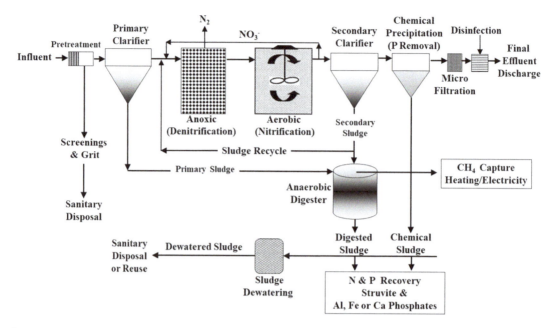

Figure 1.9 Example flow diagram of secondary wastewater treatment with nutrient removal and tertiary processes, and resource recovery in the US circa 2015. *Source:* Adapted from Metcalf and Eddy/AECOM (2007, 2014).

The final effluent is discharged to surface waters with very strict discharge requirements, or reused for high-end uses such as landscape irrigation, industrial, urban and recreational uses, groundwater discharge, and indirect or direct potable reuse (with more added processes). Agricultural reuse with domestic wastewater has historically not been a priority in the EU and the US.

Few plants have all of the unit processes shown in Figure 1.9, but the trend in the EU and the US is to incorporate more of them as effluent discharge requirements become stricter, along with increasing interest in the recovery of heat and energy, and N and P for fertilizer use.

1.2.3.2 Wastewater reuse in agriculture in the EU and the US

Wastewater reuse in the EU is not common. It is estimated that 964 million m^3/yr of urban wastewater is reused annually in the EU, which is only 2.4% of all the treated urban wastewater of 40 billion m^3/yr, most of which is discharged to surface waters (European Commission, n.d.).

Agricultural reuse is also not common in the US. Figure 1.10 presents data from the USEPA on total wastewater treatment plant flowrates and final disposition of effluents in 2008. Of the total flowrate of 121,703,040 m^3/d, less than 6,868,800 m^3/d, or 5.6% of the total flowrate, was reused for agricultural irrigation.

Because the entire infrastructure of wastewater management in both the EU and the US has been developed for surface water discharge, with wastewater treatment plants located downgrade near outfall locations, future emphasis on wastewater reuse as a result of climate change, water stress, and water scarcity will be difficult to implement.

1.2.4 Wastewater treatment and resource recovery in China: 1980–2020

1.2.4.1 Wastewater treatment and discharge of excess nitrogen to surface waters

Urban use of night soil in China, which had existed for 5000 years, ended in the 1990s as a result of widespread urbanization, with the concomitant loss of peri-urban agricultural land; the construction

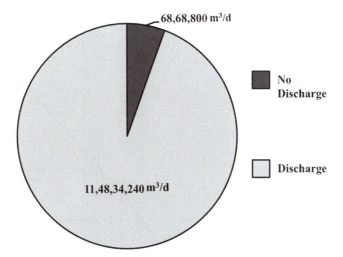

Figure 1.10 Flowrates of all domestic wastewater treatment facilities in the US in 2008 as reported by the USEPA. Of the total flowrate treated (121,703,040 m³/d), 94.4% was discharged to surface waters, and 5.6% was not discharged and disposed of by evaporation, reused in industry, and reused for agricultural irrigation. *Source*: Developed from data by Metcalf and Eddy/AECOM (2014).

of sewerage with wastewater treatment plants discharging to surface waters became the replacement (Figure 1.11). As a result of urban growth, and the development of sewerage with wastewater treatment and surface water discharge, China now has problems of excess nitrogen in surface waters from the excessive use of synthetic fertilizers. This experience is repeated in large cities worldwide where mechanized wastewater treatment is designed for surface water discharge.

Figure 1.12 shows the rapid growth of N discharges to surface waters from 1980 on as a result of urban growth and the development of wastewater treatment plants. At the same time, the use of night soil declined and disappeared by 2005. Figure 1.13 shows the deterioration of water quality in terms of N concentrations as a result of the increased use of synthetic fertilizers from 1965 to 2014. Finally, the decrease in nutrient reuse in agriculture with the growth of synthetic N fertilizer use, and subsequent discharge to surface waters, is shown in Figure 1.14.

Figure 1.11 An activated sludge wastewater treatment plant with nitrogen removal built near the site of the night soil fields in Chengdu shown in Figure 1.7, which no longer exists. Mechanized wastewater treatment has grown rapidly over the last 30 years in China. The first unit process in the left photo is an open anaerobic reactor without the capture of methane. The middle photo shows the second unit process, an anoxic denitrification basin receiving the anaerobic reactor effluent. The nitrified effluent from the activated sludge basin (right photo) is returned to the denitrification basin, where nitrate is reduced to N_2 and released into the atmosphere.

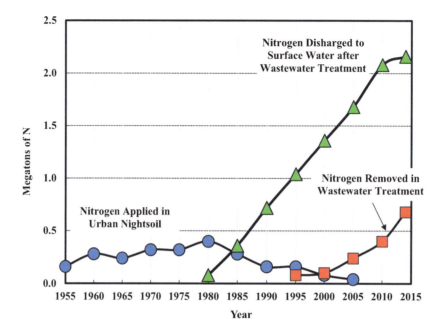

Figure 1.12 Consequences of the end-of-pipe paradigm in China. The decline of N applied in night soil coincided with the growth of synthetic fertilizer use. At the same time, urban growth and widespread coverage of sewerage caused a dramatic increase in domestic wastewater N discharges to surface waters from the 1980s to 2014. N removal in wastewater treatment began in 1995 and is increasing to the present day. *Source*: Developed from data by Yu *et al.* (2019).

An example of a resource recovery wastewater treatment plant with a goal of energy self-sufficiency is presented in the following case study from China.

1.2.4.2 Resource recovery in a Chinese 'concept wastewater treatment plant'

China has the world's largest municipal wastewater infrastructure, with a treatment capacity in 2018 of 200 million m³/d (Qu *et al.*, 2019). Most treatment plants are based on various modifications of activated sludge technology, some with the implementation of downstream unit processes for nitrogen removal (Yu *et al.*, 2019). Recently, a prototype 'New Concept Wastewater Treatment Plant' was built in Sui County to demonstrate the valorization of wastewater by realizing the following four goals (Qu *et al.*, 2019):

(1) Sustainable water supply for replenishing local surface water supplies,
(2) Energy self-sufficient operation,
(3) Resource recovery of nutrients;, and
(4) Environmental harmony with the local surroundings.

These goals are similar to those promoted in the EU and the US, and the plant design is similar to that shown in Figure 1.4. Thus, the historical use in agriculture of organic matter and nutrients in human excreta and animal manure, for thousands of years, has been replaced with urbanization and advanced secondary treatment in the last 40 years in China.

Figure 1.15 presents the operating data for energy consumption and production for the Sui County treatment plant. The goal of energy self-sufficiency has not yet been obtained, with energy production at 42% of total consumption. Although there are various wastewater treatment plants worldwide

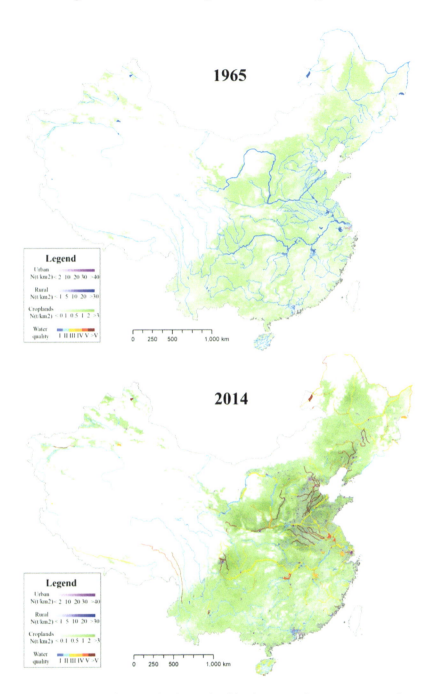

Figure 1.13 Surface water quality from total N inputs in China in 1965 and 2014. In 1965, surface waters were at Class I and II levels (≤ 0.2 to ≤ 0.5 mg/L TN). By 2014, many stretches of rivers were at Class III, IV, V, and >V levels (≤ 1.0 to >2.0 mg/L TN). Concentrations of TN >1.0 mg/L are considered above the pollution threshold (Yu et al., 2019). TN inputs in 2014 are due to extensive use of synthetic fertilizers, with 30 Mt applied in 2014, compared with 2 Mt in 1965. *Source*: Figures used with permission from Springer Nature.

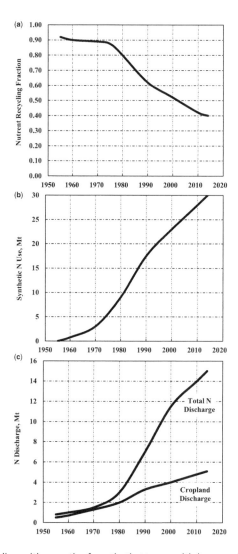

Figure 1.14 Decrease in N recycling with growth of synthetic N use, with losses of N to surface waters.
(a) Decline in the use of organic N fertilizers, both from animal manure and human excreta, began in the 1970s and continues to the present day. Causes of the decline in traditional methods include:

- Growth of the urban population;
- Human wastes increasingly disposed of in sanitary sewers rather than used as night soil fertilizer; and
- Economic growth with increased access to synthetic fertilizers.

(b) The use of synthetic N fertilizers made with the Haber–Bosch process started in 1955, increasing exponentially until 1990, then continuing with linear growth to reach 31 megatons (Mt) in 2014. (c) The growth of total N discharged to surface waters, from approximately 1.0 Mt in 1955 to 15 Mt in 2014, has paralleled the growth of synthetic fertilizer use and the urban population. Of the difference between total discharge and cropland discharge, 10 Mt, approximately 3 Mt is due to N in domestic wastewater discharged from sewers, or wastewater treatment plants, most of which do not yet have N removal processes – wastewater treatment plants removed 0.7 Mt of N in 2014 (Yu et al., 2019). It is planned the existing and future plants will remove or recover N, but at an additional increased consumption of energy of at least 0.14 kWh/m^3 (Li et al., 2015; Qu et al., 2019).
Source: Redrawn from Yu et al. (2019).

Integrated wastewater management for reuse in agriculture 23

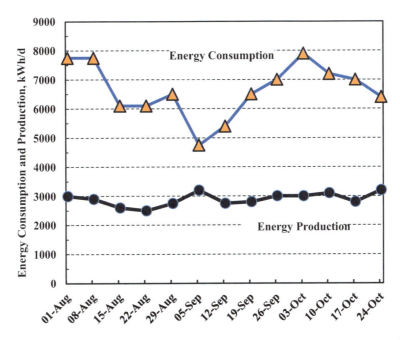

Figure 1.15 Energy production and consumption at the Chinese Concept Wastewater Treatment Plant, Sui County No. 3, in 2020. The plant was designed for a flowrate of 40,000 m³/d at a cost of US$28 million, or US$500/m³. The average energy production over the 3-month monitoring period was 42% of the total energy required to operate the plant: 0.14 kWh/m³ mean production compared to 0.33 kWh/m³ mean consumption. *Source*: www.thesourcemagazine.org/chinas-sustainable-concept-wastewatertreatment-plant-progress/.

that have met the goal of energy self-sufficiency, most have not as a result of the technology and costs required for design, construction, and operation. This approach will obviously not be feasible in resource-poor cities that have difficulty constructing sewer capacity to keep up with population growth.

1.2.5 End-of-pipe paradigm in resource-limited cities/peri-urban areas: 2000–2020

The end-of-pipe paradigm dominates as urban areas grow at rates of 3.5% per year, but the majority have inadequate wastewater treatment, or none at all. Many downstream surface waters are seriously contaminated from direct raw wastewater discharges from sewers, or overland flow when piped water is provided to dwellings before sewers are built as shown in Figure 1.16.

Figure 1.17 underscores the magnitude of the public health problem from wastewater discharges on a global scale with measured fecal coliform concentration greater than 1000 CFU/100 mL in river stretches throughout Latin America, Northern Africa, Western Asia, Central and Southern Asia, and Eastern and South-Eastern Asia (UNEP, 2017). In all of these regions coverage of wastewater treatment is low, and pathogen reduction in existing treatment plants is limited (Oakley & Mihelcic, 2019). Table 1.9 presents estimates of the millions of persons coming in contact with these surface waters in Africa, Asia, and Latin America (UNEP, 2016).

1.2.5.1 Indirect reuse of wastewater in agriculture
Indirect use of wastewater in agriculture is common worldwide, with an estimated 29.3 million ha irrigated in 2017 with surface water flows highly dependent on inadequately treated wastewater (Thebo *et al.*, 2017). Farmers and agricultural workers, however, face high risks of excreta-related

Figure 1.16 The beginnings of the end-of-pipe paradigm in peri-urban areas before sewerage is constructed. Piped-in water is provided before the construction of sewers, and wastewaters are discharged to the streets – a continuing practice since the Middle Ages. In the top left photo, raw sewage flows down the street in a new peri-urban development for a population displaced by Hurricane Mitch in Honduras. In the top right photo, raw wastewater flows by sewer pipes that have been delivered for installation, with eventual discharge to a nearby river. The bottom photo shows a middle-class housing development built without sewerage. (Top: Amarateca and Danlí, Honduras; Bottom: Ciudad Juarez, Mexico.)

infections. (Indirect reuse is wastewater diluted in surface water that is used for irrigation; direct reuse is the direct application of effluent wastewater for irrigation. In both cases, the wastewater may be well-treated, poorly treated, or untreated.)

Farmers, both small-scale and large-scale, are generally more interested in the reuse of wastewater than engineers, and local and national government officials. Figure 1.18 shows an example of large-scale indirect reuse of raw wastewater mixed with surface water on plantain and banana plantations. Figure 1.19 shows an example of small-scale indirect reuse of inadequately treated wastewater effluent mixed with river water.

1.2.5.2 Direct reuse of inadequately treated wastewater in agriculture

While direct reuse with raw wastewater in agriculture is not common, reuse with inadequately treated wastewater is. Many wastewater treatment plants have poor operation and maintenance, and many are also abandoned by the municipality after a few years of operation, or when a new government takes charge of the municipality.

Figure 1.17 Frequency (months/year) where fecal coliform concentrations exceeded 1000 CFU/100 mL in river stretches for the years 2008–2010. Red: >6 months. Orange: ≤6 months. The red and orange areas on the map are areas where wastewaters discharged into rivers are poorly treated or not treated at all. *Source*: Figure reproduced with permission, copyright CESR, University of Kassel, April 2016.

Table 1.9 Estimated number of persons coming in contact with surface waters with >1000 CFU/100 mL fecal coliform concentrations.

Africa	Asia	Latin America
31,700,000–164,300,000	30,600,000–13,370,000	8,100,000–24,800,000

Source: UNEP (2016).

Figure 1.20 presents an example of direct reuse of waste stabilization pond effluent from a pond system that had been abandoned by the municipality in charge of operating it. In this case, the farmers were more interested in the reuse of wastewater than the municipal engineers and officials responsible for the operation and maintenance of the system.

1.2.5.3 Direct reuse in agriculture with effluent wastewater meeting WHO guidelines

Direct reuse with well-treated wastewater is found in various countries worldwide, most often with natural process treatment systems, principally WSPs designed for pathogen removal and stabilization of putrescible organic matter. Some of the best examples are in Mendoza, Argentina, which has been using treated wastewater in agriculture for over 50 years, and has been using large waste stabilization pond systems since the 1980s. As of 2020, two large waste stabilization pond systems, Campo Espejo and Paramillo, with design flowrates of 128,000 and 151,475 m³/d, respectively, were irrigating a total area of 5640 ha in the summer, and 1670 ha in the winter; the combined flowrate of these two systems comprises 77% of all wastewater flows in the province of Mendoza (Rauek, 2020).

Figure 1.21 shows part of the irrigation canal system receiving the waste stabilization pond effluent from the Paramillo system.

Figure 1.18 An example of indirect irrigation with diluted raw wastewater. The Chotepe River receives the majority of raw wastewater from the city of San Pedro Sula, Honduras (top photo), and mixes with the Chamelecón River 5 km downstream. Shortly downstream of the union of the rivers, water is pumped from the river to irrigate thousands of hectares of plantain and banana plantations (bottom photo). The large plantations could have selected other sources of water, but likely prefer the river for its nutrient value from the wastewater. The population of the urban area of San Pedro Sula is approximately 1.5 million, and there is no wastewater treatment. This irrigation practice is common worldwide where wastewater treatment is non-existent or poorly managed.

Figure 1.19 Indirect reuse in agriculture. An irrigation pump is placed near the wastewater discharge of a poorly operating activated sludge wastewater treatment plant. The treatment plant was designed without nutrient removal, and it is likely the placement was intentional as farmers know well the value of wastewater as fertilizer. The wastewater treatment plant was also designed without disinfection, and the health risk is high for the farmers and consumers of their produce. The effluent enters Lake Atitlán (seen at top), which suffers from eutrophication and pathogen inputs from wastewater discharges, and which is also a major drinking water source in the lake basin for at least 100,000 persons (Panajachel, Guatemala).

In conclusion, direct or indirect reuse of inadequately treated wastewater is common as a result of the end-of-pipe paradigm, and is summarized in Figure 1.22. Excellent examples exist, however, of WSPs with agricultural reuse that should be modeled as is the case with Mendoza, Argentina discussed above.

The key issue for engineers working in resource-challenged cities worldwide is the use of sustainable wastewater treatment technologies that inactivate pathogen concentrations to levels where effluents can be directly reused in agriculture. WSPs offer the best example of sustainable treatment and reuse. It is worthwhile to quote from Feachem *et al.* (1983) on this point (Feachem *et al.*, 1983: 63–64):

> Those whose job is to select and design appropriate systems for the collection and treatment of sewage in developing countries must bear in mind that European and North American practices do not represent the zenith of scientific achievement, nor are they the product of a logical and rational design process. Rather, treatment practices in the developed countries are the product of history, a history that started about 100 years ago... Conventional sewage works were originally developed in order to prevent gross organic pollution in European and North American rivers; they were never intended to achieve high removal of excreted pathogens. Their use in tropical countries in which excreted infections are endemic is only justifiable in special circumstances, for there is an alternative treatment process much superior in obtaining low survivals of excreted pathogens – the waste stabilization pond system.

Figure 1.20 An example of farmers reusing the effluent of an abandoned waste stabilization pond system for irrigation of their crops in Punata, Bolivia. Top left: A facultative pond, which is part of an anaerobic–facultative–maturation pond system long abandoned by the nearby municipality. Top right: An irrigation canal built by local farmers carrying the inadequately treated waste stabilization pond effluent to their fields, and to a storage reservoir. Bottom left: The small reservoir built by one of the local farmers to store the stabilization pond effluent. Bottom right: One of the fields irrigated with the reservoir effluent. The abandonment of the wastewater treatment system by the municipality is a common problem with the end-of-pipe paradigm in resource-challenged areas, where municipal staff and local engineers see wastewater as a problem to avoid rather than a resource; the farmers, however, know the value of the water with nutrients and organic matter. At this site, the waste stabilization ponds had not been desludged or maintained for years, and overall performance was extremely low for pathogen reduction, posing risks for field workers and their families, and consumers, as shown in the monitoring data below (Verbyla *et al.*, 2016):

Pathogen	Influent Wastewater Concentration	$\log_{10}$ Reduction by Treatment	Final Effluent Concentration Prior to Irrigation
E. coli	$2.0 \times 10^{+07}$ CFU/100 mL	0.7	$4.0 \times 10^{+06}$ CFU/100 mL
Giardia	1300 cysts/L	0.8	326 cysts/L
Cryptosporidium	22.8 oocysts/L	0.8	5.7 oocysts/L
Helminth eggs	316 eggs/L	0.5	100 eggs/L

Integrated wastewater management for reuse in agriculture   29

Figure 1.21 An irrigation canal receiving effluent from the Paramillo waste stabilization pond system in Mendoza, Argentina. The final effluent meets the local regulatory guidelines for crops eaten raw, which are equivalent to the 2006 WHO guidelines for restricted and unrestricted irrigation. Approximately 2780 ha are irrigated with a mean flowrate of 151,475 m^3/d; the official irrigation area is called the Lavalle ACRE. Irrigated crops include grapes (right photo), olives, fruits, and vegetables. This system has been in operation for over 20 years. *Source*: Data from Rauek (2020).

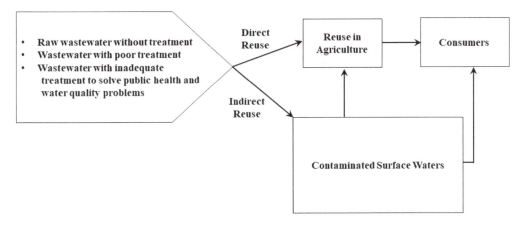

Figure 1.22 Direct and indirect reuse of inadequately treated wastewater as a result of the end-of-pipe paradigm that continues to look at wastewater as 'waste' rather than a resource.

1.3 WASTEWATER TREATMENT FOR AGRICULTURAL REUSE IN RESOURCE-LIMITED REGIONS

1.3.1 Urban population growth

Resource-limited cities typically have high urban growth rates that exacerbate the problems of infrastructure development in terms of piped water, sewerage, and wastewater treatment. Historical and projected urban growth and growth rates of the two principal UN development regions of the world are shown in Figure 1.23. From 1950 to 2018, the resource-limited regions of Africa, Asia, Latin America/Caribbean, and the Pacific Islands had an urban growth rate of 3.5%, which gives a doubling

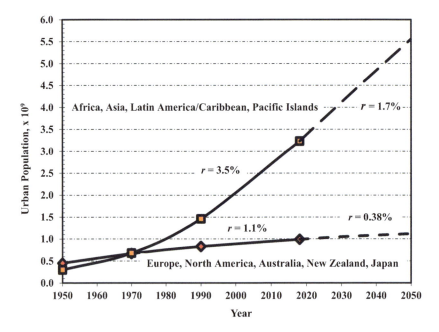

Figure 1.23 Historical and projected (dashed line) urban growth and growth rates of the two principal UN development regions of the world. *Source*: UN (2019).

time of 20 years ($t_{2x} = 0.693/0.035 = 19.8$ years). In contrast, the regions of Europe, North America, Australia, New Zealand, and Japan, with an urban growth rate of 1.1%, had a doubling time of 63 years ($t_{2x} = 0.693/0.011 = 63$ years). The so-called more developed regions had considerably more time to develop the wastewater infrastructure of sewerage and treatment, which still was not near completion for wastewater treatment, for example, until the 1980s in North America. It is worth noting that the standard design period in the US for both sewerage and wastewater treatment facilities is 20–25 years.

1.3.2 Coverage of wastewater treatment in the EU and North America

During the 17-year period for which data are available, from 2000 to 2017, wastewater treatment kept pace with the growth of the population as shown in Figure 1.24. In 2000, the coverage of wastewater treatment was 79% (population with WWT/sewered population). In 2017, the coverage increased to 85%. This is not the case with resource-limited cities as discussed in Section 1.3.3.

1.3.3 Coverage of wastewater treatment in resource-limited SDG regions

Figure 1.25 presents the urban population, and populations with piped water, sewerage, and wastewater treatment, for the years 2000 and 2017, for the SDG regions of Central and Southern Asia, Eastern and South-Eastern Asia, Latin America and the Caribbean, Northern Africa and Western Asia, and Sub-Saharan Africa. All of these regions have urban growth rates near 3.5% with an approximated doubling time of 20 years as discussed in Section 1.3.1.

The patterns of provision of piped water, sewerage, and wastewater treatment for both 2000 and 2017 follow the same, long-existing historical pattern for resource-limited regions:

- Piped water has priority, and is provided before the required sewer capacity is achieved.
- Wastewater treatment has the lowest priority, far behind sewerage, piped water, and urban development.

Integrated wastewater management for reuse in agriculture

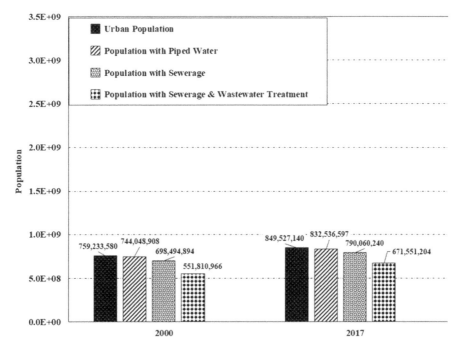

Figure 1.24 Urban population, population with piped water, population with sewerage, and population with sewerage and wastewater treatment for 2010 and 2017 for the SDG Regions of the EU and North America. *Source*: UNICEF and WHO (2019).

- Wastewater treatment capacity lags far behind sewerage, and further still for piped water, and the difference increased from 2000 to 2017.
  - The population with sewerage but without wastewater treatment increased to slightly over 1.0 billion persons in 2017.
  - The population with piped water also increased by 1.0 billion persons from 2000 to 2017.
  - The population with piped water but without wastewater treatment increased to 1.86 billion persons in 2017.

If both the piped water and the total urban population will eventually be connected to a sewer system, which is the likely scenario, then the wastewater treatment capacity as a percentage of the piped water, and total population, will be lower than that reported as a percentage of the sewered population, as shown below for the data in Table 1.10 from Figure 1.25.

The percent wastewater treatment based on total sewered, piped water, and urban populations is calculated as

$$(1) \quad \frac{\text{Population with WWT}}{\text{Sewered Population}}(100\%) = \left(\frac{764{,}719{,}006}{1{,}776{,}284{,}190}\right)(100\%) = 43.1\%$$

$$(2) \quad \frac{\text{Population with WWT}}{\text{Population w/Piped Water}}(100\%) = \left(\frac{764{,}719{,}006}{2{,}611{,}939{,}859}\right)(100\%) = 29.3\%$$

$$(3) \quad \frac{\text{Population with WWT}}{\text{Urban Population}}(100\%) = \left(\frac{764{,}719{,}006}{3{,}266{,}887{,}800}\right)(100\%) = 23.4\%$$

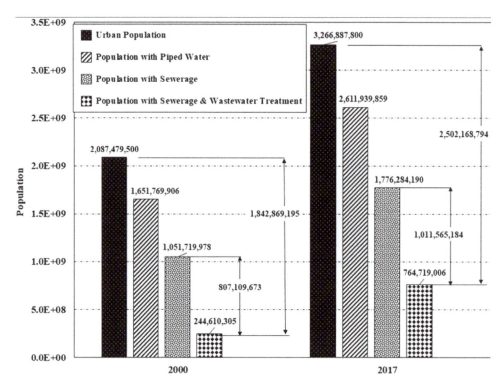

Figure 1.25 Urban population, population with piped water, population with sewerage, and population with sewerage and wastewater treatment for 2000 and 2017 for the SDG Regions of Central and Southern Asia, Eastern and South-Eastern Asia, Latin America and the Caribbean, Northern Africa and Western Asia, and Sub-Saharan Africa. *Source*: UNICEF and WHO (2019).

Table 1.10 Urban populations with piped water, sewerage, and wastewater treatment.

SDG Region	Urban Population	Population w/ Piped Water	Sewered Population	Percent of Sewered Population w/ WWT	Population w/ WWT
Central and Southern Asia	698,336,640	467,885,549	230,451,091	13	29,958,642
Eastern and South-Eastern Asia	1,332,228,100	1,105,749,323	839,303,703	55	461,617,037
Latin America and Caribbean	516,474,400	495,815,424	397,685,288	37	147,143,557
North Africa and Western Asia	310,783,060	276,596,923	239,302,956	49	117,258,449
Sub-Saharan Africa	409,065,600	265,892,640	69,541,152	13	8,741,323
Total	**3,266,887,800**	**2,611,939,859**	**1,776,284,190**		**764,719,006**

Although the percent coverage of wastewater treatment to sewered population increased to 43.1% in 2017, the population difference between the sewered and wastewater treatment populations increased by 1.0 billion persons, 200 million more than estimated in 2000. If sewerage does not keep pace with urban population growth, the provision of wastewater treatment will always lag behind as a result of the end-of-pipe paradigm, where wastewater is devalued in urban development.

1.3.4 Effectiveness of wastewater treatment in resource-challenged urban areas

The existence of a wastewater treatment plant does not guarantee it is operating correctly, or that it was designed and constructed to address the relevant public health, water quality and environmental problems, and the valorization possibilities of agricultural reuse. The following examples of wastewater treatment in Bolivia, Honduras, Burkina Faso, and Argentina are typical of the global situation in resource-challenged urban areas where reuse in agriculture is important.

1.3.4.1 Bolivia: waste stabilization ponds and wastewater reuse

A study of wastewater treatment and agricultural reuse in Bolivia, with an inspection of 84 wastewater treatment plants throughout the country, was performed in 2012 (Marka, 2012). The technologies included WSPs, Imhoff tanks, UASBs, tricking filters, and constructed wetlands, several installations with the combination of technologies (e.g., UASB followed by a facultative pond). The following conclusions and recommendations were reported:

(1) Of the 84 wastewater treatment plants inspected:
  - 34 were abandoned.
  - Of the remaining 53 in operation, 26 plants had less than 50% removal efficiencies for BOD_5 and TSS.
  - Disinfection of effluents was not practiced at most installations, and there were serious risks to public health from effluent discharges and reuse in agriculture.
  - Key problems encountered included:
    - Lack of training of operators on the understanding and control of treatment processes,
    - Deterioration of the physical structures from corrosion and lack of maintenance,
    - Insufficient budget to adequately operate the treatment plant,
    - Insufficient number of personnel, and
    - Existing personnel lack adequate training.
(2) Wastewater reuse with both treatment plant effluent and raw wastewater is a common practice in the arid and semi-arid water-scarce regions of Bolivia. This wastewater is an important resource as a supplementary source of water for agriculture (see Figures 1.26–1.28).
(3) Wastewater reuse in agriculture is common in the principal urban centers of Bolivia. It is estimated that approximately 5000 ha are irrigated with wastewater, with 53% of the area in the arid Department of Cochabamba.
(4) A technical evaluation of the potential for wastewater reuse in agriculture in Bolivia based on a technical and economic analysis of two detailed case studies in Cochabamba and Tarija demonstrated the following:
  (a) A great potential exists for (i) the safe reuse of wastewater as a solution to the problems of water scarcity that exist in the different regions of the country and (ii) as a driver for economic development.
  (b) Waste stabilization pond effluents can easily meet the necessary water quality requirements for unrestricted irrigation, maximizing the cultivated land area, and simplifying the operation and maintenance requirements for plant personnel.
  (c) It will be necessary to overcome certain existing barriers that would endanger the long-term sustainability of the waste stabilization pond-reuse system (e.g., generalized dissatisfaction of the population living in the vicinity of the country's wastewater treatment plants that were not well-operated and created foul odors, etc.).

Figure 1.26 An abandoned waste stabilization pond system in Tarata, Bolivia. The effluent discharges to a small river used for irrigation by farmers (indirect reuse). The health risks are high for farmers, their families, and the consumers of produce. Waste stabilization ponds are the recommended system for treatment and reuse in agriculture, but even the most appropriate, sustainable technologies require operation and maintenance.

Figure 1.27 A farmer pumps wastewater-contaminated water from the Rio Rocha near the city of Cochabamba for crop irrigation. This indirect irrigation with raw wastewater is common throughout the arid climates of Bolivia. Once again, the health risks are high for farmers, their families, and the consumers of produce.

Figure 1.28 An abandoned facultative pond at the Albarrancho wastewater treatment facility in the city of Cochabamba, Bolivia. The original waste stabilization pond system built in the 1980s was designed for agricultural reuse and discharge. The growth of the city and lack of pond desludging overloaded the system in the 1990s, and a new treatment system using UASBs with trickling filters is planned. The final effluent of the new plant will discharge to the contaminated Rio Rocha. No integrated reuse in agriculture is planned, although downstream farmers could have better quality water for irrigation if the final effluent of the new plant was used directly for irrigation.

1.3.4.2 Honduras: pathogen reduction in waste stabilization ponds

A monitoring study of 10 waste stabilization pond systems in Honduras, with emphasis on pathogen reduction and sustainability issues, was reported in 2005 (Oakley, 2005). Most of the pond systems consisted of a facultative pond followed by a maturation pond as shown in Figure 1.29. The results of the monitoring program showed the following:

(1) Helminth egg concentrations were significant in all pond influents, ranging from 9 to 744 eggs/L, with a mean concentration of 80.8 eggs/L for all systems. These values underscore the high prevalence of helminth infections throughout the country.
(2) No helminth eggs were detected in any pond effluent, and the mean $\log_{10}$ reduction was >1.9 assuming a detection limit of 1 egg/L.
(3) Helminth eggs were concentrated in facultative pond sludges, ranging from 1 to 4473 eggs/g total solids.
(4) *E. coli* $\log_{10}$ reduction ranged from 1.87 to 4.45, with a mean value of 2.97. Lower values were found in ponds with shorter hydraulic retention times as a result of bad design with a low hydraulic retention time, or long-term sludge accumulation causing hydraulic short-circuiting.
(5) All ponds met the WHO guidelines for restricted irrigation as a result of helminth egg removal.
(6) Two ponds met the WHO guidelines for unrestricted irrigation as a result of >4.0 $\log_{10}$ reduction in *E. coli* concentrations.
(7) In two systems, farmers used a portion of the final effluent for irrigation of crops in an informal manner without the support of the municipality.

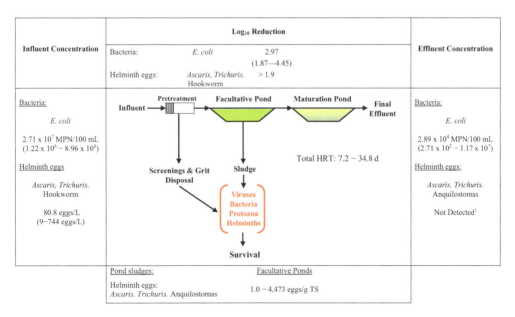

Figure 1.29 *E. coli* and helminth egg removal and survival in effluent and sludges from 10 wastewater stabilization pond systems in Honduras. Influent and effluent data are mean values (geometric for *E. coli* and arithmetic for helminths) for the 10 systems, with ranges in parentheses. Helminth egg concentrations were not detected in effluent samples, and $\log_{10}$ reduction was calculated as >1.9 assuming a detection limit of 1 egg/L. Pond sludges, however, had high helminth egg concentrations, up to 4473 eggs/g TS, and should never be used in agriculture but buried onsite at the treatment plant.

Table 1.11 Physical condition, operation and maintenance, sustainability of Honduras WSP systems.

Physical Condition	• Most systems are hydraulically and organically overloaded, without accurate flow measurement devices and grit chambers. • Several systems are in urgent need of desludging.
Routine Monitoring	• None of the systems measure flowrates or have sampling programs. • Only a few systems have attempted to monitor the accumulation of sludge in primary ponds.
Maintenance	• Most of the systems have adequate physical maintenance of the installation.
Personnel	• While most systems have permanent operators assigned to operate and maintain the installation, all lack training in the measurement of flowrates, sampling, and measurement of sludge accumulation.
Plans for Expansion and Sludge Removal	• None of the municipalities have plans for expansion, even though many are arriving at their hydraulic and organic limits. • No municipality has planned, let alone prepared a budget, for the desludging of primary ponds.
Sustainability	• Most installations have technical and financial support for maintenance, and most have public acceptance. • Wastewater reuse in agriculture has not been seriously considered as most systems were designed with the end-of-pipe paradigm for discharge to surface water. • The major problem in all municipalities is long-term planning for agricultural reuse, plant expansion, and sludge removal.

Table 1.11 presents a summary of physical conditions, operation and maintenance, and sustainability issues for the pond systems.

At the time of the monitoring study, many of the ponds were relatively new, with only a few years of operation, but various still had problems as shown in Table 1.11 because of hydraulic overloading, mostly due to under-design of the system as a result of assuming flowrates rather than measuring them. Unfortunately, at the present time (2022), several of the systems have been abandoned by the municipalities, or given the lowest priority for operation and maintenance (L. Eveline; C. Flores, personal communications) – a common occurrence in resource-limited cities where wastewater treatment has the lowest of priorities.

The monitoring data in Figure 1.29 for the 10 systems, however, show that waste stabilization pond systems can easily meet the WHO guidelines for restricted irrigation in spite of design and operational problems in resource-limited municipalities. As a result, WSPs are always recommended as the first choice in integrated wastewater reuse projects (WHO, 2006).

1.3.4.3 Ouagadougou, Burkina Faso: protozoan cyst and helminth egg removal in the WSP system

Parasite removal with WSPs in resource-limited communities in the Sahel region of West Africa is important to protect public health in a region where the prevalence of protozoan and helminth infections is high. At the same time, it is important to produce a valuable resource with treated wastewater for reuse in agriculture in a water-scarce and nutrient-poor environment, with high levels of poverty and malnutrition. To address these issues, a study on the protozoan cyst and helminth egg removal in a waste stabilization pond system on the campus of the International Institute of Water and Environmental Engineering (2ie) in Ouagadougou was reported by Konaté *et al.* (2013), with the following objectives:

(1) Determine the concentrations in raw wastewater, and the percent removal, of protozoan cysts and helminth eggs in an existing waste stabilization pond system treating domestic wastewater that is also reused for agriculture.

(2) Determine the concentrations of protozoan cysts and helminth eggs in pond sludges. This is particularly important since fecal and wastewater sludges are commonly used in agriculture in West Central Africa, which contributes to the high prevalence of parasite infections.
(3) Determine the viability and distribution of helminth egg species in pond sludge.

A photo of the anaerobic–facultative–maturation waste stabilization pond system on the campus of the International Institute for Water and Environmental Engineering is shown in Figure 1.30.

Figure 1.31 shows the results of the monitoring study, which are summarized below:

(1) The mean influent combined protozoan concentrations of *Entamoeba coli*, *Entamoeba histolytica*, and *Giardia lamblia* were high at 111 cysts/L, with a range from 4 to 327 cysts/L.
(2) No protozoan cysts were detected in the final effluent, and the $\log_{10}$ reduction was estimated at >1.74.

Figure 1.30 The anaerobic–facultative–maturation waste stabilization pond system at the International Institute for Water and Environmental Engineering (2ie), Ouagadougou, Burkina Faso. The system treated domestic wastewater for a population equivalent of 448 persons, with a mean flowrate of 55 m³/d, and a theoretical hydraulic retention time of 18 days.

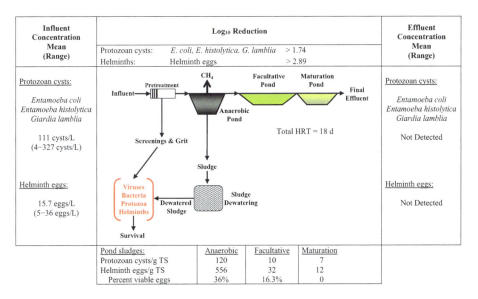

Figure 1.31 Protozoan cyst and helminth egg removal in effluent and sludges at a wastewater stabilization pond system in Ouagadougou, Burkina Faso (Konaté et al., 2013). Protozoan cyst and helminth egg concentrations were not detected in any effluent samples from the maturation pond. The $\log_{10}$ removal shown is based on mean values of measured concentrations from the facultative pond effluent: protozoan cysts = 2 cysts/L; helminth eggs = 0.02 eggs/L. Q = 55 m³/d. Population equivalent = 448 persons.

(3) The mean influent concentration of helminth eggs was 15.7 egg/L, with a range of 5–36 eggs/L.
(4) No helminth eggs were detected in the final effluent, and the $\log_{10}$ reduction was estimated at >2.89.
(5) High concentrations of protozoan cysts were found in all three pond sludges, with the highest value of 120 cysts/g TS found in the anaerobic pond sludge.
(6) Helminth eggs were also found in all pond sludges, with the highest concentration of 556 eggs/g TS found in the anaerobic pond sludge.
(7) The percent of viable eggs in the anaerobic and facultative ponds were 36 and 15.3%, respectively.

The data from this waste stabilization pond study show that WSPs can produce effluents without measurable concentrations of protozoan cysts and helminth eggs, which are major public health problems in the Sahel region of West Africa. Equally important, the effluent from well-designed WSPs can be used for restricted wastewater reuse in agriculture, where the effluent will have value for its water, nutrients, and organic matter, greatly helping low-income farmers. Pond sludges, however, have high concentrations of protozoan cysts and helminth eggs, and should be permanently buried onsite at the wastewater treatment facility.

This case study is an excellent example of the IWWM paradigm, of how a well-designed, sustainable waste stabilization pond system can solve a major public health problem, while at the same time creating valuable resources for agriculture.

A common configuration where effluent is to be reused for unrestricted irrigation is a facultative pond followed by two maturation ponds in series ($F/M_1/M_2$ system). The design strategy is to maximize pathogen removal through extended hydraulic retention time for sedimentation of particles, and exposure to solar radiation. The following two examples from Lima, Peru, and Mendoza, Argentina, show the successful application of pathogen reduction and agricultural reuse.

1.3.4.4 Lima, Peru: Vibrio cholera reduction in the San Juan de Miraflores WSP-reuse system

Figure 1.32 shows the results of a study where an actual bacterial pathogen, *Vibrio cholera* 01, was monitored throughout an $F/M_1/M_2$ system along with fecal coliforms. The monitoring took place during the 1991 cholera epidemic in Lima. At the peak of the epidemic in March 1991, *V. cholera* 01 was detected at concentrations as high as 4.3×10^5 MPN/100 mL in one of the principal wastewater collectors in Lima (Castro de Esparza *et al.*, 1992).

The monitoring study at San Juan de Miraflores, where the final effluent was being reused in aquaculture and agriculture, was implemented from June to August 1991. During the monitoring period, raw wastewater influent and effluent from each pond in series were monitored for fecal coliform and *V. cholera* 01 concentrations; the aquaculture ponds were also monitored for fecal coliforms and *V. cholera* 01, and tissue from tilapia cultivated in the ponds was monitored for the presence of *V. cholera*.

The results in Figure 1.20 show a 4.26 $\log_{10}$ reduction of *V. cholera* 01, and a 4.89 $\log_{10}$ reduction for fecal coliforms, with geometric mean effluent concentrations of 0.1 MPN/100 mL and 2.11×10^4 MPN/100 mL, respectively. *V. cholera* 01 was found in the aquaculture ponds at very low concentrations (0.03 MPN/100 mL), and the presence of *V. cholera* was detected in the tilapia tissue (skin, gills, intestines). Figure 1.33 shows the serial removal of fecal coliforms and *V. cholera* 01 through each pond in series during the monitoring period.

1.3.4.5 Mendoza, Argentina: Campo Espejo waste stabilization ponds with reuse in agriculture

Another example of an F/M/M system with agricultural reuse that has operated for decades is the Campo Espejo system in Mendoza, Argentina (Figure 1.34). The system was built in 1976 and upgraded in 1996 to meet the WHO guidelines for wastewater use in agriculture. The system consists of 12 parallel batteries of F/M/M ponds in series; the design flowrate was 146,620 m^3/d, with a theoretical hydraulic retention time of 21.1 days, and an irrigated area of approximately 3000 ha

Integrated wastewater management for reuse in agriculture

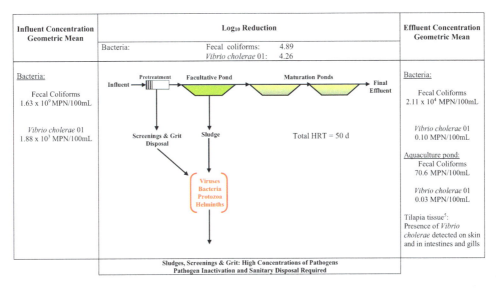

Figure 1.32 Reduction of fecal coliforms and *Vibrio cholerae* 01 in the waste stabilization pond system at San Juan de Miraflores, Lima, Peru, from June 9 to August 10, 1991 (Castro de Esparza *et al.*, 1992). Influent and effluent values, and values in the aquaculture pond, are geometric mean concentrations during the monitoring period.

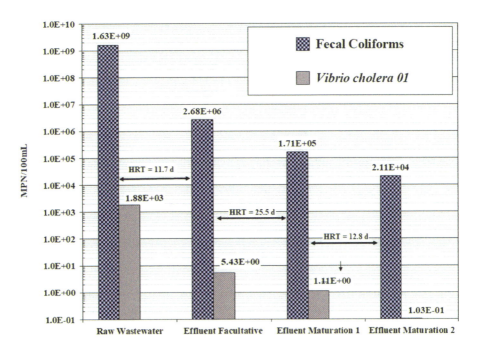

Figure 1.33 Reduction of fecal coliform and *Vibrio cholerae* 01 concentrations at the San Juan wastewater stabilization pond system in Lima, Peru, during the cholera epidemic in the months of June–August 1991. The influent flowrate was controlled during the monitoring period to maintain a hydraulic retention time of 50 days. The mean water temperature in the system averaged 17.5°C. *Source*: Data from Castro de Esparza *et al.* (1992).

40   Integrated Wastewater Management for Health and Valorization

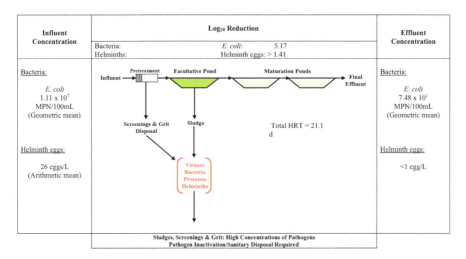

Figure 1.34 Reduction of *E. coli* and helminth eggs at the Campo Espejo waste stabilization pond system, Mendoza, Argentina. Mean values are from daily monitoring for 1 year (January 1, 2000 to December 31, 2000). The detection limit of helminth eggs was not mentioned by the author and the $\log_{10}$ removal of >1.41 assumes the detection limit was 1 egg/L. Approximately 3000 ha are irrigated with effluent from this system. *Source*: Data from Barbeito Anzorena (2001).

(Barbeito Anzorana, 2001; Vélez *et al.*, 2002). The system has been continuously monitored for fecal coliforms and helminth eggs to meet the WHO guidelines for agricultural reuse. Figure 1.34 shows the mean results for daily monitoring during 1 year (January–December 2000). Mean fecal coliform concentrations were reduced from 1.1×10^7 MPN/100 mL to 75 MPN/100 mL, a 5.14 $\log_{10}$ reduction, and helminth eggs from a mean of 26 to <1 egg/L, thus meeting the WHO effluent guidelines for unrestricted irrigation.

The examples in this section show that well-designed, constructed, and operated waste stabilization pond systems have been proven to reduce pathogen concentrations to produce effluents meeting the WHO guidelines for restricted and unrestricted reuse in agriculture. These systems have been recommended by government authorities for agricultural reuse in various countries, including Argentina, Bolivia, Burkina Faso, and India. The combined problems of public health and excreta-related infections, water scarcity and irrigation water supply, fertilizers, and sustainable agriculture, can be addressed in resource-limited urban areas as the paradigm shifts from end-of-pipe to IWWM.

1.4 THE SUSTAINABLE DEVELOPMENT GOALS AND INTEGRATED WASTEWATER MANAGEMENT

1.4.1 The 2030 Agenda for Sustainable Development.

The 2030 Agenda for Sustainable Development, adopted by the United Nations General Assembly in 2015, came with a set of interlinked goals designed to end poverty, protect the environment, and ensure prosperity for all (UN, 2019). The Agenda includes 17 Sustainable Development Goals (SDGs), with each having specific outcome targets with measurable indicators, to be achieved by 2030.

1.4.2 Sustainable development goals relevant for integrated wastewater management

Table 1.12 lists the SDG goals, targets, and indicators most relevant for IWWM with reuse in agriculture.

Table 1.12 Sustainable development goals relevant to integrated wastewater management with valorization (IWWM).

Goals and Targets	Indicators	IWWM Alternative
Goal 2. End hunger, achieve food security, improve nutrition, promote sustainable agriculture		
2.4 By 2030, ensure sustainable food production systems and implement resilient agricultural practices that increase productivity and production, that help maintain ecosystems, that strengthen capacity for adaptation to climate change, extreme weather, drought, flooding and other disasters, and that progressively improve land and soil quality	2.4.1 Proportion of agricultural area under productive and sustainable agriculture	Wastewater effluent treated in waste stabilization ponds (WSPs) with water and embedded nutrients used in productive agriculture
Goal 3. Ensure healthy lives and promote well-being for all at all ages		
3.3 By 2050, end the epidemics of AIDS, tuberculosis, malaria, and neglected tropical diseases and combat hepatitis, water-borne diseases, and other communicable diseases	3.3.5 Number of people requiring interventions against neglected tropical diseases	Pathogen reduction in WSPs lowers disease transmission of excreta-related infections
Goal 6. Ensure availability and sustainable management of water and sanitation for all		
6.2 By 2030, achieve access to adequate and equitable sanitation and hygiene for all and end open defecation, paying special attention to the needs of women and girls and those in vulnerable situations	6.2.1 Proportion of population using (a) safely managed sanitation services and (b) a hand-washing facility with soap and water	Pathogen reduction in WSPs with reuse in agriculture lowers excreta-related disease risks from surface waters
6.3 By 2030, improve water quality by reducing pollution, eliminating dumping and minimizing the release of hazardous chemicals and materials, halving the proportion of untreated wastewater, and substantially increasing recycling and safe reuse globally	6.3.1 Proportion of domestic and industrial wastewater flows safely treated 6.3.2 Proportion of bodies of water with good ambient water quality	Use of waste stabilization ponds designed for pathogen reduction for domestic wastewater treatment Treated wastewater reused in agriculture and not discharged to surface waters
6.4 By 2030, substantially increase water-use efficiency across all sectors and ensure sustainable withdrawals and supply of freshwater to address water scarcity and substantially reduce the number of people suffering from water scarcity	6.4.1 Change in water-use efficiency over time 6.4.2 Level of water stress: freshwater withdrawal as a proportion of available freshwater resources	Wastewater used for irrigation in place of groundwater or surface waters
6.5 By 2030, implement integrated water resources management at all levels, including through transboundary cooperation as appropriate	6.5.1 Degree in integrated water resources management	Wastewater treated in WSPs is considered a valuable water resource for agriculture
		Current Status
6.a By 2030, expand international cooperation and capacity-building support to developing countries in water- and sanitation-related activities and programs, including water harvesting, desalination, water efficiency, wastewater treatment, recycling, and reuse technologies	6.a.1 Amount of water- and sanitation-related official development assistance that is part of a government-coordinated spending plan	Rare in the form of IWWMV in resource-limited cities and peri-urban areas
6.b Support and strengthen the participation of local communities in improving water and sanitation management	6.b.1 Proportion of local administrative units with established and operational policies and procedures for participation of local communities in water and sanitation management	Rare for IWWMV, with a few good examples (e.g., Mendoza, Argentina)

Source: United Nations (2015) (https://sdgs.un.org/goals).

1.4.2.1 Goal 2: end hunger, achieve food security, improve nutrition, promote sustainable agriculture

The reuse of wastewater in agriculture can significantly contribute to the achievement of this SDG. Irrigation with wastewater produces higher crop yields than irrigation with freshwater, even with freshwater plus added nutrients (Figure 1.5). Higher yields promote improved food availability, which can lead to lower prices (WHO, 2006), enabling food security and improved nutrition.

Target 2.4 outcomes in Table 1.12 are to (i) ensure sustainable food production systems and (ii) implement resilient agricultural practices that increase productivity, maintain ecosystems, and strengthen capacity for adaptation to climate change. IWWM using WSPs for pathogen reduction, followed by reuse in agriculture, would ensure sustainable food production with a constant supply of water, nutrients, and organic matter for crops and soils. This would also foster the capacity to adapt to climate change.

1.4.2.2 Goal: 3 ensure healthy lives and promote well-being for all ages

The Target 3.3 outcome relevant to IWWM is to end the epidemics of neglected tropical diseases and combat water-borne diseases. Schistosomiasis, the soil-transmitted helminth infections of *Ascaris*, *Trichuris*, Hookworm, and excreta-related bacteria, protozoa, and viruses, can all be reduced in waste stabilization pond systems to meet the WHO guidelines for unrestricted or restricted irrigation (WHO, 2006). Thus, sustainable agriculture with wastewater irrigation would easily be combined with the reduction of the major excreta-related infections in resource-limited areas worldwide.

1.4.2.3 Goal 6: ensure availability and sustainable management of water and sanitation for all

The following targets are relevant to IWWM with reuse in agriculture:

- Target 6.2 Achieve access to adequate and equitable sanitation and hygiene for all
  The collection of wastewaters in sewerage, with treatment in waste stabilization pond systems, lowers excreta-related disease risks from surface waters.

- Target 6.3 Improve water quality and increase safe reuse globally
  Wastewaters are treated in waste stabilization pond systems for pathogen reduction; and all wastewater is used in agriculture and not discharged to surface waters, protecting water quality.

- Target 6.4 Increase water-use efficiency, ensure sustainable withdrawals
  Wastewater is used for irrigation in place of groundwater and surface water.

- Target 6.5 Implement integrated water resources management at all levels
  Wastewater effluents treated in waste stabilization pond systems are considered a valuable water resource for agriculture.

doi: 10.2166/9781789061536_0043

Chapter 2
Selection of natural systems for wastewater treatment with reuse in agriculture

2.1 INTRODUCTION

Domestic wastewater is the water supply of an urban area after it has been used and discarded as waste at the point of generation. In poor cities and peri-urban areas, the principal components of fresh wastewater are putrescible solid and dissolved organic matter deriving from excreta, microorganisms (including excreta-related pathogens), and nutrients (nitrogen and phosphorus) originating largely from urine. In large cities, organic matter from food wastes and nutrients from soaps and detergents are added to the waste stream, increasing its strength somewhat in terms of biochemical oxygen demand (BOD) and nutrient mass loadings.

The end-of-pipe paradigm has focused on wastewater treatment for surface water discharge, emphasizing constituent removal to protect water quality (e.g., BOD, total suspended solids (TSS), nitrogen (N) and phosporous (P) removal, etc.). The development of treatment evolved to what has been termed 'high-rate engineered processes', characterized by high flow rates with short hydraulic retention times (hours) (WHO, 2006), which includes primary treatment for TSS removal, secondary aerobic treatment for BOD removal, secondary sedimentation, and anaerobic digestion for primary/secondary sludge digestion. As discussed in Chapter 1, many add-on processes have been gradually inserted to the treatment train to remove specific constituents of concern, with disinfection for pathogen reduction being one of the last.

The integrated wastewater management paradigm focuses first on wastewater treatment for pathogen reduction to protect public health, and second, on the productive reuse of treated effluent in agriculture. (An excellent example of work on this theme is the detailed study of wastewater pollution, treatment, and reuse throughout Latin America published by the Pan American Center for Sanitary Engineering and Environmental Science (Egocheaga & Moscoso, 2004). The study concluded that the proper management of domestic wastewater in Latin America should focus on public health as a first priority, with the removal of pathogens as the principal objective of wastewater treatment. The study also concluded that in order to resolve the problem of agricultural demand for water and the sustainability of wastewater treatment in impoverished cities, the treatment of

© 2022 The Author. This is an Open Access book chapter distributed under the terms of the Creative Commons Attribution Licence (CC BY-NC-ND 4.0), which permits copying and redistribution for noncommercial purposes with no derivatives, provided the original work is properly cited (https://creativecommons.org/licenses/by-nc-nd/4.0/). This does not affect the rights licensed or assigned from any third party in this book. The chapter is from the book *Integrated Wastewater Management for Health and Valorization: A Design Manual for Resource Challenged Cities*, Stewart M. Oakley (Author).

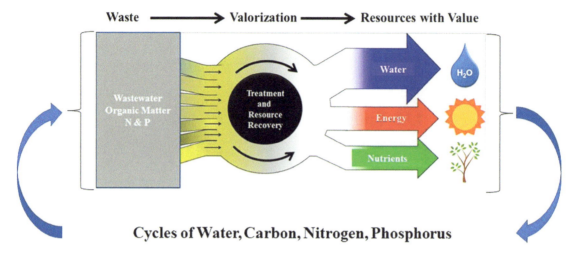

Figure 2.1 Integrated wastewater management incorporates the natural cycles of water, carbon, nitrogen, and phosphorus into the design with the objective of applying treated wastewater effluent to agricultural fields.

wastewater focusing on pathogen removal should be integrated with the productive reuse of the treated wastewater (CEPIS/OPS, 2000; Egocheaga & Moscoso, 2004). The study concluded that the best available technology for accomplishing this goal is wastewater stabilization pond systems, which can most easily meet the World Health Organization guidelines for wastewater reuse in agriculture than any other technology.) Natural system concepts used for the wastewater treatment and reuse include:

(1) Physical processes such as adsorption, sedimentation, and ultraviolet radiation from sunlight to enhance pathogen inactivation.
(2) Biological processes such as aerobic decomposition of organic matter by bacteria, with oxygen produced by algal photosynthesis, and the anaerobic decomposition of organic matter with the production of methane.
(3) The natural cycles of water, carbon, nitrogen, and phosphorous are incorporated into the design.
(4) Treatment is part of a larger system and is concerned with pathogen removal for agricultural reuse, not surface water discharge.
(5) Nitrogen and phosphorus are valorized as fertilizers for agricultural reuse.
(6) The energy characteristics of wastewater for the production and use of methane are also valorized under the appropriate conditions.

Figure 2.1 presents the circular concept of integrated wastewater management with valorization.

2.2 WASTEWATER CHARACTERISTICS AND TRADITIONAL LEVELS OF TREATMENT

2.2.1 Characteristics of domestic wastewater

Table 2.1 lists the main constituents of concern in raw domestic wastewater from small cities and peri-urban areas in order of importance for design. (It is assumed that there are no industrial waste discharges – this should be verified in the planning stages of a wastewater valorization project.)

Table 2.1 Constituents of concern in raw wastewater of small cities/peri-urban areas.

Constituent	Importance
Screenings and grit	Screenings are coarse solids in raw wastewater, which includes rags, paper, plastic, rubber, and vegetable matter that are removed by bar screens as influent enters a treatment plant; grit is comprised of solids with a specific gravity greater than putrescible organic matter (e.g., sand, gravel, coffee grounds) and is removed in grit chambers following the bar screens.Screenings and grit are not treatable and cause maintenance and operational problems at treatment plants if not removed; they are also highly contaminated with pathogens and must be buried onsite in small treatment plants, or buried offsite in sanitary landfills in large installations.
Pathogens (excreta-related)	Disease causing microorganisms present in wastewater that must be inactivated prior to reuse or, if necessary, discharge to surface waters. Includes pathogens in the broad categories of bacteria, helminths, protozoa, and viruses.Excreta-related infections are widespread in Africa, Asia, and Latin America, and pathogen removal or inactivation should be the primary purpose of treatment for wastewater reuse in agriculture or aquaculture.
TSS	Particulate organic matter, including microorganisms, that is highly putrescible. Historically removed by sedimentation, forming unstable primary sludge that is typically stabilized with anaerobic digestion.
Biodegradable organics (soluble and particulate)	Dissolved and particulate organic matter derived from excreta, soaps and detergents, and food wastes. Highly putrescible with foul odors. The biodegradable fraction is measured as the BOD, the principle parameter for the design and operation of biological processes used in wastewater treatment.If anaerobic processes are used as part of the valorization of wastewater, the capture of methane can be an important sustainable energy source.
Nutrients: Total N and P (TN, TP)	Nitrogen and phosphorus, deriving primarily from excreta, soaps, and detergents, can valorize the wastewater if it reused in agriculture and aquaculture for its fertilizer value, which can also contribute carbon offsets for substituting synthetic fertilizers.If wastewater is discharged to surface waters, nutrients can cause eutrophication, hypoxia, and ammonia toxicity in aquatic organisms. In North America and the EU, nutrient removal as a requirement for discharge has become common.

Source: Adapted from Metcalf and Eddy/AECOM (2014).

2.2.1.1 Screenings and grit

Screenings and grit always enter the collection system and must be removed as the first unit process, called preliminary treatment, and disposed daily by burial onsite or offsite using sanitary protocols because of their pathogen content (Oakley, 2018). Treatment plant designs should include onsite disposal of screenings and grit, and proper training of operators and maintenance workers. This is especially important in resource-challenged cities and peri-urban areas where open dumps are common, and where it cannot be assumed that materials will be buried offsite in a sanitary landfill that does not exist. It also cannot be assumed that all wastewater treatment plants are designed and built with consideration of screening and grit removal processes (Figure 2.2).

2.2.1.2 Pathogens

Excreted pathogens are the major constituent of concern if wastewater is to be valorized for reuse in agriculture. The urban areas of the world that can benefit most from wastewater valorization are also among the ones having the highest prevalence of excreta-related infections. Pathogen reduction was not introduced in the development of wastewater treatment in the US and the EU until the 1960s–1970s, when disinfection was borrowed from the water treatment industry and placed at the

Figure 2.2 Examples of screenings and grit accumulation in primary waste stabilization ponds designed and built without considering preliminary treatment. Screenings and grit contain high concentrations of excreted pathogens and must be disposed in a sanitary manner onsite or offsite. Top: Punata, Bolivia. Bottom: Chinendega, Nicaragua.

end of the wastewater treatment train (Feachem et al., 1983), where it still lies to this day. Designs using natural processes, however, such as waste stabilization ponds, can be designed specifically for pathogen reduction as an integral part of the treatment process.

2.2.1.3 Total suspended solids

TSS are particulate matter, approximately 80% organic, that are retained on a standard glass fiber filter (pore size $\approx 2.0\ \mu m$). After sedimentation in treatment processes, TSS form putrescible anaerobic sludges high in pathogen content that must be stabilized. Elevated concentrations of TSS discharged to surface waters can form anaerobic sludge deposits with deleterious effects on water quality and aquatic organisms.

2.2.1.4 Biodegradable organics

Dissolved and particulate organic matter can be biodegraded by three principle processes in wastewater treatment: aerobic oxidation, sulfate reduction, and methanogenesis. Table 2.2 ranks these processes by the standard Gibbs free energy yield per electron equivalent.

Aerobic oxidation yields the most free energy, and reacts more rapidly, than any other biochemical pathway, as long as oxygen is present. As a result, aerobic processes have traditionally been the process of choice for wastewater treatment. A key issue for valorization and sustainability is the type of process used to supply the oxygen, and the energy required to supply it.

In sulfate reduction, SO_4^{2-} serves as the electron acceptor rather than O_2, forming hydrogen sulfide, H_2S, which can be oxidized to sulfate to form sulfuric acid, H_2SO_4, by *Acidithiobacillus* if exposed to air. The formation of H_2S and H_2SO_4 is common in sewers, causing corrosion problems at the crown and inspection ports. In areas where high sulfate concentrations exist in the water supply, significant concentrations of SO_4^{2-} can be present in the raw wastewater; if an anaerobic process such as an upflow

Table 2.2 Principle biological processes in biodegradation of organics in wastewater.[1]

	ΔG^0 (kJ/e$^-$ eq)
Aerobic *Oxidation* $0.02C_{10}H_{19}O_3N + 0.25O_2 \rightarrow 0.18CO_2 + 0.02HCO_3^- + 0.02NH_4^+ + 0.14H_2O$ Organic Matter in Wastewater	−111.52
Anoxic *Sulfate Reduction* $0.02C_{10}H_{19}O_3N + 0.125SO_4^{2-} + 0.187H^+ \rightarrow 0.18CO_2 + 0.06H_2S + 0.06HS^- + 0.02NH_4^+ + 0.02HCO_3^-$ Organic Matter in Wastewater	−11.95
Anaerobic *Methanogenesis* $0.02C_{10}H_{19}O_3N + 0.11H_2O \rightarrow 0.125CH_4 + 0.055CO_2 + 0.02HCO_3^- + 0.02NH_4^+$ Organic Matter in Wastewater	−9.29

Source: Developed from Rittman and McCarty (2001) and Sawyer et al. (2002).
[1]In biological degradation of organic matter by heterotrophic bacteria, part of the carbon in the substrate goes to energy, and another to form biomass; the fraction transferred to biomass is not shown here.

anaerobic sludge blanket (UASB) is used as the first unit process in a treatment plant under these conditions, H_2S and HS^- will be produced instead of CH_4, although BOD_L removal will still occur.

Methanogenesis yields the least amount of energy of any heterotrophic pathway. Where did the original energy go? The majority of it is still contained in the CH_4 formed in methanogenesis. If this methane were completely oxidized, it would yield almost as much energy as that released by aerobic oxidation:

$$0.125CH_4 + 0.25O_2 \rightarrow +0.125CO_2 + 0.25H_2O$$

$\Delta G^0, kJ/e^-eq$
-102.24

Methanogenesis can potentially be an important valorization process for sustainable energy production for cooking in small treatment plants, and heating and electricity generation in large ones. Wastewater treatment can theoretically be a net producer of energy rather than a consumer, depending on the processes chosen for treatment and valorization (McCarty *et al.*, 2011). Unfortunately, the results in practice demonstrate for resource-poor urban areas it is better to emphasize reuse in agriculture with wastewater treated in aerobic natural systems, such as waste stabilization ponds, that avoid all of the problems of operating and maintaining anaerobic treatment systems.

2.2.1.5 Nutrients

Raw wastewater generally has total N concentrations ranging from 20 to 80 mg/L and total P concentrations from 3 to 12 mg/L. The conventional wastewater treatment was not developed to address nutrient removal and at best removes <50% of total nitrogen and phosphorus. Extensive monitoring of wastewater treatment plants designed for TSS and BOD removal in Brazil, for example, including activated sludge, waste stabilization ponds, and UASB reactors followed by polishing ponds, showed removal efficiencies ranging from 24 to 50% for total nitrogen (54 plants) and 23–46% for total phosphorus (76 plants) (Oliveira & von Sperling, 2008).

When treated wastewater is discharged to surface waters, increased nitrogen and phosphorus loadings contribute to eutrophication, with ammonia nitrogen also having toxicity effects on aquatic organisms – all of which are now happening as more urban areas worldwide build their first sewerage systems and subsequent wastewater treatment plants. In North America and the EU nutrient removal, or at minimum nitrification to eliminate ammonia toxicity, is often a requirement for discharge permits. At the same time, there is increasing interest in resource recovery of nutrients in wastewater by the chemical precipitation of phosphorus in the liquid stream, and production of struvite (magnesium ammonia phosphate) from digested sludges and sidestreams as discussed in Chapter 1.

The integrated wastewater management paradigm for reuse in agriculture avoids discharge to surface waters, along with all the complex processes of nutrient removal and/or recovery.

2.2.2 Levels of wastewater treatment

Traditionally, levels of wastewater treatment have been grouped together as preliminary, primary, secondary, tertiary, and advanced (see Table 2.3), with disinfection and anaerobic digestion included separately. In spite of the persistent use of these terms, it has been argued they are of little value, and that a better approach would be to establish the necessary degree of treatment for discharge or reuse, and then select the specific unit processes required to obtain it (Metcalf & Eddy/AECOM, 2014). Nevertheless, the terms primary, secondary, and tertiary/advanced treatment are still commonly used in textbooks, design manuals, and regulatory standards worldwide.

Table 2.3 Historical classification of levels of wastewater treatment.

Treatment Type	Description
Preliminary	Removal of screening and grit in raw wastewater with final disposal in landfills or onsite burial.
Primary	Removal of a portion of suspended solids ($\approx$60%) and organic matter ($\approx$40% as BOD_5) by sedimentation, producing primary sludge.
Secondary	Removal of biodegradable organic matter (soluble and particulate), and suspended solids, with aerobic processes and secondary sedimentation, producing secondary sludge. Commonly an activated sludge process.
Secondary with nutrient removal	Removal of biodegradable organic matter, suspended solids, and nutrients (N or P, or both), and secondary sedimentation, producing secondary sludge, and a chemical sludge if chemical precipitation of P is used. Commonly an activated sludge process.
Tertiary	Removal of residual suspended solids by filters or microscreens. Can also include nutrient removal and disinfection.
Advanced	Removal of dissolved and suspended solids after normal biological treatment when required for high-quality wastewater reuse applications.
Disinfection	Disinfection with chlorine, ultraviolet radiation, or ozone before effluent discharge or reuse. The last unit process in the treatment train.
Anaerobic digestion	Stabilization of primary and secondary sludges by the anaerobic process of methanogenesis; methane production is flared, or used as an alternative energy source for heating and electricity generation in large plants. Digested sludge is dewatered and disposed in landfills, or reused in agriculture, or as a soil conditioner, after pathogen reduction.

Source: Adapted from Metcalf and Eddy/AECOM (2014).

2.3 PATHOGEN REDUCTION IN WASTEWATER TREATMENT PROCESSES

2.3.1 High-rate treatment processes

High-rate wastewater treatment systems are engineered processes characterized by high flow rates and low hydraulic retention times on the order of hours rather than days; they also often require energy inputs, such as diffused air aeration for activated sludge processes. High rate processes include primary treatment for TSS removal, secondary treatment for BOD removal, secondary sedimentation, and finally disinfection as the last unit process before final discharge (Figure 2.3).

While the term 'high-rate treatment processes' implies engineering achievement, this is not necessarily true for the disinfection unit processes. As discussed in Chapter 1 in the quote by Feachem *et al.* (1983: 35), the conventional wastewater treatment was developed to remove organic matter and was never intended to achieve high removal of excreted pathogens.

The problem of pathogen removal continues to the present day as seen in the pathogen reduction data in Table 2.4 for various high-rate wastewater treatment processes. The ranges of $\log_{10}$ reductions for primary sedimentation, UASB treatment, activated sludge, and trickling filters range from 0 to 2, which is insufficient to meet the WHO guidelines for wastewater reuse in agriculture. The $\log_{10}$ reduction data for chlorine, ozone, and UV disinfection, which all require a very high-quality effluent entering the disinfection chamber, are also low, especially for helminth egg reduction, which is an essential requirement for wastewater reuse in agriculture.

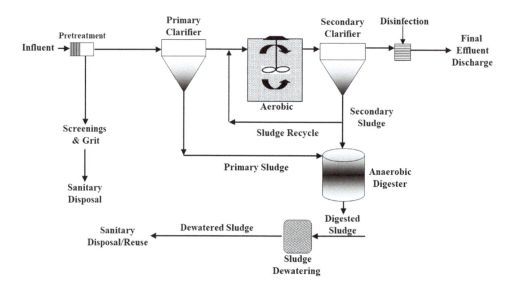

Figure 2.3 The conventional treatment flow diagram of activated sludge is a classic example of high-rate treatment processes. The primary clarifier, aeration basin, secondary clarifier, and disinfection chamber have hydraulic retention times ranging from 1 to 6 h. The disinfection unit process was added in the 1970s in Europe and the US but is still not common in treatment plants in resource-limited urban areas.

Table 2.4 Pathogen reduction for select wastewater treatment processes.

Treatment Process	Log$_{10}$ Reduction			
	Viruses	Bacteria	Protozoan (Oo) cysts	Helminth Eggs
High-rate processes[1]				
Primary sedimentation	0–1.8	0.1–0.4	0–0.1	0–1
UASB	0–1	0.5–1.5	0–1	0.5–1
Activated sludge with secondary sedimentation	1–2	0–2	1–2	0–1
Trickling filters with secondary sedimentation	0–0.5	0–1.4	0–0.8	0–1
Chlorine disinfection[2] (free available chlorine)	0–4	2–6	0–2[3]	0
Ozone disinfection[2]	3–6	2–6	1–2	No data
UV disinfection[2]	1 to >3	2 to >4	>3	0
Natural system processes				
Constructed wetlands	1–2	0.5–3	0.5–2	1–3
Waste stabilization ponds in series[4]	1–4	1–6	2–4	1–4
Batch stabilization reservoirs[4]	1–4	1–6	1–4	1–3

Sources: Feachem *et al.* (1983), Oakley (2018), Oakley and von Sperling (2017), Verbyla *et al.* (2017), and WHO (2006).
[1]High-rate treatment systems are engineered processes characterized by high flow rates and low hydraulic retention times, and usually include primary treatment for TSS removal followed by secondary treatment for BOD removal, and anaerobic digestion for primary and secondary sludges.
[2]Disinfection of a high-quality effluent from an activated sludge treatment plant.
[3]Protozoan cysts are exclusively *Giardia*. *Cryptosporidium* is resistant to free available chlorine.
[4]High-end values for each group of pathogens are for maximum reduction in optimally designed, functioning, and well-maintained systems.

2.3.2 Pathogen reduction data from operating high-rate treatment systems

The following case studies illustrate the continued problems with pathogen reduction in high-rate treatment systems, especially when there are high concentrations of pathogens in raw wastewaters, as is the case in much of the world outside of North America and Europe.

2.3.2.1 Activated sludge treatment plants without disinfection in Tunisia

Measured protozoan cyst (*Entamoeba coli*, *Entamoeba histolytica*, *Giardia*) and helminth egg (*Ascaris*, *Enterobius vermicularis*, *Taenia*) concentrations in four activated sludge plants in Tunisia not using disinfection are shown in Figure 2.4 (Oakley & Mihelcic, 2019). The results show high concentrations of cysts and eggs both in influents and effluents, with total $\log_{10}$ reductions of only 0.73 and 0.90, respectively. The data demonstrate that $\log_{10}$ reduction can vary greatly among different species of protozoa and helminths: the mean reduction of *E. coli* (1.28) was more than double that of *E. histolytica* (0.50) and *Giardia* (0.57), and reductions for both *Ascaris* (0.90) and *E. vermicularis* (1.13) were much greater than *Taenia* (0.64). The study concluded that wastewater treatment efficiency for parasite removal needs to be improved to protect public health in Tunisia, where the prevalence of protozoan and helminth infections is high (Ben Ayed *et al.*, 2009).

2.3.2.2 Activated sludge treatment plant with chlorine disinfection in the US

A conventional activated sludge plant with chlorine disinfection was monitored for removal of select pathogens that included *Giardia* and *Cryptosporidium* (Oakley & Mihelcic, 2019). The treatment plant had a mean flow rate of 155,000 m³/d and served a population of approximately 500,000; influent

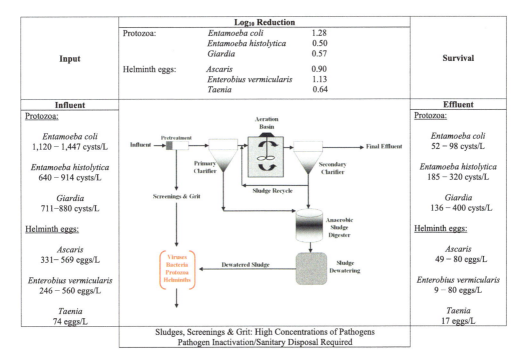

Figure 2.4 Protozoan cyst and helminth egg reduction at four conventional activated sludge treatment plants without disinfection in Tunisia. *Taenia* data are for one system only; all other data are from the four treatment systems. Influent/effluent ranges are from mean values for each treatment plant; mean values of $\log_{10}$ reductions are calculated from the geometric mean influent/effluent concentration for each plant. *Source*: Data from Oakley and Mihelcic (2019).

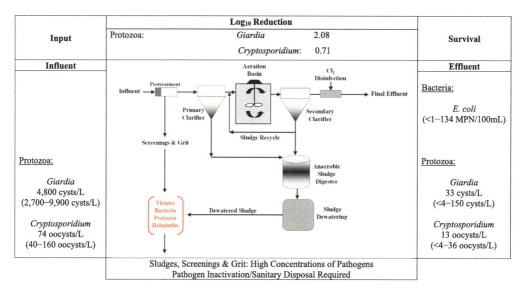

Figure 2.5 Reduction of *Giardia* and *Cryptosporidium* at an activated sludge treatment plant using chlorine disinfection in Arizona, US. Influent and effluent concentrations are mean values with ranges in parentheses. *E. coli* was only monitored in effluent samples and met the regulatory effluent requirement of <200 MPN/100 mL. *Source*: Data from Oakley and Mihelcic (2019).

and final effluent samples were collected once a month for 1 year. The results are shown in Figure 2.5 for influent and effluent concentrations, and $\log_{10}$ reduction for each (oo)cyst. Effluent *E. coli* concentrations are also shown; although influent concentrations were not monitored, *E. coli* was monitored as an effluent discharge requirement. The $\log_{10}$ reductions of *Giardia* and *Cryptosporidium* were 2.08 and 0.71, respectively. While effluent concentrations of *E. coli* were below 150 MPN/100 mL (<1–134 MPN/100 mL) for the study period, meeting the effluent discharge requirement, *Giardia* and *Cryptosporidium* concentrations in the final effluent were measured as high as 150 cysts/L and 36 oocysts/L, respectively.

2.3.2.3 Activated sludge treatment plants with microfiltration and disinfection in Spain

Figure 2.6 presents the results of the study of Ramo *et al*. (2017) for three activated sludge plants using microfiltration and disinfection, with two plants using UV disinfection, and one chlorine disinfection. The authors did not specify the microfilter pore size used, which can range from 0.05 to 2.0 μm, and which should easily filter protozoan cysts if operating properly. The results in Figure 2.6 show that even with microfiltration and disinfection, high concentrations of protozoan (oo)cysts can pass in the final effluent, with concentrations as high as 266 cysts/L for *Giardia*, and 47 oocysts/L for *Cryptosporidium*.

These brief case studies clearly demonstrate that high-rate wastewater treatment processes cannot guarantee adequate pathogen reduction for protozoan and helminth pathogens, especially when influent concentrations are high as encountered in resource-limited urban areas around the globe.

2.3.3 Natural system treatment processes

The most reliable and recommended natural system treatment processes are waste stabilization pond systems, consisting of facultative ponds followed by maturation ponds in series as shown in Figure 2.7. Several case studies on pathogen reduction in these systems were discussed in Chapter 1, Section 1.3.4. Well-designed and maintained pond systems should consistently remove 100% of helminth eggs,

Selection of natural systems for wastewater treatment with reuse in agriculture

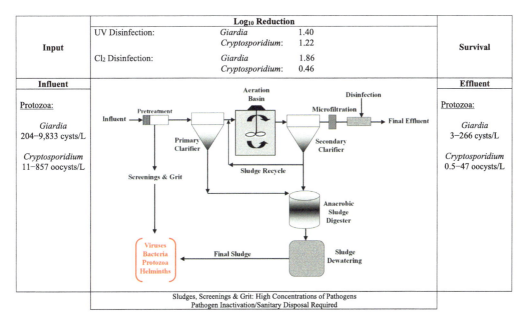

Figure 2.6 *Giardia* and *Cryptosporidium* reductions at three wastewater treatment plants in Spain using activated sludge, microfiltration, and disinfection (two plants used UV and one chlorine). Influent and effluent values are ranges, and $\log_{10}$ reductions are mean values. *Source*: Data from Oakley and Mihelcic (2019).

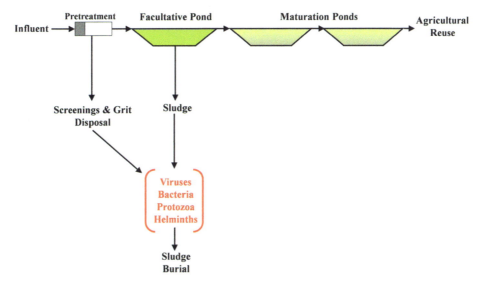

Figure 2.7 A waste stabilization pond system consisting of a facultative pond followed by two maturation ponds in series is an excellent example of natural system treatment processes capable of high pathogen reductions, If well-designed, with a total hydraulic retention time of $\geq$30 d in tropical and semi-tropical climates, the final effluent should have 100% removal of helminth eggs, and a 5.0–6.0 $\log_{10}$ reduction of *E. coli*, meeting the WHO guidelines for restricted irrigation and unrestricted irrigation, especially if a waiting period is used after the last irrigation (Chapter 7).

and a 4.0–6.0 $\log_{10}$ reduction of *E. coli*, and easily meet the WHO guidelines for restricted wastewater reuse in agriculture and unrestricted reuse if $\log_{10}$ reductions are at the high end for *E. coli*.

2.4 NATURAL SYSTEM TREATMENT PROCESSES FOR INTEGRATED WASTEWATER MANAGEMENT

The recommended natural systems for wastewater treatment designed for reuse in agriculture are presented in Table 2.5. Facultative ponds followed by maturation ponds in series are considered the

Table 2.5 Recommended natural systems for wastewater reuse in agriculture.

Natural System	Description
Facultative pond	Facultative ponds range from 1.5 to 2.0 m in depth, with hydraulic retention times ranging from 15 to 45 days. As raw wastewater enters a pond, organic suspended solids settle to the bottom, forming an anaerobic sludge layer. Soluble and colloidal organic solids in the wastewater are oxidized by bacteria in the aerobic and facultative zone in the first 1.0–1.5 m, where oxygen is produced by algae during the day. Carbon dioxide produced by bacterial decomposition is, in turn, used by algae as a carbon source. Anaerobic breakdown of solids in the sludge layer results in the production of CO_2, H_2S, and CH_4, which can be oxidized by aerobic bacteria as the gases rise to the aerobic zone. Facultative ponds combine three levels of wastewater treatment in one pond: primary sedimentation, secondary treatment (BOD removal), and anaerobic sludge digestion.

Maturation pond	Shallow (0.4–1.0 m) aerobic ponds following facultative ponds that are designed for pathogen reduction. Usually are baffled, or have high length to width ratios, to promote plug flow. Design hydraulic retention times are based on *E. coli* reduction kinetics and helminth egg removal overflow rates. Environmental conditions within maturation ponds contributing to pathogen reduction include:
	
	    Bacteria and viruses:
	      • Ultraviolet radiation from sunlight;
	      • High pH during daylight hours as a result of algal photosynthesis.
	    Protozoa and helminths:
	      • Sedimentation of protozoan (oo)cysts and helminth eggs.

Anaerobic pond	Deep (3–5 m) ponds with short hydraulic retention times (3–6 days), capable of removing 50–70% influent BOD, mostly by sedimentation of suspended solids. If used, anaerobic ponds should be covered for methane capture and flaring, for cooking, or for gas motors to generate electricity in large systems. Anaerobic ponds are not recommended for small cities, as most are abandoned after filling with sludge, with methane escaping to the atmosphere for years afterwards.
UASB reactor	UASB reactors can be operated by gravity with no energy input, with methane captured for its energy value. The risk of failure is high, however, with many systems abandoned as with anaerobic ponds.

Batch stabilization reservoir	Waste stabilization pond system final effluent is stored for 30 to >60 days in a batch mode stabilization reservoir to increase pathogen reduction prior to crop irrigation. Surface organic loadings ≤50 kg BOD_5/ha d are used to maintain aerobic conditions throughout reservoir depth, which ranges from 3 to 6 m. The following pathogen reductions have been reported:
	
	    • Fecal coliforms: 4.0–5.0 $\log_{10}$
	    • *Giardia* and *Cryptosporidium*: 4.0 $\log_{10}$
	    • Helminth eggs: 100% removal (3.0–4.0 $\log_{10}$).

Source: Adapted from Tchobanoglous and Schroeder (1985), von Sperling (2007), and WHO (2006).

best alternative for treatment, and ease of operation and maintenance. When land area limitations limit the use of facultative ponds, UASB reactors are preferred over anaerobic ponds for their ease of desludging and methane capture. Finally, the use of batch stabilization reservoirs can significantly increase pathogen reduction while wastewater is in storage during the non-irrigation season.

2.4.1 Facultative→maturation pond systems

Facultative ponds followed by maturation ponds in series are among the most common wastewater treatment systems worldwide. They have long been promoted as the most sustainable option for resource-limited cities and peri-urban areas, especially when wastewater effluent is to be used in agriculture. The arguments in favor of facultative/maturation systems include:

- Simplicity and low cost of construction, and operation and maintenance;
- Low sludge production and minimal handling of sludges;
- Minimal training of personnel, and minimal process monitoring requirements;
- Zero energy requirements for process operation;
- High resilience to process perturbations;
- Ability to consistently meet the WHO guidelines for wastewater reuse in agriculture.

2.4.1.1 Simplicity

Facultative/maturation pond systems are simpler to design, build, operate, and maintain than any other wastewater treatment system. Excavation for 2.0 m water depths in facultative, and 1.0 m in maturation ponds, is the main earth-moving activity. Minor concrete construction is used for preliminary treatment, inlet/outlet structures, revetments, and influent/effluent canals.

Operation and maintenance typically consist of routine tasks such as cutting vegetation on embankments, removal of floating scum and solids, daily flowrate measurement, and periodic monitoring of key influent and effluent constituents. Microbiological monitoring for pathogen reduction is best done by private laboratories on a routine basis.

2.4.1.2 Land requirements

The main disadvantage of waste stabilization pond systems is the required area. Table 2.6 shows estimates of the area required for the various treatment processes.

2.4.1.3 Low cost

Waste stabilization pond systems cost much less than any other treatment process for both construction, and operation and maintenance. Table 2.6 shows cost estimates for three different treatment systems in Bolivia as developed by Wagner (2010). The waste stabilization pond system is lowest in construction,

Table 2.6 Area required for various treatment processes.

Treatment Process	Required Area (m^2/person)
Activated sludge with primary and secondary sedimentation, and anaerobic sludge digestion	0.20–0.25
Extended aeration activated sludge with secondary sedimentation	0.15–0.2
Trickling filter with primary and secondary sedimentation, and anaerobic sludge digestion	0.2–0.3
UASB/secondary facultative/maturation	1.2–3.0
Waste stabilization pond system (facultative/maturation)	1.0–6.0[1]

Source: Adapted from Arceivala and Asolekar (2007).
[1] Area required depends on climate and level of treatment (e.g., number of ponds in series).

Table 2.7 Costs of select wastewater treatment systems in Bolivia.[1]

Wastewater Treatment System	Construction Cost	Operation and Maintenance Cost
Activated sludge with primary and secondary sedimentation	US$92/person	US$4.46/person yr
UASB/secondary facultative	US$30/person	US$1.22/person yr
Anaerobic/facultative/maturation	US$19/person	US$0.67/person yr

Source: Developed from Wagner (2010).
[1]Costs in 2008 dollars do not include the price of land for each system.

but the costs do not include the price of land, which can vary greatly, and pond systems will require much more land than activated sludge systems, for example. Mara (2003) has argued, however, that land purchase is an investment with stabilization pond systems, while operation and maintenance are annual costs that must be paid. Operation and maintenance for mechanized treatment plants can cost over six times that of waste stabilization pond systems as shown in Table 2.7.

2.4.1.4 Minimal sludge handling

Often the highest cost in the operation of secondary wastewater treatment plants, with primary and secondary sedimentation and anaerobic sludge digestion, is the management of process-produced sludges. An important advantage in the use of waste stabilization pond systems is the low sludge production, lower than any other treatment process as shown in Table 2.8. Sludges produced in mechanized treatment plants with anaerobic digestion have total solid concentrations ≤5% and must be dewatered in centrifuges or belt presses. Facultative pond sludges, however, remain in the pond for 10–20 years, where they decompose anaerobically, and gradually consolidate to total solid concentrations from 11 to 15% (Nelson *et al.*, 2004; Oakley, 2005).

As wastewater treatment-produced sludges will be contaminated with pathogens, helminth eggs, protozoan (oo)cysts, and bacterial and virus pathogens, an additional advantage of facultative ponds is the minimal handling of sludge, with lower concentrations of pathogens in the sludge due to its age. An activated sludge plant with primary and secondary sedimentation, and anaerobic digestion, would have to remove sludge with a frequency of at least once per month, with all the risks of handling, drying and disposing of the highly contaminated sludge. In contrast, sludge removal from a primary facultative pond is necessary only once every 10–15 years, when the pond is drained and the sludge removed with an excavator after drying *in situ* in the pond.

2.4.1.5 Process complexity and operation and maintenance requirements

Table 2.9 presents the level of complexity of various processes, and the requirements for staff training and process monitoring. Waste stabilization ponds have a low level of complexity because they

Table 2.8 Quantities of sludge produced by various unit processes.

Treatment Process	Sludge Production m³ of Wet Sludge per 1000 m³ of Treated Wastewater
Primary sedimentation	2.1–3.3
Activated sludge	1.4–1.9
Primary and secondary treatments with sludge digestion	2.6–3.9
Facultative waste stabilization ponds	0.4–0.6

Sources: Metcalf and Eddy/AECOM (2014) and Mara and Pearson (1998).

Table 2.9 Complexity, training, and process monitoring requirements.

Treatment Process	Level of Complexity	Training Level of Personnel	Process Monitoring Requirements
Activated sludge with primary and secondary sedimentation, and anaerobic digestion	High	High	High
Tricking filter with primary and secondary sedimentation, and anaerobic digestion	Medium–High	Medium–High	Medium–High
UASB waste stabilization ponds (UASB/secondary facultative/maturation)	Medium	Medium	Medium
Waste stabilization ponds (facultative/maturation)	Low	Low	Low

function autonomously as a natural process. As a result, personnel training and process monitoring requirements are minimal.

2.4.1.6 Energy consumption

Table 2.10 presents the energy requirements for operation of conventional activated sludge, extended aeration activated sludge, facultative/maturation ponds, and UASB facultative/maturation ponds. Many activated sludge plants around the world have failed as a result of the high costs of electricity for aeration (Figure 2.8), while natural system treatment plants operate within natural cycles such as photosynthesis and heterotrophic aerobic/anaerobic decomposition.

2.4.1.7 Process stability and resilience

Waste stabilization ponds, as a result of their large volumes, with long hydraulic retention times measured in days (15 to >35) rather than hours, have much more resilience to high organic and hydraulic loads, and to high concentrations of toxic compounds (Table 2.11). As a result, waste stabilization ponds are frequently used to treat high-strength industrial wastewaters such as those from pulp and paper mill wastes, brewery and winery wastes, dairy wastes, and meat-packing and feedlot wastes (Nemerow & Dasgupta, 1991). Thus, waste stabilization pond systems have a built-in safety factor that is especially valuable to resource-limited cities that do not have the resources to trouble-shoot problems of organic or hydraulic overloads, or of sensitivity to toxic compounds, that frequently occur in high-rate treatment processes.

2.4.2 Anaerobic→secondary facultative→maturation pond systems

Figure 2.9 shows the preferred design of a covered anaerobic pond for methane capture, followed by secondary facultative and maturation ponds in series for pathogen reduction.

Table 2.10 Energy requirements for wastewater treatment plants.

Treatment Process	Energy Consumption (kWh/person yr)
Activated sludge with primary and secondary sedimentation, and anaerobic digestion	12–15
Extended aeration activated sludge with secondary sedimentation	16–20
Waste stabilization pond system (facultative/maturation-1/maturation-2)	0
UASB waste stabilization pond system (UASB/secondary facultative/maturation)	0

Source: Adapted from Arceivala and Asolekar (2007).

Figure 2.8 An abandoned activated sludge treatment plant in an urbanization within the city of San Salvador, El Salvador. New developments often require wastewater treatment to be approved, and activated sludge is chosen to save land area for more housing. The cost of operation and need for skilled personnel, however, is normally beyond the resources of the urbanization, and treatment plants like this one quickly fail, if they were ever in operation to begin with.

Table 2.11 Resilience of select wastewater treatment processes.

Treatment Process	Sensitivity to High Organic Loads	Sensitivity to High Hydraulic Loads	Sensitivity to Toxic Compounds
Activated sludge with primary and secondary sedimentation, and anaerobic digestion	High	High	High
Tricking filter with primary and secondary sedimentation, and anaerobic digestion	Medium	Medium	Medium
UASB/waste stabilization ponds (UASB/ secondary facultative/maturation)	Medium	Medium	High
Waste stabilization ponds (facultative/ maturation)	Low	Low	Low

Anaerobic ponds are designed to be 3–5 m deep, with hydraulic retention times of 3–6 days. They are often recommended as the first ponds to use in a waste stabilization system, primarily as a means to remove the majority of influent BOD and suspended solids, allowing the secondary facultative pond to be smaller than a primary facultative pond as a result of a lower BOD loading (Mara, 2003; von Sperling, 2007). Anaerobic ponds do not significantly reduce pathogens, however, and a shorter retention time in the secondary facultative pond also reduces pathogen reduction potential, which is counter to the goal of wastewater reuse in agriculture. All anaerobic ponds should also be covered for methane capture, not only as an alternative energy source, but also to prevent methane, which has a global warming potential of 25, to escape to the atmosphere.

The most serious problem with anaerobic ponds is the removal of 2–3 m of sludge every 2–5 years. Figures 2.10 and 2.11 are typical examples the problems encountered in large systems with covered ponds and small systems in resource-limited municipalities.

Selection of natural systems for wastewater treatment with reuse in agriculture

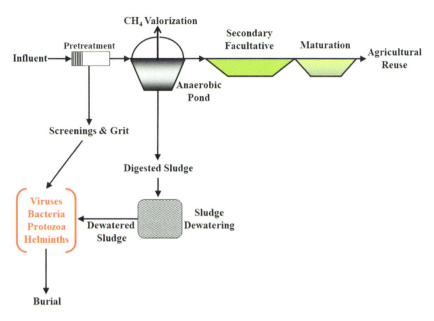

Figure 2.9 A covered anaerobic pond for methane capture, followed by a secondary facultative and maturation pond in series for pathogen reduction.

Figure 2.10 Top photo: An anaerobic pond/trickling filter wastewater treatment plant designed for an urban population of 60,000 in Sicuani, Peru. The four geomembrane-covered anaerobic ponds are operated in parallel, with the generated biogas burned in a flare rather than used as an alternative energy source. The anaerobic ponds were chosen to save area since the available land for wastewater treatment was limited – a common problem in small cities where wastewater treatment is not taken seriously by local governments. Bottom photo: After 5 years of operation, the geomembrane on one pond was removed for desludging. At the time of this writing, the pond had been uncovered and out of service for more than 6 months as plant personnel searched for an affordable solution for desludging and final disposal, neither of which were included in the original design or operations manual. *Source*: Photos courtesy of CONASIN SRL, Cusco, Peru.

Figure 2.11 Common examples of abandoned anaerobic ponds without methane capture, without a desludging plan, and without maintenance (top to bottom: Arani, Punata, and Tarata, Bolivia).

2.4.3 UASB→secondary facultative→maturation pond systems

Figure 2.12 shows the UASB/pond system monitored in Brazil over a 10-year period with excellent results for pathogen reduction (see Chapter 6, pp. 134–140). This system could be a good prototype for projects where land area is limited. Figure 2.13 is an example of a well-constructed UASB in Colombia.

UASBs can have serious problems with corrosion as shown in Figure 2.14. The best corrosion-resistant materials must be used in construction to avoid this potentially serious problem.

Selection of natural systems for wastewater treatment with reuse in agriculture

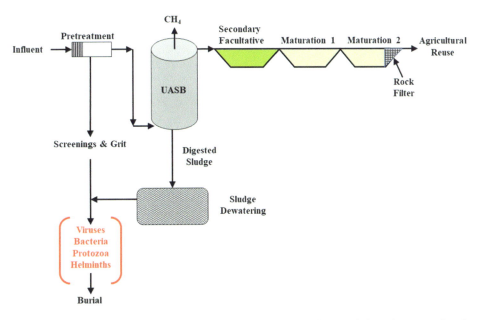

Figure 2.12 A UASB followed by a secondary facultative and two maturation ponds in series as used at the 10-year pilot project of the University of Minas Gerais-Copasa, Brazil (Dias *et al.*, 2014). The rock filter at the effluent end of the last maturation pond enabled an additional 0.59 $\log_{10}$ reduction of *E. coli*, for a total $\log_{10}$ reduction of 5.0 in the three ponds in series.

Figure 2.13 A well-designed and constructed UASB system for a population of 125,000. The methane is flared and not used as an alternative energy source (Rionegro, Colombia).

2.4.4 UASB→trickling filter→batch stabilization reservoir

Where land area is not available, or where there are steep slopes, a UASB/trickling filter could be used for BOD and TSS removal, with a stabilization reservoir used for pathogen reduction to meet restricted irrigation requirements, as shown in Figure 2.15.

Figure 2.14 Serious corrosion problems can arise with the production of H_2S and H_2SO_4 in UASBs as shown in this unit in Aracaju, Brazil. Raw wastewater monitoring for sulfate concentrations, and the use of corrosion-resistant materials throughout the UASB, must be specified by design engineers to avoid these problems.

Selection of natural systems for wastewater treatment with reuse in agriculture

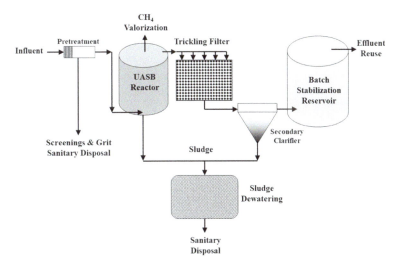

Figure 2.15 A UASB/trickling filter system with a batch stabilization reservoir to increase pathogen reduction to meet restricted irrigation recommendations.

Figure 2.16 A well-designed and built UASB/trickling filter wastewater treatment plant that has operated for over 25 years in Sololá, Guatemala. The produced biogas is used for cooking in various houses near the plant, and a maximum of 24 ha is irrigated with the effluent during the growing season. Unfortunately, a batch stabilization reservoir was never built, and the final effluent has concentrations of *E. coli* measured at 1.2×10^7 MPN/100 mL, posing serious health risks to the farmers, their families, and local consumers (see Section 7.2.2, pp. 205–211).

Two excellent examples of this system that have operated for over 25 years exist in Sololá, Guatemala, and one is shown in Figure 2.16. Both systems capture methane, which is used for cooking by residents livening near the treatment plants. The effluents from both plants are used for agricultural irrigation 6 months a year, with 100 farmers irrigating a total of 26 ha (see Section 7.2.2, pp. 205–2011).

Unfortunately, batch stabilization reservoirs were never built for these systems, and effluent concentrations of *E. coli* have ranged from 1.2 to 2.2×10^7 MPN/100 mL, posing serious health risks to the farmers, their families, and local consumers.

doi: 10.2166/9781789061536_0065

Chapter 3
Wastewater flows, design flowrate, and flow measurement

3.1 SOURCES OF WASTEWATER

The wastewater generated from small- to medium-sized cities, generally called domestic wastewater, has three important components as defined in Table 3.1.

The major component of wastewater in a collection system should, in theory, be domestic wastes from toilets, showers, laundry, and kitchens. These types of waste are also derived from commercial, institutional, public, and industrial facilities. Together, they comprise domestic wastewater.

Infiltration and inflow are waters that derive from rainfall and groundwater entering the sanitary sewer system from various sources. These waters can range from minor to major components from dry season to wet season and can be a major one year-round if there are many illegal connections between stormwater and surface drainage, which unfortunately is common in many parts of the world.

Industrial wastewaters are usually a small percentage of total flows in small- to medium-sized cities, but they can be significant depending on the water use of a particular industry. Even industries with small flows can have a negative impact on municipal wastewater treatment if the wastewater is very strong in terms of organic loads, such as in cheese processing plants or slaughterhouses.

3.2 WASTEWATER FLOWS
3.2.1 Domestic wastewater flow and urban water consumption

Wastewater flow is a direct result of water consumption and the fraction discharged to the sewer system. The distribution of water consumption gives insight into understanding the complexities of wastewater flows. As an example, Table 3.2 presents the consumption of water in the US for the key water uses excluding industry.

The total water consumption for typical values equals 550 L/cap d. This value is considerably higher than the per capita water consumption used for design in various countries as shown in Table 3.3.

Historically in the US, average daily and per capita water consumption, in combination with measured or estimated flow data for infiltration/inflow and industrial wastewater, were used to

© 2022 The Author. This is an Open Access book chapter distributed under the terms of the Creative Commons Attribution Licence (CC BY-NC-ND 4.0), which permits copying and redistribution for noncommercial purposes with no derivatives, provided the original work is properly cited (https://creativecommons.org/licenses/by-nc-nd/4.0/). This does not affect the rights licensed or assigned from any third party in this book. The chapter is from the book *Integrated Wastewater Management for Health and Valorization: A Design Manual for Resource Challenged Cities*, Stewart M. Oakley (Author)

Table 3.1 Components of wastewater from small- to medium-sized cities.

Component	Description
Domestic wastewater	Wastewaters discharged from residences, commercial, institutional, public, and industrial facilities
	• Domestic wastewater is comprised of human excreta flushed from toilets/urinals, bathing and laundry wastewaters, and food preparation wastes
Infiltration/Inflow	Water that enters the collection system through direct and indirect means
	• Infiltration is extraneous water that enters the collection system through leaking joints, cracks, or breaks in sewer pipes
	• Inflow is stormwater entering the collection system from storm drain or catch basin connections, foundation/basement drains, and sewer access covers
	• In many parts of the world, storm drainage is illegally connected to the sanitary sewer system causing excess flows that can damage the collection system and treatment plants
Industrial wastewater	Wastewater produced from industrial process wastes
	• In small to medium cities, this is often a minor component compared to domestic wastewater and stormwater inflow but can be a major problem if certain industries are present (e.g., slaughterhouses)

Source: Metcalf and Eddy/AECOM (2014).

Table 3.2 Municipal water uses and consumption in the US.

Use	L (cap d) Range	L (cap d) Typical Value
Domestic	150–300	250
Commercial	40–300	150
Public	60–100	75
Loss and waste	60–100	75
Total:		**550**

Source: Metcalf and Eddy/AECOM (2014).

estimate wastewater flows (WPCF, 1969). Design per capita flowrates developed for 35 cities and regions across the US, from 1937 to 1965, ranged from 348 to 1134 L/cap d (WPCF, 1969), underscoring the effects of local conditions (e.g., infiltration/inflow, shallow groundwater, and use of basements) on wastewater flows. While water consumption is an important factor in wastewater generation, it is not the only one, and flows can be much greater than estimates as a result of local conditions. Monitoring of wastewater flows, rather than estimating them by using national or regional guidelines, has long been recommended as the most reliable method (ASCE, 1959).

Today in the US and EU essentially all design flows are developed from detailed flow monitoring of collection systems or existing wastewater treatment plants. In the case of new developments, flowrates are derived from measured flowrates, population data, and per capita flowrates from similar nearby cities (Metcalf & Eddy/AECOM, 2014). In other regions of the world, engineers, development specialists, and government authorities, both local and foreign, do not require verified flow monitoring with design documents, even though it is usually specified in design standards (but never enforced). This significantly contributes to inadequate designs resulting in widespread underperformance and failure.

Table 3.3 Water consumption design values for urban areas in various countries.

Country/Organization	Water Consumption Classification		Design Values L (cap d)
Bolivia	Population:	2000–5000 5000–20,000 20,000–100,000 >100 000	50–120 80–180 100–250 150–350
Brazil	Population:	Small >100,000	80 150–350
Honduras	City size:	Small–intermediate Large	100–150 200
India	City size:	Megacities Non-metropolitan	$\geq$150–200 135
Mexico	Socioeconomic class:	Residential Middle class Lower class	300 205 130
Peru	Climate:	Cold Temperate/Tropical	120–200 150–250
WHO	Optimal consumption for minimal health risks. Drinking, cooking, and hygiene needs are met		$\geq$100–300

Source: CNA (2007); DIGESBA (2001); Howard and Bartram (2003); Mendonça (2000); Shaban and Sharma (2007).

3.2.2 Infiltration and inflow

Infiltration and inflow can cause significant increases in wastewater flows in old sewer systems, and in poorly constructed ones, during dry and especially during wet seasons. This wastewater flow increases the impact of wastewater treatment plant processes and causes sanitary sewer overflows, contaminating the environment and posing public health risks. The USEPA has estimated that between one-third and two-thirds of the US sanitary sewer system have problems with sanitary sewer overflows and that approximately 40,000 sewer overflows occur per year (Lai, 2008). Most of the infiltration/inflow problem is due to the aging sewer infrastructure built with vitrified clay, brick and concrete, and materials still used widely around the world for new sewer construction.

Infiltration and inflow during the wet season can be much higher than the dry weather flow and volumes should be determined by flow measurement in the sewer system or at the wastewater treatment plant (a task seldom if ever performed in cities worldwide where missing access port covers is a common sight). When flow measurement is not possible design procedures have been developed using infiltration allowances per length of pipe or per service area of collection (WPCF, 1969). Figure 3.1 and similar figures are used for infiltration design per unit service area for new and old sewer systems, but results are susceptible to large errors due to site-specific conditions. Flow measurement is always preferable, especially so in cities where illegal sewer connections of wastewater and stormwater are common, and sewer maintenance is lacking.

3.2.3 Industrial wastewater flows

Industrial wastewater flowrates vary greatly with the type and size of industry. In small- to medium-sized cities, industrial wastewater is usually not a significant part of the collection system wastewater flow. In cases where various industries exist, flowrates and constituent loadings need to be monitored, ideally with the help of the industry. Small flows from some industries, however, can have high BOD_5 and TSS loadings that can affect treatment processes. Examples commonly found in small cities include slaughterhouses and cheese processing plants (Figure 3.2).

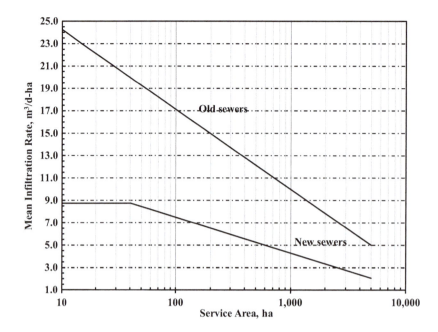

Figure 3.1 Mean infiltration allowances for new and old sewers. New sewers have precast access ports and pipe joints with flexible gaskets, while old sewers have cement mortar joints and access ports constructed of brick masonry. Flow measurement is always preferable to the use of infiltration allowances per service area or per length of pipe. Adapted from Metcalf & Eddy (1981).

Figure 3.2 A facultative pond receiving domestic wastewater from the municipality of Huata, Peru (Population ~1600) overloaded by discharges of cheese processing wastewaters. In this case, a small cheese processing plant had much stronger wastewater than the design population.

Table 3.4 Flowrate parameters for design of wastewater treatment facilities.

Parameter	Description	Purpose
Mean dry weather flow	Mean of daily flow data for dry periods	Sizing of unit processes for dry/wet flow (e.g., parallel units)
Mean wet weather flow	Mean of daily flow data for wet periods	
Mean daily flow	Mean daily flow over a 24 h period based on annual data	Development of flow ratios: (Q_{max}/Q_{mean}); (Q_{min}/Q_{mean})
Maximum daily flowrate	Maximum daily flow over a 24 h period based on annual data	Determine the need for an equalization basin
Peak hourly flowrate	Peak sustained hourly flowrate during a 24 h period	Sizing of physical unit processes: grit chambers, sedimentation basins; filters. Sizing of channels and weirs
Maximum sustained flow or load	Flow or mass loading rate sustained for a given time (4, 8, and 12 h)	Design of biological processes
Minimum daily flowrate	Maximum daily flow over a 24 h period based on annual data	Sizing of influent channels and grit chambers to control solids settling

Source: Adapted from Metcalf & Eddy/AECOM (2014).

3.3 DESIGN FLOWRATE

3.3.1 Design flowrate from wastewater flow data: the ideal case

The key flowrate design parameters that need to be developed from flow data are listed in Table 3.4. These parameters are necessary for the proper sizing and hydraulic design of unit processes, piping, pumping, and various flow appurtenances. Figure 3.3 shows several of the design parameters measured over a 24 h period: mean daily flow, peak hour flow, minimum hour flow, and maximum sustained flow for a 12 h period. In areas with inflow during wet seasons, there will be wet weather

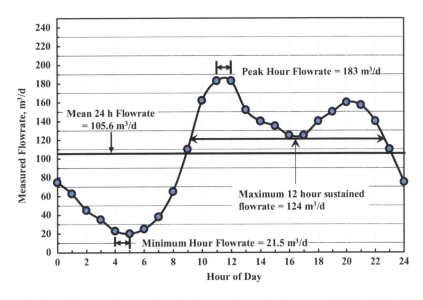

Figure 3.3 Example diurnal dry weather wastewater flow pattern showing mean (24 h), minimum hour, peak hour, and maximum sustained (12 h) flowrates. Adapted from Metcalf & Eddy/AECOM (2014).

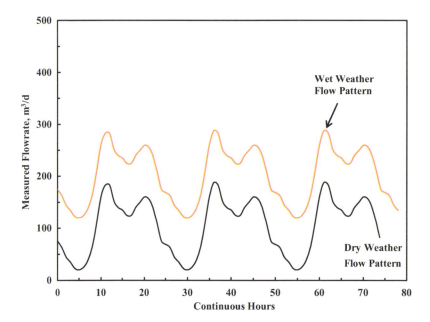

Figure 3.4 The wet weather flow pattern, if there is considerable inflow, will typically mirror the dry weather pattern with increased flows. Treatment facilities should be designed for wet weather flows under these conditions.

and dry weather flow patterns as shown in Figure 3.4. A treatment facility in this case needs to be designed for the wet weather flows.

Figure 3.5 is an example of direct inflow effects from rainfall on wastewater flows. The dry weather flow is assumed to have steady infiltration and inflow, and a major rainfall event causes an instantaneous peak inflow over three times the dry weather peak flowrate. Without continuous flow monitoring during storm events, this effect may go unnoticed. This peak inflow from stormwater is common in cities where sewers are poorly maintained and access port covers are missing; when the day comes when a wastewater treatment plant is built in such a city without flow monitoring data, the plant will likely fail as the ones shown in Figures 3.6 and 3.7.

For accurate engineering design, flowrates should be measured continuously, for long periods during dry and wet seasons, to develop the parameters shown in Table 3.4. Flowrates should be measured directly in the collection system or at a wastewater treatment plant if one already exists.

All large cities have sewer systems that must be maintained (adequately or poorly), and the means, if not the wherewithal, should be available to monitor flows, especially at existing wastewater treatment facilities. In the case where flow data are not available and not easily attainable, data from wastewater treatment facilities in similar cities could be used – if conditions are similar (Metcalf & Eddy/AECOM, 2014).

Flowrates measured directly from the collection system should use 24 h monitoring during different days of the week, and during wet and dry seasons, to have sufficient measured data to estimate mean, peak hour, maximum sustained, and minimum flowrates. Most country wastewater treatment design standards require flow monitoring to determine design flows, and an example from Peru is shown in Table 3.5. Unfortunately, as discussed previously, this requirement is not enforced.

In all but the very largest cities flow monitoring is never performed for design, and even flowrates at existing treatment plants typically are not monitored (Oakley, 2004). Monitoring is a difficult task in small- to medium-sized cities where the collection system lacks proper maintenance, and even

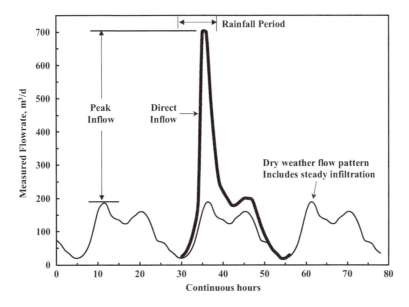

Figure 3.5 Graphic identification of direct inflow effects from rainfall on wastewater flows by continuous flow monitoring. In this example, the dry weather flow is assumed to have steady infiltration and inflow and a major rainfall event causes a peak inflow over three times the dry weather peak flowrate. Adapted from Metcalf & Eddy/AECOM (2014).

Figure 3.6 A secondary clarifier in an activated sludge treatment plant filled with activated sludge from the aeration basin that was washed out due to excessive inflow into the collection system from Hurricane Stan (Panajachel, Guatemala). The plant had no bypass channel and no pump to return the sludge to the aeration basin after the storm event; as a result, it was discharged to the anaerobic digester. Re-startup of aeration basins with raw wastewater can take many days to several weeks if there is no source of activated sludge from a nearby plant, which was the case here as this plant was the only one in the country. This problem is unlikely to happen in waste stabilization ponds with their large volumes and long retention times of weeks rather than 6 h in activated sludge plants.

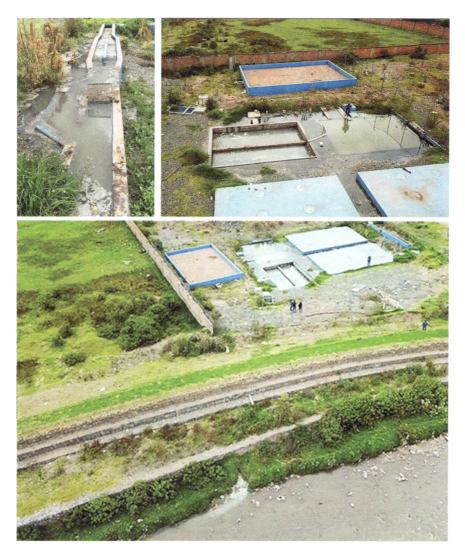

Figure 3.7 This wastewater treatment plant was designed and built without flow monitoring using population and per capita estimates for the design flowrate. Within a year after construction, the plant, which never had an operator, was inundated: first, the grit chamber filled and overflowed; next, the wastewater flowed over the anaerobic reactors and filled and clogged the media filters; after overflowing the filters, it drained into a sump with a gravity discharge pipe to the river – a case of wastewater flowing over the treatment facility rather than through it. Photos courtesy of CONASIN SRL, Cusco, Peru.

an existing wastewater treatment plant may not have flow measuring capabilities; local engineers additionally lack experience in flow monitoring and availability of equipment such as area velocity meters that can be placed in sewers or access ports for 24 h monitoring. Under these conditions, the design flowrate is estimated by assuming per capita wastewater flow, infiltration by pipe length and diameter or service area, and industrial flows by local knowledge.

Table 3.5 Design flowrate monitoring requirements for Peru.

Monitoring Objective	Number of Monitoring Events	Monitoring Days	Duration per Event	Outcomes
Initial wastewater characterization for select sewer discharges	5	Different days of the week	24 h	• Preliminary mean flowrates • Mass loading of select parameters
Wastewater characterization for design flowrates	5	Representative days with the highest flowrates	24 h	• Mean and peak hourly flow • Mean flowrate per capita • I/I flowrate • Industrial wastewater flows

Source: Ministerio de Vivienda, Construcción, y Saneamiento (n.d.)

3.3.2 Design flowrate by equation: the non-ideal case (but most common)

When flowrates are not monitored they must be estimated from domestic water consumption, assumed infiltration rates per length or area of the collection system, and from known industrial wastewater discharges. Typical design equations for estimating mean daily wastewater flowrates are shown in Equations (3.1) and (3.2) (modified from studies by DIGESBA, 2001; Mendonça, 2000):

$$Q_{mean} = C \cdot P \cdot q + (q_{inf,L})L + \sum_{i}^{n} Q_{ind,i} \tag{3.1}$$

$$Q_{mean} = C \cdot P \cdot q + (q_{inf,A})A + \sum_{i}^{n} Q_{ind,i} \tag{3.2}$$

Q_{mean} is the mean daily domestic wastewater flowrate, m³/d; C is the return coefficient for daily water consumption to sewer (usually 0.8 assumed); P is the design population; q is the per capita water consumption, m³/cap d; $q_{inf,L}$ is the infiltration into sewer system based on collector length, m³/d m; L = total length of pipe in collection system, m; $q_{inf,A}$ is the infiltration into sewer system based on total service area, m³/d ha; A is the service area of the collection system, ha; and $Q_{ind,i}$ is the flowrate of industry i, m³/d.

Equations such as these are oftentimes used without checking data on actual water consumption since those data may not exist. Many small cities (and some large ones) charge a flat rate per month without metering consumption, and as a result, consumption is high with much wastage. This is especially true for cities in the mountains, where water sources from springs do not pass through a water treatment plant, where flow monitoring should occur (and still may not). It is also true in rapidly growing urban and peri-urban areas where connections, legal or illegal, occur faster than the ability to meter them.

Values for sewer infiltration based on diameter and length of pipe, $q_{inf,L}$, or service area, $q_{inf,A}$, are published in national and regional design standards, with data for pipe infiltration the most common. Design engineers will need detailed plans of the collection system, however, to use published values of $q_{inf,L}$, and in older systems, they may not be available.

Without flow measurement, none of the flowrate parameters listed in Table 3.4 can be determined empirically and must be estimated from published data that relate calculated mean flowrate and population to maximum, minimum, and peak flows. Once the mean design flowrate is calculated, the maximum and minimum daily flowrates, and the peak hourly flow, are estimated from published charts such as those shown in Figures 3.8 and 3.9.

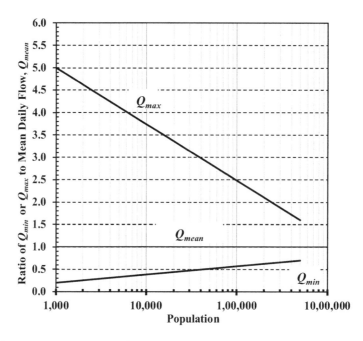

Figure 3.8 Ratio of minimum and maximum flows to mean daily wastewater flow for prevailing dry weather flow conditions. Graphs were developed from various cities and regions in the US and recommended for design by various authorities. Redrawn from WPCF (1969).

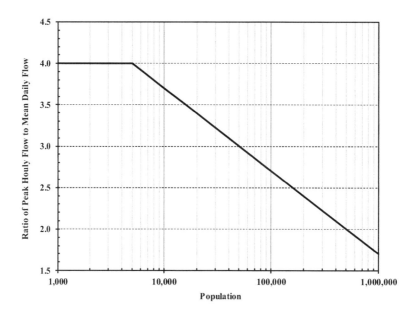

Figure 3.9 Ratio of peak hourly flow to mean daily flow for domestic wastewater in the US. This figure was developed from an analysis of records of numerous cities throughout the US and is based on mean residential flowrates with small contributions of commercial and industrial wastewaters; infiltration/inflow flowrates were excluded. Redrawn from Metcalf & Eddy/AECOM (2014).

3.4 DESIGN EXAMPLE: DESIGN FLOWRATES FOR THE CITY OF TRINIDAD, HONDURAS

The design flowrate for the city of Trinidad, Honduras, must be determined for the design of a waste stabilization pond system. It is assumed that no flow monitoring of the existing collection system has been performed, as is typical for most small cities; as a result, the design engineer must use Equation (3.2) to estimate mean, maximum, and minimum daily flowrates, and peak hourly flowrate. The following conditions are assumed to apply to Equation (3.2):

- $P = 6108$;
- $C = 0.8$;
- $q = 0.100$ m³/cap-d from Table 3.3 for Honduras; this is a small city and water consumption is considered to be low per capita;
- $A = 53.3$ ha calculated from Figure 3.8; and
- It is assumed the collection system is relatively new and in good condition (Figure 3.10).

Using Equation (3.2),

$$Q_{mean} = C \cdot P \cdot q + (q_{inf,A})A$$

$C = 0.8$;
$P = 6108$ (design population); $q = 0.100$ m³/cap d;
$A = 532\,793$ m² $= 53.3$ ha; $q_{inf,A} = 8.0$ m³/d-ha; (Figure 3.1 for new sewer);

Q_{mean} = (0.8)(6108)(0.100 m³/cap-d) + (8.0 m³/ha d)(53.3 ha)
= 488.6 m³/d + 426.4 m³/d = 915 m³/d

From Figure 3.8 for $P = 6108$,

$Q_{max}/Q_{mean} = 4.0$
$Q_{min}/Q_{mean} = 0.8$

$Q_{max} = (4.0)(915$ m³/d$) = 3660$ m³/d
$Q_{min} = (0.8)(915$ m³/d$) = 732$ m³/d

Figure 3.10 The service area for Trinidad is estimated to be 53.3 ha using Google Earth. In practice, the municipal plans of the city should be used to calculate service areas and lengths of sewer pipes for use in Equations (3.1) or (3.2).

Using different assumptions for per capita flowrate ($q = 0100, 0.150, 0.200$ m³/cap d), and a new or old collection system ($q_{inf, A} = 8.0$ or 20.0 m³/d ha from Figure 3.1), the engineer could have arrived at a value of Q_{mean} ranging between 915 and 2043 m³/d, Q_{max} from 3660 to 8173 m³/d, and Q_{min} from 732 to 1635 m³/d, as shown in the table below.

Q (m³/cap d)	C·P·q (m³/d)	$q_{inf, A}$ (m³/ha d)	$(q_{inf, A})(A)$ (m³/d)	Q_{mean} (m³/d)	Q_{max} (m³/d)	Q_{min} (m³/d)
0.100	488.6	8.00	426	915	3660	732
0.100	488.6	20.0	1066	1555	6219	1244
0.150	733.0	8.00	426	1159	4637	927
0.150	733.0	20.0	1066	1799	7196	1439
0.200	977.3	8.00	426	1404	5615	1123
0.200	977.3	20.0	1066	2043	8173	1635

Measured flowrates for the city of Trinidad over a continuous 3-day period with an area velocity flow meter are shown in Figure 3.11. The measured mean flowrate of 1816 m³/d is double the value first calculated above and higher than all calculated values in the table except for the last assumption assuming the highest per capita flowrate and highest infiltration rate (0.200 m³/cap d and 20 m³/ha d). The use of Equations (3.1) and (3.2) in obtaining the best design flowrate without flow monitoring requires either specific local knowledge or good luck, which is unlikely.

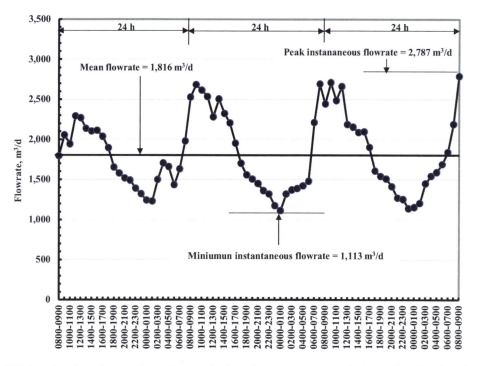

Figure 3.11 Results of continuous three-day monitoring of raw wastewater discharge for the city of Trinidad, Honduras. The flow was measured with an area velocity meter with ring inserts such as that shown in Figure 2.12 (*Source*: ECOMAC (2004)).

3.5 CASE STUDY: DESIGN FLOWRATE FOR SAYLLA, PERU

Saylla is a rapidly growing peri-urban district adjacent to the city of Cuzco. It is divided into three sub-districts in which three wastewater treatment plants have been proposed for each district. For one district a design engineer, lacking any information on measured flowrates, assumed a population of 1910 in 2015 with an urban growth rate of 1.9%. The engineer's design flowrates used for the wastewater treatment plant design up to 2035 were as follows:

Year	Population	L (ha d)	Q_{mean} (m³/d)
2015	1910	100	191
2035	2738	100	274

Subsequent to the submission of the design, a consulting engineering firm with flow monitoring capabilities was contracted to monitor flowrates in the principal collector that would feed the proposed wastewater treatment plant (Figure 3.12). The results of continuous 8 h monitoring of the outfall are

Figure 3.12 Measuring flowrate with an area velocity digital flow meter, which has ring inserts for different diameter sewers, for the design of a wastewater treatment plant for a small city, Saylla, Peru. Note the second discharge pipe that was not part of the original sewer design and was clandestinely added more recently as a result of rapid population growth. Photos courtesy of CONASIN SRL, Cusco, Peru.

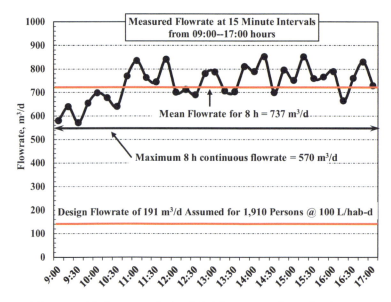

Figure 3.13 Measured flowrate with the area velocity digital flow meter shown in Figure 2.12 compared with the assumed flowrate for the design.

shown in Figure 3.13. The measured mean flowrate was 737 m³/d, 3.8 times greater than the design flowrate and 2.7 times greater than the projected flowrate in 2035. The designer's assumption of an urban growth rate of 1.9% was very low when the rate in this rapidly growing peri-urban area was 6.9% as reported by the Instituto Nacional de Estadística e Informática in Peru (Figure 3.14).

Without monitoring data of actual wastewater flowrates, very large errors in the estimation of design flowrates will occur. Designs using estimated flowrates rather than measured ones must be evaluated with extreme caution.

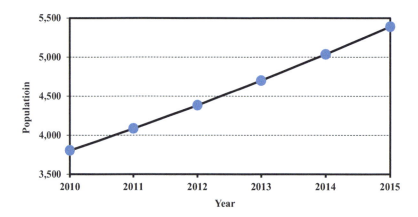

Figure 3.14 Population growth in the District of Saylla, Peru from census data by the Instituto Nacional de Estadística e Informática. The growth rate was 6.9% during this period.

Unfortunately, in resource-challenged cities around the world most designs, such as this one, are performed without field measurements, without having accurate population data, and without knowledge of commercial and industrial discharges. Even if the number of connections is accurately known, high urban growth rates can result in a rapid increase in connections in the short term with clandestine outfalls such as happened in Saylla. Additionally, if flows are not measured, the actual effects of infiltration/inflow and industrial discharges will remain unknown.

doi: 10.2166/9781789061536_0081

Chapter 4
Preliminary treatment

4.1 INTRODUCTION

Pretreatment is necessary in municipal wastewater treatment so materials that cannot be treated biologically can be removed in the first unit processes. These materials, termed screenings and grit, consist of the following (Departamento de Sanidad del Estado de Nueva York, 1993; Mara, 2003; MOPT, 1991).

(1) Coarse solids such as paper, plastics, rags, and cloth that float or are suspended in raw wastewater. Depending on bar spacing, human excreta can also be stopped by screens.
(2) Inorganic solids such as sand and gravel that have entered the sewerage system; these inorganic solids are collectively known as grit. Grit enters through sewer connections and inspection covers, varies according to local soil characteristics, and has a specific gravity much greater than the organic solids in wastewater.

In small- to medium-sized cities, the most appropriate method to remove coarse solids and grit is with manual bar screens and horizontal flow grit chambers, where the horizontal velocity is controlled with a Parshall flume; the flume is also used to measure flowrates. Figures 4.1 and 4.2 show a typical installation of a bar screen with a double-channeled grit chamber and Parshall flume.

4.2 REMOVAL OF COARSE SOLIDS: BAR SCREENS

For the removal of coarse solids, bar screens are inserted transversely to the flow direction. As the water passes through the bars, coarse solids are retained. The material must be manually removed with a rake and buried daily. The amount of material retained varies depending on the spacing between the bars. Studies in Brazil and Peru have found amounts of coarse solids retained between 0.008 and 0.038 m^3/1000 m^3 in screens with openings between 20 and 50 mm (Mendonça, 2000; Minsterio de Vivienda, Construcción y Saneamiento (n.d.)). Using these ranges, and assuming a flow per person of 120 L/capita d, and a population of 10,000 inhabitants, it would be possible to have a production of retained material between 0.01 and 0.05 m^3/d. Designers should verify the quantity retained through field measurements of systems in operation near the design site.

© 2022 The Author. This is an Open Access book chapter distributed under the terms of the Creative Commons Attribution Licence (CC BY-NC-ND 4.0), which permits copying and redistribution for noncommercial purposes with no derivatives, provided the original work is properly cited (https://creativecommons.org/licenses/by-nc-nd/4.0/). This does not affect the rights licensed or assigned from any third party in this book. The chapter is from the book *Integrated Wastewater Management for Health and Valorization: A Design Manual for Resource Challenged Cities*, Stewart M. Oakley (Author).

Figure 4.1 Bar screen and grit chamber, followed by a pre-fabricated Parshall flume to control horizontal velocity in the grit chamber and measure flow rates. Above the bar screen is a drainage platform to drain the screenings before removal and burial. The grit chamber has two canals: the flow is diverted to one, while the other is drained to remove solids. This grit chamber was installed because of excessive grit load to the downstream stabilization pond, which was filling prematurely with inorganic sludge. Note the piles of the screenings and the grit removed by the operator, and the wheelbarrow is used to carry the solids to their burial site (León, Nicaragua).

4.2.1 Design of bar screens

Table 4.1 and Figure 4.3 show recommended design parameters and details of bar screen design, and Figure 4.4 is an example of a typical installation. Bar screens should have rectangular bars with 5–15 mm width and thicknesses from 25 to 40 mm. The bar screen installation should also have a drainage platform to drain the solids before removal for final sanitary disposal; coarse solids can have a water content greater than 80% and need to be drained after removal from the bar screen (Mendonça, 2000). The spacings between bars should be 50 mm in small installations, so human feces will pass through without being retained (see Figure 4.5). The approach channel to the bar screen should have a bypass in case the bar screen becomes blocked (Figure 4.5). The angle of the bar screen should be 45–60° from the horizontal, so material can be easily removed with a rake. The bars should

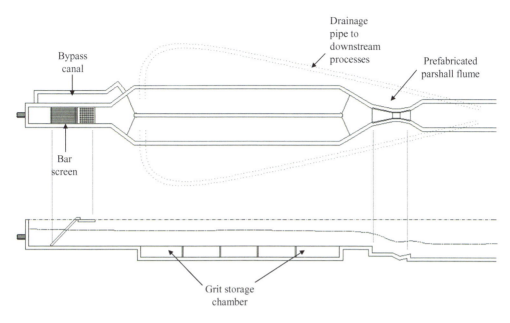

Figure 4.2 Recommended design layout of bar screen, horizontal grit chamber with two canals, and a Parshall flume to control velocity and measure the flowrate.

Table 4.1 Design specifications for manual bar screens.

Parameter	Recommendation
Shape	Rectangular
Width	5–15 mm
Thickness	25–40 mm
Spacing between bars	25–50 mm 50 mm recommended so Human feces pass through the bars
Inclination from horizontal	45–60°
Drainage platform	Made of corrosion-resistant material and sufficient size for adequate drainage before daily disposal
Bypass channel	Sufficient capacity to bypass maximum flow during an emergency
Bar screens and drainage platform materials	Stainless or galvanized steel; aluminum
Approach velocity	0.45 m/s
Hydraulic retention time in approach channel	≥ 3 s
Approach channel length	≥ 1.35 m
Velocity through bars	≤ 0.6 m/s for average flow ≤ 0.9 m/s for maximum flow
Maximum headloss through bars	0.15 m
Quantities of screenings (volume/flow)	0.008–0.038 m^3/1000 m^3
Final disposal of screenings	Disposed onsite with daily cover

Source: Adapted from Reynolds and Richards (1996), Mendonça (2000), and Viceministerio de Vivienda y Construcción (1997).

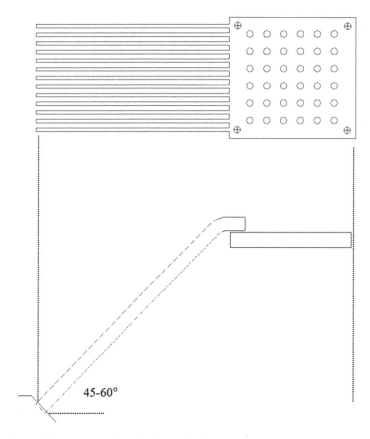

Figure 4.3 Detail of a metal bar screen with a drainage platform.

be made of corrosion-resistant metals such as stainless steel, galvanized steel, and aluminum (Figures 4.4, 4.6 and 4.7).

4.2.2 Design equations for bar screens and approach canal

The approach channel and bar screen are designed with the following equation adapted from Mara (1976):

$$a_{canal} = \frac{Q_{max}}{0.6 P_{max}} \cdot \left[\frac{a_b + e_b}{e_b}\right] \tag{4.1}$$

a_{canal} = the width of approach channel, m
Q_{max} = the maximum flowrate, m³/s
0.6 = the maximum velocity through the bars, m/s
P_{max} = the maximum depth of water in the channel when $Q = Q_{max}$, m
a_b = the width of bars, mm
e_b = the spacing between bars, mm

The maximum depth in the channel, P_{max}, is determined in the design of the grit chamber and will be shown later in the design example using Equation (4.16).

Preliminary treatment

Figure 4.4 A well-designed bar screen should have rectangular bars with widths of 5–15 mm and thicknesses of 25–40 mm, with a drainage platform to drain the retained coarse solids – which have a water content of ≥80% – before sanitary disposal. A spacing between the bars of 50 mm is recommended, so that human feces pass through the grate without being retained (see Figure 4.5). The inclination with the horizontal should be between 45 and 60°, so that the retained material is easily removed with a rake. The construction material of the bars and the drainage platform should be corrosion-resistant metal such as stainless steel, galvanized steel, or aluminum (Alta Vista Urbanization, San Salvador, El Salvador).

The velocity in the approach channel is calculated with the following equation:

$$v = \frac{0.6}{((a_b + e_b)/e_b)} \tag{4.2}$$

$v =$ the velocity in the approach channel, m/s.

Equation (4.2) assumes that the maximum velocity through the bars is 0.6 m/s and, as a result, the velocity in the approach channel should be close to 0.45 m/s if the typical dimensions of a_b and e_b shown in Table 4.1 are used.

Approach channels need a minimum hydraulic retention time of 3 s, and a minimum length of 1.35 m to ensure uniform velocity through the bars. If the hydraulic retention time and the channel length are less, it is likely the channel will have turbulence through the bars as seen in Figure 4.8.

Headloss through the bar screen is calculated with the following equation (Metcalf & Eddy, 1991):

$$h_f = \frac{1}{0.7} \cdot \left[\frac{v_R^2 - v_a^2}{2g} \right] \tag{4.3}$$

Figure 4.5 The timely removal of screenings is a fundamental operational task to maintain uninterrupted flow through the grit chamber. A bar screen has to be cleaned daily or even hourly depending on the raw wastewater characteristics. The top photo shows the water level 24 hours after cleaning the bar screen, while the lower photo shows the level right after cleaning the bars. This bar screen should have a bypass channel (Figure 4.6). The retained solids are human feces that should pass through the bars to be treated in the wastewater treatment plant (Urbanización Alpes Suiza, San Salvador, El Salvador).

h_f = the headloss through the bar screen, m
v_R = the velocity through the bar screen, m/s
v_a = the velocity in the approach channel, m/s
g = the acceleration due to gravity, 9.81 m/s^2

Equation (4.3) is valid only when the bar screen is clean (Metcalf & Eddy, 1991).

Preliminary treatment

Figure 4.6 The approach channel to the bar screen should have a bypass as shown above to divert the flow during an emergency when the operator is not available to clean the screen. Note in the photo that the bars are made of aluminum. The influent bar screen should have a drainage platform as shown in Figure 4.4 (Sanarate, Guatemala).

4.2.3 Final disposal of screenings

Screenings are highly contaminated with pathogens, unsightly, and highly odiferous. They should be buried daily with the minimum of handling by the facility operator. The design of the pretreatment facility should include a reserved area near the bar screen where the operator can bury screenings with minimal handling as shown in Figure 4.9.

Figure 4.7 The bar screen on the left, while having an appropriate opening to retain the coarser solids and allow human feces to pass, and also a suitable drainage platform, should not be constructed of rebar. Reinforcing bars are not corrosion resistant and deteriorate rapidly in a wastewater environment where corrosion conditions are favored as shown in the right photo. The screening solids in the left photo should be buried as soon as possible to protect public health (Left: Granada, Nicaragua; Right: Choloma, Honduras).

Figure 4.8 The approach channel leading to the bar screen must have a velocity of 0.45 m/s, so the grit solids do not settle and the velocity through the bars does not exceed 0.6 m/s. The channel also must have a minimum hydraulic retention time of 3 s and a minimum length of 1.35 m to ensure a uniform velocity through the bars. The channel to the left has significant turbulence; under these conditions, the bar screen will not function properly. The channel on the right is long enough to ensure a uniform velocity without turbulence (left photo: Urbanization in Guatemala City; right photo: Trinidad, Honduras).

Figure 4.9 An excavation next to the bar screen and the grit chamber to bury the solids collected daily. The design of the pretreatment facility should include a reserved area close-by where the operator can bury screenings and grit solids with minimal handling. Pretreatment solids will always be contaminated with pathogens, and screening solids are unsightly with noxious odors like bad smells and bad looks. Bar screen solids need to be buried daily, while grit solids require burial only when cleaning the grit chamber (Urbanización Alta Vista, San Salvador, El Salvador).

4.3 GRIT REMOVAL: DESIGN OF GRIT CHAMBERS

Wastewater contains significant quantities of inorganic solids such as sand, gravel, and cinders, and organic solids such as eggshells, seeds, and coffee grounds, that together have a specific gravity from 1.5 to 2.65 and thus settle more rapidly than organic matter. Grit is produced in the sewer system and the quantity produced is highly variable, depending on factors such as sewer infiltration rate, condition of the collection system, topography, soil types, and the percentage of paved or unpaved streets. The quantity of grit can also vary significantly during the wet and dry seasons (AECOM/Metcalf & Eddy, 2014; ASCE/WPCF, 1977; Mendonça, 2000).

Table 4.2 shows the quantities of grit in sewer systems reported in several studies in Latin America, India, and the US. During the wet season, and especially during a storm period, grit production increases significantly. Studies in the US have shown the ratio of maximum daily production to mean daily production during a storm can rise to 1800 (ASCE/WPCF, 1977). Unfortunately, there are few data for many regions of the world on grit production, especially during the wet and dry seasons. Designers must estimate production from the few data that exist, or from data collected at large wastewater treatment facilities where the number of hauled truckloads of grit and screenings is known.

If large quantities of grit enter a small wastewater treatment system such as a facultative pond, they will cause many of the problems mentioned previously. The problem would be much worse if grit were to enter a closed reactor such as a UASB, where it would very difficult to remove.

4.3.1 Free-flow Parshall flume equations for the design of grit chambers

Figure 4.10 shows the details of a Parshall flume connected at the end of a rectangular, horizontal grit chamber.

The flowrate equation for a Parshall flume is defined as follows (Gloyna, 1971; Marais & van Haandel, 1996):

$$Q = 2.27 \cdot W(H_a)^{1.5} \tag{4.4}$$

Q = the flowrate, m³/s
W = the throat width in the Parshall flume, m
H_a = the depth of water (hydraulic head) at point 2/3 A (Figure 4.10) measured from the base of the Parshall flume, m

Table 4.2 Quantities of grit measured in wastewater in Latin America, India, and the US.

Country	Quantities (m³/1000 m³)	Ratio: Daily Maximum / Daily Mean
Brazil (1970)		
Dry season	0.015–0.029	
Rainy season	0.030–0.040	
Honduras (2003) (Estimated)	0.010–0.085	
India (1970)		
Daily mean	0.026–0.090	
Peak load	0.370–0.740	
(during 1–2 hours)		
US		
Mean (separate sewer)	0.004–0.04	
Mean (combined sewer)	0.004–0.20	
Daily maximum	0.006–3.90	1.0–1800
(during a storm)		

Source: Adapted from Arceivala *et al.* (1970), Oakley (2004); Mendonça (2000), ASCE/WPCF (1977), and Metcalf and Eddy/AECOM (2014).

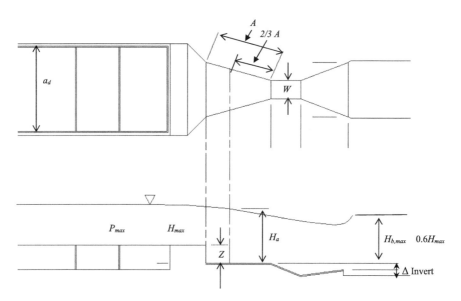

Figure 4.10 A Parshall flume at the effluent end of a grit chamber.

The hydraulic head upstream of the Parshall flume in the grit chamber channel is defined as (Gloyna, 1971):

$$H = 1.1 H_a \tag{4.5}$$

H = head in the grit chamber channel (Figure 4.10) measured with the reference to the base of the Parshall flume, m.

Combining Equations (4.4) and (4.5) yields the following equation:

$$Q = 2.27 \cdot W \left[\frac{H}{1.1}\right]^{1.5} \tag{4.6}$$

Rearranging Equation (4.6), the following relation for the hydraulic head in the grit chamber canal is obtained (Figure 4.10):

$$H = \left[\frac{1.1 \cdot Q}{2.27 \cdot W}\right]^{0.667} \tag{4.7}$$

Rearranging Equation (4.7) for the maximum flowrate, Q_{max}, gives the following equation:

$$Q_{max} = 2.27 \cdot W \left[\frac{H_{max}}{1.1}\right]^{1.5} \tag{4.8}$$

H_{max} = the maximum head in the grit chamber channel when $Q = Q_{max}$, m.

Rearranging Equation (4.8) gives the following relationship for the maximum head in the grit chamber channel:

$$H_{max} = \left[\frac{1.1 \cdot Q_{max}}{2.27 \cdot W}\right]^{0.667} \tag{4.9}$$

Equations (4.4) through (4.9) assume there is free flow through the Parshall flume. For free-flow conditions to exist, the hydraulic head downstream of the Parshall flume must be ≤60% of the head in the grit chamber channel ($H_{b,max}$ in Figure 4.10). To satisfy this condition for design, the value of $H_{b,max}$ is calculated with the following equation:

$$H_{b,max} = 0.6 H_{max} = 0.6 \left[\frac{1.1 \cdot Q_{max}}{2.27 \cdot W} \right]^{0.667} \tag{4.10}$$

It is recommended to use a safety factor in design by lowering the invert elevation of the downstream channel from the Parshall flume by the value, Δ Invert (Figure 4.10), with reference to the invert base of the Parshall flume, to ensure free-flow.

Figures 4.11 and 4.12 show good examples of operating, pre-fabricated Parshall flumes with downstream free-flow to satisfy Equation (4.10). Figures 4.13 and 4.14 are examples of poorly designed and constructed Parshall flumes that should never have been built.

Figure 4.11 An example of a pre-fabricated Parshall flume installed for a horizontal flow grit chamber. The designer should always specify pre-fabricated and calibrated Parshall flumes in the design to avoid the construction problems shown in Figures 4.13 and 4.14 (Masaya, Nicaragua).

Figure 4.12 Examples of free-flow Parshall flumes used with horizontal grit chambers. To ensure free flow, the downstream head at the maximum flowrate, $H_{b,\mathrm{max}}$, must be $\leq 0.60\, H_{\mathrm{max}}$. In the examples above the invert elevations of the downstream canals are much lower than the Parshall flume invert and thus give an added safety factor ($H_{b,\mathrm{max}} \ll 0.60\, H_{\mathrm{max}}$) (left photo: León, Nicaragua; right: Masaya, Nicaragua).

Figure 4.13 Examples of Parshall flumes that never should have been built. The dimensions are wrong, the construction is poor, and there is no headloss through the structure. For these reasons, pre-fabricated Parshall flumes that have certified calibration should always be used (left photo: Catacamas, Honduras; right photo: Villanueva, Honduras).

Figure 4.14 Another example of a poorly designed and constructed Parshall flume without free-flow (Choloma, Honduras). Pre-fabricated Parshall flumes are always recommended to avoid this common problem.

The relation between Q and H for several common throat widths for free-flowing Parshall flumes is shown in Figure 4.15. Flowrate ranges most likely to be encountered in cities up to 100,000 inhabitants are shown in Table 4.3.

4.3.2 Design of rectangular grit chambers

The invert, Z, which is the difference in elevation between the Parshall flume invert and the grit chamber canal invert (Figure 4.10), is calculated with the following equations (Babbitt & Baumann, 1958; Gloyna, 1971; Marais & van Haandel, 1996):

$$Z = \left[\frac{R^{1/3}-1}{R-1}\right] \cdot 1.1 \left[\frac{Q_{max}}{2.27W}\right]^{2/3} \tag{4.11}$$

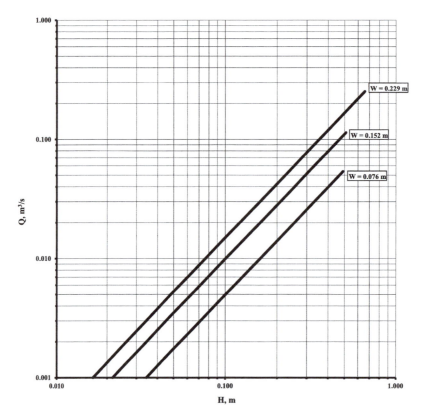

Figure 4.15 Flowrate, Q, versus hydraulic head, H, in the grit chamber canal upstream of a free-flow Parshall flume with throat width W.

Table 4.3 Flowrate ranges for free-flow Parshall flumes.

Throat Width	Q_{min}		Q_{max}		Q_{max}/Q_{min}
W (m)	(m³/s)	(m³/d)	(m³/s)	(m³/d)	
0.076	0.0008	69	0.0538	4648	67.3
0.152	0.0015	130	0.1104	9539	73.6
0.229	0.0025	216	0.2519	21,764	100.8
0.305	0.0031	268	0.4556	39,364	146.9
0.457	0.0042	363	0.6962	60,152	165.8
0.610	0.0119	1028	0.9367	80,931	78.7
0.915	0.0176	1521	1.4263	123,232	81.0
1.220	0.0368	3180	1.9215	166,018	52.2
1.525	0.0628	5426	2.422	209,261	38.6
1.830	0.0744	6428	2.9290	253,066	39.4
2.135	0.1154	9971	3.4400	297,216	29.8
2.440	0.1307	11,292	3.9500	341,280	30.2
3.050	0.2000	17,280	5.6600	489,024	28.3

Source: Marais and van Haandel (1996).

$$Z = \left[\frac{R^{1/3}-1}{R-1}\right] \cdot H_{max} \qquad (4.12)$$

$$Z = C_r \cdot H_{max} \qquad (4.13)$$

$$C_r = \frac{R^{1/3}-1}{R-1} \qquad (4.14)$$

$$R = \frac{Q_{max}}{Q_{min}} \qquad (4.15)$$

Z = difference in elevation from the Parshall flume invert to the grit chamber canal invert (Figure 4.10), m.

Figure 4.16, developed from Equations (4.14) and (4.15), shows the relationship of C_r to R for the ranges most likely to be encountered in small to large cities.

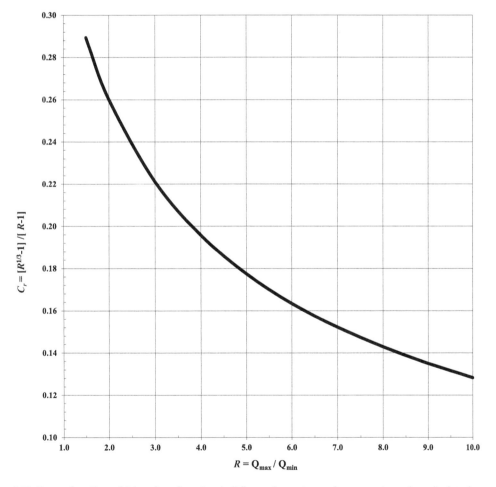

Figure 4.16 C_r as a function of R in a free-flow Parshall flume downstream from a rectangular grit chamber.

After determining the invert drop, Z, the maximum depth of water in the grit chamber canal with reference to the canal invert (not the Parshall flume invert) is determined with Equation (4.16)

$$P_{max} = H_{max} - Z \qquad (4.16)$$

P_{max} = the maximum depth of water in the grit chamber canal measured from the invert of the canal, m.

The width of the grit chamber canal is calculated with Equation (4.17):

$$a_d = \frac{Q_{max}}{P_{max} \cdot v_{max}} = \frac{Q_{max}}{P_{max} \cdot (0.3)} \qquad (4.17)$$

a_d = the width of grit chamber canal, m
v_{max} = the maximum horizontal velocity in the canal, 0.3 m/s

The length of the grit chamber channel is a function of water depth, horizontal velocity, and particle settling velocity (Mara, 2003; Mendonça, 2000) as shown in the following equation:

$$L = \frac{P_{max} v_{max}}{v_{settling}} \qquad (4.18)$$

L = the length of channel, m
P_{max} = the maximum depth of water in channel, m
v_{max} = the máximum horizontal velocity, m/s
$v_{settling}$ = the settling velocity of grit particles, m/s

For design, the maximum horizontal velocity recommended, v_{max}, is 0.3 m/s and the settling velocity, $v_{settling}$, most commonly used is that for a 0.20 mm particle with $v_{settling}$ = 0.02 m/s (AECOM/Metcalf & Eddy, 2014; Mara, 2003; Mendonça, 2000). Using these values, Equation (4.18) reduces to the following:

$$L = \frac{P_{max}(0.3\,\text{m/s})}{0.02\,\text{m/s}} = 15 P_{max} \qquad (4.19)$$

Using a safety factor of 1.7, Equation (4.20) should be used for design (Mendonça, 2000):

$$L = 25 P_{max} \qquad (4.20)$$

The volume of grit that accumulates during a given time period can be estimated with Equation (4.21):

$$V_{grit} = t_{op} \cdot Q_{mean} \cdot C_{grit} \qquad (4.21)$$

V_{grit} = the grit volume, m³
t_{op} = the operational period, d
Q_{mean} = the mean flowrate, m³/d
C_{grit} = the volumetric loading of grit, m³/1000 m³

The volumetric loading can initially be estimated with data from Table 4.2, or from local data if available, which is unlikely except for larger treatment plants. Experienced operators at small plants, however, will know grit cleaning frequencies and quantities, assuming they are in fact operating and maintaining the plant.

Table 4.4 summarizes the principal design parameters recommended for horizontal grit chambers.

Table 4.4 Design parameters for horizontal grit chambers.

Parameter	Design Recommendations
Horizontal velocity	$v_{max} = 0.3$ m/s $v_{min} \geq 0.24$ m/s
Sedimentation velocity	0.02 m/s (particles with 0.2 mm diameter)
Transverse section in canal	Rectangular (with hydraulic drop from the base of canal to the base of Parshall flume)
Hydraulic retention time	≤ 60 s for v_{min} ≥ 45 s for v_{max}
Length of canal	$L = 25 P_{max}$
Velocity control	Pre-fabricated Parshall flume with downstream free-flow
Hydraulic head in the canal downstream from the Parshall flume to ensure free-flow	$\leq 60\%$ hydraulic head in the grit chamber canal
Number of canals	Two in parallel with drainage to downstream treatment (one in operation with second for cleaning)

Source: Adapted from Marais and van Haandel (1996), Reynolds and Richards (1996), Mendonça (2000), and Viceministerio de Vivienda y Construcción (1997).

4.4 BYPASS CHANNEL DESIGN

A bypass channel is always necessary to protect the preliminary treatment works (and the treatment plant) during extreme flow events. The simplest design uses a rectangular channel sized for the maximum peak hourly flow plus a safety factor.

The Manning equation is most often used for the design of rectangular open channels. The figure below presents a cross-section of a rectangular open channel of depth, d, width w, and area A_{flow}

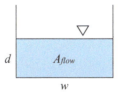

The cross-sectional area of the channel at the maximum flow is calculated with the following equation:

$$A_{flow} = w \cdot d \tag{4.22}$$

A_{flow} = the cross-sectional flow area, m²
w = the channel width, m
d = the water depth at maximum flow, m

Equation (4.23) relates the maximum flowrate to the cross-sectional area, the hydraulic radius, and the channel slope:

$$Q_{max} = A_{flow} \cdot \left(\frac{1}{n} (R)^{2/3} (S)^{1/2} \right) \tag{4.23}$$

R = the hydraulic radius, m
S = the slope of channel, m/m
n = the coefficient of roughness for channel surface ($n = 0.012$ for concrete)

The hydraulic radius is calculated with the following equation:

$$R = \frac{A_{\text{flow}}}{wp} \tag{4.24}$$

wp = the wetted perimeter at Q_{max}, m.

$$wp = 2d + w \tag{4.25}$$

The minimum velocity in the channel is calculated with Equation (4.26):

$$v = \frac{1}{n}(R)^{2/3}(S)^{1/2} \tag{4.26}$$

v = the minimum velocity in channel, $v \geq 0.6$ m/s.

Equation (4.26) is rearranged to solve for the design slope of the channel at the maximum flowrate:

$$S = \left[\frac{v}{(1/n)(R)^{2/3}}\right]^2 \tag{4.27}$$

4.5 PROCEDURE FOR PRELIMINARY TREATMENT DESIGN WITH THE PARSHALL FLUME

(1) Determine the maximum, minimum, and mean flowrate for design in m³/s. It is important that flow measurements are made during the dry and wet seasons and over 24 hour periods.
(2) Select the throat width of the Parshall flume using Table 4.3.
(3) Calculate the maximum hydraulic head in the grit chamber canal with reference to the base of the Parshall flume, H_{max}, using Equation (4.9):

$$H_{\text{max}} = \left[\frac{1.1 \cdot Q_{\text{max}}}{2.27 \cdot W}\right]^{0.667} \tag{4.9}$$

(4) Calculate R and C_r from Equations (4.14) and (4.15), or use Figure 4.16:

$$C_r = \frac{R^{1/3} - 1}{R - 1} \tag{4.14}$$

$$R = \frac{Q_{\text{max}}}{Q_{\text{min}}} \tag{4.15}$$

(5) Calculate, Z, from Equation (4.13):

$$Z = C_r \cdot H_{\text{max}} \tag{4.13}$$

(6) Calculate the maximum depth of water in the grit chamber canal, P_{max}, measured from the bottom of the canal, using Equation (4.16):

$$P_{\text{max}} = H_{\text{max}} - Z \tag{4.16}$$

(7) Calculate the width of the grit chamber using Equation (4.17):

$$a_d = \frac{Q_{\text{max}}}{P_{\text{max}} \cdot v_{\text{max}}} = \frac{Q_{\text{max}}}{P_{\text{max}} \cdot (0.3)} \tag{4.17}$$

(8) Calculate the length of the grit chamber from Equation (4.20):

$$L = 25 P_{\text{max}} \tag{4.20}$$

Preliminary treatment

(9) Estimate the volume of grit produced during a selected time interval from Equation (4.21).

$$V_{grit} = t_{op} \cdot Q_{mean} \cdot C_{grit} \tag{4.21}$$

(10) Select the invert elevation in the canal downstream from the Parshall flume to have a safety factor for free flow in the case that $H_{b,max}$ was exceeded with high flows ($H_{b,max} > 0.6H_{max}$) (see Figure 4.10).

(11) Select the thickness, width and spacing, a_b and e_b, of the bars for screening (see Table 4.1).

(12) Using Equation (4.1), calculate the width, a_{canal}, of the approach channel to the bar screen.

$$a_{canal} = \frac{Q_{max}}{0.6P_{max}} \cdot \left[\frac{a_b + e_b}{e_b}\right] \tag{4.1}$$

(13) Calculate the velocity in the approach channel and headloss through the bar screen using Equations (4.2) and (4.3). If results are not within design norms, redesign.

$$v = \frac{0.6}{((a_b + e_b)/e_b)} \tag{4.2}$$

$$h_f = \frac{1}{0.7} \cdot \left[\frac{v_R^2 - v_a^2}{2g}\right] \tag{4.3}$$

(14) Complete the pretreatment design with additional parameters shown in Tables 4.1 and 4.4.
(15) Design a bypass channel using Equations (4.22)–(4.27).

4.5.1 Case study design: preliminary treatment, WSP system, Catacamas, Honduras

A preliminary treatment system of bar screen, grit chamber with Parshall flume, and bypass channel, is to be designed for the existing waste stabilization pond system at East Catacamas, Honduras. Figure 4.17 shows the results of continuous flowrate monitoring with an area–velocity meter for 72 h (Oakley, 2005).

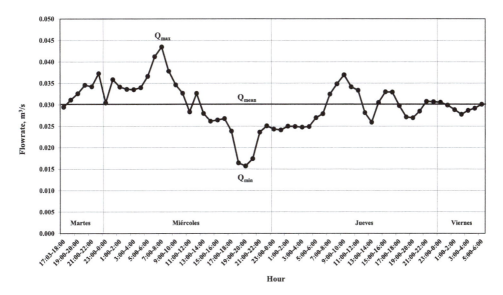

Figure 4.17 Influent flowrate measurement at the East Catacamas waste stabilization pond system, September 2–5, 2003.

(1) Determine the mean, maximum, and minimum flowrate for design.
    From Figure 4.16, $Q_{mean} = 0.03$ m³/s; $Q_{max} = 0.045$ m³/s; $Q_{min} = 0.015$ m³/s

$$\frac{Q_{max}}{Q_{min}} = \frac{0.045}{0.015} = 3.0$$

The ratio Q_{max}/Q_{min} is low and is increased to 5.0 to have an increased safety factor.

Therefore, $Q_{max}/Q_{min} = 5.0$; $Q_{max} = 0.075$ m³/s for design.

(2) Select the throat width for the Parshall flume.

    $W = 0.152$ m (Table 4.3)

(3) Calculate the maximum head, H_{max}, in the grit chamber channel using Equation (4.9).

$$H_{max} = \left[\frac{1.1 \cdot Q_{max}}{2.27 \cdot W}\right]^{0.667} = \left[\frac{1.1 \cdot (0.075)}{2.27 \cdot (0.152)}\right]^{0.667} = 0.385 \text{ m}$$

(4) Calculate R and C_r from Equations (4.13) and (4.14) or Figure 4.16:

$$R = \frac{Q_{max}}{Q_{min}} = \frac{0.075}{0.015} = 5; \quad C_r = \frac{R^{1/3} - 1}{R - 1} = \frac{(5)^{1/3} - 1}{5 - 1} = 0.18$$

(5) Calculate the invert drop, Z, from Equation (4.13):

    $Z = C_r \cdot H_{max} = (0.18) \cdot (0.385 \text{ m}) = 0.07$ m

(6) Determine the maximum depth of water in the grit chamber channel, P_{max}, measured from the invert of the channel, using Equation (4.16).

    $P_{max} = H_{max} - Z = 0.385 \text{ m} - 0.07 \text{ m} = 0.315$ m

(7) Determine the width of the grit chamber channel with Equation (4.17).

$$a_d = \frac{Q_{max}}{P_{max} \cdot v_{max}} = \frac{0.075}{(0.315) \cdot (0.3)} = 0.8 \text{ m}$$

(8) Calculate the length of the grit chamber channel with Equation (4.20).

    $L = 25 P_{max} = 25(0.315 \text{ m}) = 7.875$ m

(9) Estimate the volume of the grit produced during a given time period from Equation (4.21) and data from Table 4.2 if no measured data or local knowledge are available.

    Assumptions:
    - $C_{grit} = 0.085$ m³/1000 m³ (worst-case scenario from data in Table 4.2 for Honduras)
    - $T_{op} = 7$ d

    $Q_{mean} = (0.03 \text{ m}^3/\text{s})(86400 \text{ s/d}) = 2592$ m³/d

    Therefore, $V_{grit} = t_{op} \cdot Q_{mean} \cdot C_{grit} = (7 \text{ d})(2592 \text{ m}^3/\text{d})(0.085 \text{ m}^3/1000 \text{ m}^3) = 1.5$ m³ / week

(10) Select the invert elevation in the canal downstream from the Parshall flume to have a safety factor for free-flow in the case that $H_{b,max}$ was exceeded with high flows ($H_{b,max} > 0.6 H_{max}$).

    $H_{b,max} = 0.6 H_{max} = (0.6)(0.428 \text{ m}) = 0.257$ m

(11) Determine the width, a_b, and the opening between bars, e_b, for the bar screen.

$a_b = 10$ mm; $e_b = 50$ mm (Table 4.1)

(12) Using Equation (4.1), calculate the width, a_{canal}, of the approach channel to the bar screen.

$$a_{canal} = \frac{Q_{max}}{0.6 P_{max}} \cdot \left[\frac{a_b + e_b}{e_b}\right] = \frac{0.075 \, m^3/s}{0.6 \cdot (0.315 \, m)} \cdot \left[\frac{10 \, mm + 50 \, mm}{50 \, mm}\right] = 0.476 \, m$$

(13) Calculate the horizontal velocity in the approach channel and the headloss through the bar screen using Equations (4.2) and (4.3):

$$v_a = \frac{0.6}{((a_b + e_b)/e_b)} = \frac{0.6 \, m/s}{((10 \, mm + 50 \, mm)/50 \, mm)} = \frac{0.6 \, m/s}{1.2} = 0.5 \, m/s$$

$$h_f = \frac{1}{0.7} \cdot \left[\frac{v_R^2 - v_a^2}{2g}\right] = \frac{1}{0.7} \cdot \left[\frac{(0.6 \, m/s)^2 - (0.5 \, m/s)^2}{2 \cdot (9.81 \, m/s^2)}\right] = 0.008 \, m$$

(14) Complete the design with the parameters in Tables 4.1 and 4.4.
(15) Design a bypass canal for extreme flow events.

Assume the ratio of $Q_{max,extreme}/Q_{mean} = 10$.

$Q_{max,extreme} = 10(0.045 \, m^3/s) = 0.45 \, m^3/s$

$$Q_{max} = A_{flow} \cdot \left(\frac{1}{n}(R)^{2/3}(S)^{1/2}\right) = A_{flow} v; \quad v = 0.6 \, m/s$$

$$A_{flow} = \frac{Q_{max}}{v} = \frac{0.45 \, m^3/s}{0.6 \, m/s} = 0.75 \, m^2$$

$A_{flow} = w \cdot d; \quad w = 2d$

$A_{flow} = 2d^2 = 0.75 \, m^2$

$d = \sqrt{\frac{0.75}{2}} = 0.61 \, m$

$w = 2(0.61 \, m) = 1.2 \, m$

$wp = 2d + w = 2(0.61) + 1.2 = 2.4 \, m$

$$R = \frac{A_{flow}}{wp} = \frac{0.75 \, m^2}{2.4 \, m} = 0.3125 \, m$$

$$S = \left(\frac{v}{(1/n)(R)^{2/3}}\right)^2 = \left(\frac{0.6 \, m/s}{(1/0.012)(0.3125)^{2/3}}\right)^2 = 0.00025 \, m/m$$

The channel should be designed with a width of 1.2 m, a maximum liquid depth of 0.61 m, and a slope of 0.00025 m/m.

4.6 FINAL DISPOSAL OF SCREENINGS AND GRIT

Screenings and grit should be disposed onsite at the wastewater treatment facility, which should have a designated area as part of the original design. For larger systems, a small backhoe can be used to construct a trench that will last several months to 1 year, with daily screenings and periodic grit quantities covered by hand with the excavated soil. Figure 4.18 shows an example of this method used for municipal solid waste. See also Chapters 9 and 10 for examples of preliminary treatment solids handling and disposal.

Figure 4.18 Municipal solid waste is deposited in this trench at Villanueva, Honduras. When full, the trench is backfilled and excess soil is used for other purposes onsite. The same procedure can be used at larger wastewater treatment facilities for preliminary treatment solids and should be included in the original design of the system.

doi: 10.2166/9781789061536_0103

Chapter 5
Theory and design of facultative ponds

5.1 NATURAL PROCESSES AS THE DRIVING FORCE IN FACULTATIVE PONDS

Facultative waste stabilization ponds are the original natural system, or nature-based solution, for sustainable wastewater treatment with valorization for reuse. Facultative ponds combine the unit processes of primary sedimentation, anaerobic sludge digestion, and aerobic removal of soluble biochemical oxygen demand (BOD). In addition, pathogen reductions can reach 100% for helminth eggs, 1.0–2.0 $\log_{10}$ for bacterial pathogens, 0–1.6 $\log_{10}$ for viruses, and 1.0–1.8 $\log_{10}$ for protozoan cysts (Verbyla et al., 2017). All of this is accomplished without an external input of energy using natural physical and biological processes: gravity sedimentation (total suspended solids (TSS), helminth eggs, and protozoan cysts), anaerobic methanogenesis (sludge digestion), aerobic heterotrophic BOD removal using oxygen from photosynthesis, and partial inactivation of pathogenic bacteria and viruses by pH increases caused by photosynthesis and by UV solar radiation. Digested sludge needs removal only once every 10–20 years, minimizing sludge handling–and potential pathogen exposure–more than any other wastewater treatment process.

Facultative ponds are characterized by an aerobic zone in the upper 0.5–1.0 m, where there is a symbiosis between algae and heterotrophic bacteria, and an anaerobic zone in the lower 1.0 m, where the sludge formed from settled suspended solids undergoes anaerobic digestion (Figure 5.1). In the aerobic zone, soluble and colloidal organic compounds are oxidized by heterotrophic bacteria using oxygen produced by the algae, which in turn use the carbon dioxide produced by bacteria as their carbon source. Anaerobic digestion in the sludge layer produces dissolved organic compounds and the gases CO_2, CH_4, H_2S, and NH_3 that diffuse to the aerobic zone: CO_2 and NH_3 can be taken up algae, while CH_4 and H_2S can be oxidized by bacteria to CO_2 and SO_4^{2-}. A small to large fraction of all the gases can also diffuse into the atmosphere.

Primary facultative ponds receive raw wastewater after preliminary treatment. Secondary facultative ponds receive effluent from an upstream biological process, usually an anaerobic pond or upflow anaerobic sludge blanket (UASB) and as a result have lower BOD_L and suspended solids loadings than primary ponds.

© 2022 The Author. This is an Open Access book chapter distributed under the terms of the Creative Commons Attribution Licence (CC BY-NC-ND 4.0), which permits copying and redistribution for noncommercial purposes with no derivatives, provided the original work is properly cited (https://creativecommons.org/licenses/by-nc-nd/4.0/). This does not affect the rights licensed or assigned from any third party in this book. The chapter is from the book *Integrated Wastewater Management for Health and Valorization: A Design Manual for Resource Challenged Cities*, Stewart M. Oakley (Author).

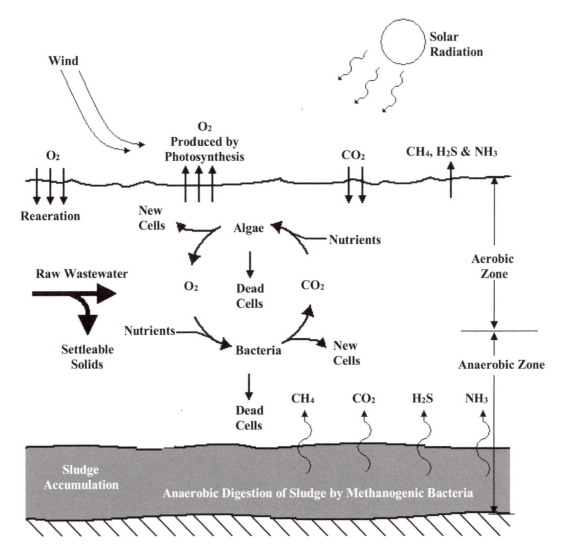

Figure 5.1 The interactions of aerobic heterotrophic bacteria and algae in the aerobic zone, and anaerobic heterotrophic bacteria (methanogens) in the anaerobic sludge zone, in a facultative stabilization pond. The depth of the aerobic zone changes during the day as motile algae move upward or downward in the photo zone in relation to incident light intensity, and at night when photosynthesis ends and algal respiration begins, hence the term facultative. In the anaerobic zone, H_2S gas can be oxidized by purple and green sulfur bacteria. In the aerobic zone, H_2S can be oxidized by autotrophic bacteria and CH_4 by methanotrophic bacteria and both may also diffuse to the atmosphere (*Source*: adapted from Tchobanoglous and Schroeder (1985).)

5.1.1 Algal and bacterial processes in the aerobic zone

Photosynthesis and respiration. The oxygen produced by algae in facultative ponds can be estimated by the following balanced photosynthesis equation modified from the original developed by Oswald and Gotaas (1957) (Rittmann & McCarty, 2001):

$$106CO_2 + 65H_2O + 16NH_3 + H_3PO_4 \xrightarrow{\text{Solar radiation}} C_{106}H_{181}O_{45}N_{16}P + 118O_2 \quad (5.1)$$

```
  |              |        |              Algal cells
1272 mg C    224 mg N  31 mg P          2428 mg        3776 mg
                                         1.0 mg         1.55 mg
```

Equation (5.1) shows that 1.0 mg of algae produced by photosynthesis produces 1.55 mg of O_2, which could satisfy a BOD_L of up to 1.55 mg. Algae must therefore be continuously produced to supply oxygen for soluble and colloidal BOD_L removal by heterotrophic bacteria. Algal concentrations in the effluent of well-operating facultative ponds can range from 40 to 100 mg/L (Oakley, 2005; Von Sperling, 2007).

The equation of photosynthesis is reversed at night as algae use oxygen for respiration as shown in Figure 5.2. In a well-designed pond, the diurnal cycle photosynthesis should produce sufficient oxygen to maintain aerobic conditions throughout the night, when both algae and bacteria consume oxygen through aerobic respiration. An algal cell concentration of 1.0 mg/L can exert a BOD_L of 0.40–0.65 mg/L (Von Sperling, 2007); thus, a typical in-pond concentration of algae of 75 mg/L would have a BOD_L of 34–45 mg/L, in addition to the wastewater oxygen demand. The rate of BOD_L removal in a facultative pond, however, is slow, on the order of days, and a well-designed pond that is not overloaded should easily maintain aerobic conditions day and night (see Section 5.2.3).

During the day, rapidly growing algae can reduce dissolved carbon dioxide below its equilibrium value with air, causing an increase in pH. As the forms of alkalinity change with the increasing pH, algae extract carbon dioxide from both bicarbonate and carbonate alkalinity according to the following equilibrium equations (Sawyer et al., 2002):

$$2HCO_3^- \rightleftarrows CO_3^{2-} + H_2O + CO_2 \rightleftarrows\underset{\text{Algae}}{} CH_2O + O_2 \quad (5.2a)$$

$$CO_3^{2-} + H_2O \rightleftarrows 2OH^- + CO_2 \rightleftarrows\underset{\text{Algae}}{} CH_2O + O_2 \quad (5.2b)$$

The algae-driven production of CO_3^{2-} and OH^- alkalinity can raise the pH to values greater than nine with the following effects (Von Sperling, 2007):

- Conversion of NH_4^+ to NH_3, which is toxic but typically released into the atmosphere
- Conversion of H_2S diffusing from the anaerobic zone to HS^-. H_2S can cause bad odors if it reaches the pond surface, but HS^- is odorless.
- High pH values contribute to pathogen reduction, especially for bacteria (Verbyla et al., 2017).

During the night when respiration occurs, Equations (5.2a) and (5.2b) reverse themselves, and the pH decreases with the production of CO_2 from respiration, causing the alkalinity to convert back to CO_3^{2-} and HCO_3^-.

A well-functioning facultative pond typically is bright green with no odors and often produces excess oxygen such that the aerobic zone becomes supersaturated (Figure 5.3). At night, as long as the pond is not overloaded, aerobic conditions should be maintained to ≥ 0.5 m depths.

Heterotrophic removal of organic matter (BOD_L). The removal of BOD_L in a facultative pond can be expressed by the following unbalanced equation for the oxidation of organic matter:

$$C_{10}H_{19}O_3N + O_2 \rightarrow CO_2 + NH_4^+ + C_5H_7O_2N$$

Organic matter in wastewater   Bacterial cells

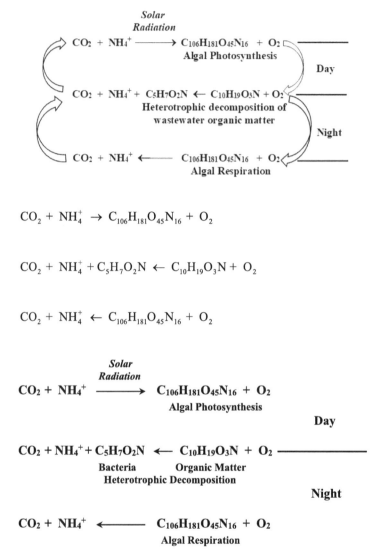

Figure 5.2 The diurnal cycle of algal photosynthesis and respiration, and bacterial respiration, in the aerobic zone of a facultative pond. Empirical formulas from Rittmann and McCarty (2001): $C_{106}H_{181}O_{45}N_{16}$ (algae); $C_5H_7O_2N$ (bacteria); and $C_{10}H_{19}O_3N$ (organic matter in wastewater).

The aerobic heterotrophic bacteria use a portion of the wastewater carbon for energy to convert the remaining portion into cellular material (Figure 5.4). When cells decay, part of their cellular material is also converted to energy and another part to cell residue.

The partitioning of carbon between energy production and cell synthesis depends on the age of the bacteria: in young rapidly growing culture, a much higher fraction is used for the cell synthesis than for energy; in a slow-growing or old culture, however, a larger fraction is used for cell maintenance than for the synthesis (Sawyer et al., 2002).

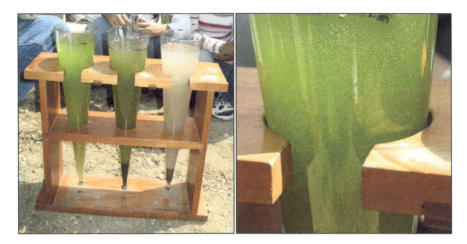

Figure 5.3 Algal growth in a well-functioning facultative/maturation pond system. The top photo shows, right to left, raw settled wastewater, settled facultative effluent, and settled maturation pond effluent. The facultative pond has high concentrations of algae that settle rapidly when confined to an Imhoff cone; the maturation pond effluent has lower concentrations that do not settle. The bottom photo shows oxygen bubbles formed as a result of supersaturation during photosynthesis. Dissolved oxygen concentrations can rise greater than 25 mg/L in facultative ponds, where the saturation concentration is 7–9 mg/L. (Aurora II Waste Stabilization Pond System, University of San Carlos, Guatemala.)

Methane oxidation. The methanotrophs are specialized bacteria that use methane diffusing from the anaerobic zone as a carbon and energy source as shown in Equation (5.3):

$$CH_4 + O_2 + NH_4^+ + HCO_3^- \xrightarrow{\text{Methanotrophs}} CO_2 + H_2O + \underset{\text{Bacterial cells}}{C_5H_7O_2N} \tag{5.3}$$

Methane derived from anaerobic natural processes is found extensively in aquatic environments, and the methanotrophs play an important role in the global carbon cycle by oxidizing it to carbon dioxide (Hanson & Hanson, 1996). Methanotrophs live in close association with methane-producing bacteria (methanogens) and are always present in the aerobic zone of well-operating facultative ponds, most commonly at the oxic–anoxic interface (Hanson & Hanson, 1996). They have been found in a facultative pond in the United Kingdom in both winter and summer (van der Linde, 2009). In a pilot-scale facultative pond in South Africa, methanotrophs averaged concentrations of 190 organisms/mL, with a measured methane-oxidizing potential of 190 mg CH_4/L d; the author concluded that although methane oxidation was significant, large quantities of methane still escaped to the atmosphere (Hoefman, 2013).

More research is needed to determine if simple design changes to facultative ponds, such as the incorporation of a porous sludge blanket for methanotrophic growth at the oxic–anoxic interface, similar to a UASB, could improve the performance of methane oxidation (Miguez *et al.*, 1999).

Chemoautotrophic oxidation of H_2S. Similar to the methanotrophs, the autotrophic bacteria that oxidize sulfide to elemental sulfur to sulfate exist at the oxic–anoxic interface between the aerobic and anaerobic zones. Hydrogen sulfide is typically oxidized in two steps by *Acidothiobacillus* as shown in Equations (5.4a) and (5.4b):

$$H_2S + 0.5O_2 + CO_2 + NH_4^+ \xrightarrow{\text{Acidothiobacillus}} S^0 + H_2O + \underset{\text{Bacterial cells}}{C_5H_7O_2N} \tag{5.4a}$$

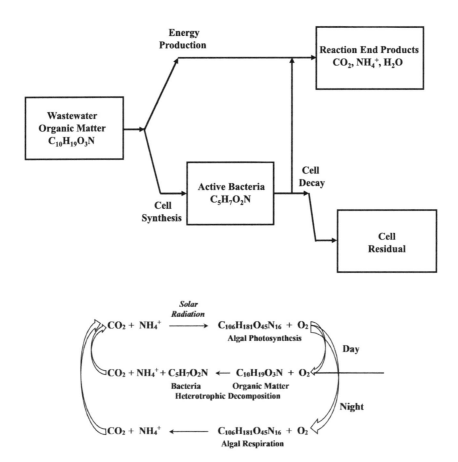

Figure 5.4 Aerobic heterotrophic utilization of carbon for energy production and synthesis (after Rittmann and McCarty, 2001).

$$\text{Acidothiobacillus} \\ S^0 + O_2 + CO_2 + H_2O + NH_4^+ \rightarrow SO_4^{2-} + 2H^+ + C_5H_7O_2N \quad (5.4b) \\ \text{Bacterial cells}$$

Hydrogen sulfide can also be oxidized chemically to sulfur when dissolved oxygen is present:

$$H_2S + 0.5O_2 \rightarrow S^0 + H_2O \quad (5.4c)$$

In Equations (5.4a) and (5.4b), the energy from the oxidation of hydrogen sulfide and sulfur is used to fix carbon dioxide for cell growth. The sulfur-oxidizing bacteria require both sulfide and oxygen at the same time (Equation (5.4a)), and sulfide is usually abundant only in a reduced environment in the absence of oxygen; as a result, they have developed an ability to grow under conditions of low-oxygen concentrations and are classified as microaerophilic bacteria (Pepper et al., 2015).

In Equation (5.4b), sulfuric acid is formed, and *Acidothiobacillus* has an optimum growth rate at a pH of 2 (Pepper et al., 2015), and under the right conditions, serious corrosion problems in wastewater collection and treatment works can develop. This is especially a problem in anaerobic reactors.

The chemoautotrophic oxidation of hydrogen sulfide is an important natural process that controls odor problems in well-designed and operated facultative ponds that do not have excessive sludge accumulation. As with methanotrophic methane oxidation, it is not a process that can be controlled operationally with the exception of sludge removal at the appropriate time intervals to prevent excessive accumulation.

5.1.2 Bacterial processes in the anaerobic zone

Methanogenesis. The final process in the biodegradation of organic matter under anaerobic conditions is the production of methane, where a portion of the original carbon from the organic substrate is reduced to its lowest oxidation state, CH_4, and another portion is oxidized to CO_2 and bicarbonate (Equation (5.5)).

$$C_{10}H_{19}O_3N + H_2O \rightarrow CH_4 + CO_2 + HCO_3^- + NH_4^+ + C_5H_7O_2N \qquad (5.5)$$

Organic matter   Bacterial cells
in wastewater

Methanogenesis occurs in the anaerobic sludge zone (Figure 5.1), which is formed by the settling of the raw wastewater suspended solids; dead algal and bacterial biomass also contribute to the sludge zone to a much lesser degree. As discussed in Chapter 4, methanogenesis yields the least amount of energy of any heterotrophic pathway for the breakdown of organic matter by converting the majority of the original energy to methane.

Equation (5.6) shows a balanced equation for the methanogenesis of wastewater organic solids for the transfer of one electron equivalent again using the method of Rittmann and McCarty (2001). Here, it is assumed that the fraction of the carbon used for the cell synthesis is 0.11, with 0.89 converted to energy in the form of CH_4 (Rittmann & McCarty, 2001).

$$0.02C_{10}H_{19}O_3N + 0.09H_2O \rightarrow 0.11CH_4 + 0.05CO_2 + 0.015HCO_3^- + 0.015NH_4^+ + 0.006C_5H_7O_2N$$

Organic matter				
Bacterial cells				
in wastewater				
0.02(201)	0.11(16)	0.015(61)		0.006(113)
4.0 mg	1.76 mg CH_4	0.91 mg		0.68 mg
(8.0 mg BOD_L	(7.04 mg BOD_L	0.75 mg $CaCO_3$		
equivalent)	equivalent)			
1.0 mg BOD_L	0.22 mg CH_4	0.11 mg $CaCO_3$		0.08 mg
equivalent	(0.88 BOD_L			
	equivalent)			

(5.6)

Cell yield

$$Y = \frac{0.68 \text{ mg new cells}}{8.0 \text{ mg } BOD_L \text{ Removed}} = \frac{0.08 \text{ mg new cells}}{\text{mg } BOD_L \text{ Removed}}$$

Equation (5.6) shows that for every 1.0 mg of BOD_L degraded, 0.88 mg is transferred to CH_4. Thus, 88% of the original energy has been converted to methane. In an anaerobic reactor, this methane can be captured and used as an alternative energy source, but in a facultative pond, it diffuses upward where it can be oxidized by the methanotrophs or escapes to the atmosphere. Of the original 0.20 mg of carbon, only 0.03 mg (15%) is transferred to bacterial cells. Methanogenesis maximizes the saving of energy by converting the majority of carbon to methane and, as a result, minimizes the biomass production.

In the past, little attention was given to methane emissions from facultative ponds, but now this must be addressed by the design engineer as methane, and other greenhouse gas emissions from wastewater treatment are a global issue. The International Panel on Climate Change has published the 2019 Refinement to the 2006 IPCC Guidelines for National Greenhouse Gas Inventories that includes a revised chapter on wastewater treatment (IPCC, 2019). This revised chapter introduces new greenhouse gas emission factors for wastewater treatment processes, including facultative ponds. An analysis of potential methane emissions in a facultative pond is given in the following example.

5.1.3 Process analysis: methane emissions from facultative pond, Catacamas, Honduras

The potential methane emissions for the facultative pond in Catacamas, Honduras, are to be estimated using modified equations from the 2019 Refinement of the IPCC Guidelines (IPCC, 2019). As a first start, it is assumed that only settled influent TSS will undergo anaerobic digestion.

The facultative pond in Catacamas was monitored with the following results (Oakley, 2005):

Estimated population served	3400
Mean flowrate	902 m³/d
Influent BOD_L	441 mg/L
Area of facultative pond	13 800 m²
Hydraulic retention time	24.1 d
Pond depth	2.0 m
Water temperature	20–25°C

(a) Calculation of methane emissions.

The 2019 Revision of the IPCC Guidelines uses the following equations to estimate methane emissions from wastewater treatment processes (IPCC, 2019):

$$CH_{4,\text{Emissions,FP}} = [(TOW_{WW} - S_{Sol}) \cdot EF_{FP}] \tag{5.7}$$

$CH_{4,\text{Emissions,FP}}$ is the methane emissions (kg CH_4/d); TOW_{WW} is the influent wastewater BOD_L (kg/d); S_{Sol} is the BOD_L that does not undergo methanogenesis (soluble/nonsettleable) (kg/d); and EF_{FP} is the emission factor for facultative pond (kg CH_4/kg $BOD_{L,\text{Removed}}$).

The IPCC Guidelines use estimated per capita BOD_L loadings to calculate TOW_{WW}, but measured flowrates and BOD_L (calculated from BOD_5 data) concentrations are much more accurate. Equation (5.8) is then used:

$$TOW_{WW} = 0.001(Q_{mean})(BOD_L) \tag{5.8}$$

Q_{mean} is the mean influent wastewater flowrate (m³/d); BOD_L is the influent BOD_L concentration (mg/L); and S_{Sol} can be estimated by assuming that the mechanism for settling of TSS in a well-functioning facultative pond is similar to that in primary sedimentation. Figure 5.5 shows historical data for TSS and BOD_L removal in conventional primary sedimentation (Metcalf & Eddy/AECOM, 2014). BOD_L removal asymptotically approaches 45%; therefore, it is assumed that 45% of influent BOD_L settles to the sludge zone, and S_{Sol} can be estimated from Equation (5.9):

$$S_{Sol} = 0.55 TOW_{WW} \tag{5.9}$$

$0.55 TOW_{WW}$ assumes that 55% of influent BOD_L is soluble/nonsettleable; EF_{FP} is determined from Equation (5.10):

$$EF_{FP} = B_0 \cdot MCF_{FP} \tag{5.10}$$

Theory and design of facultative ponds

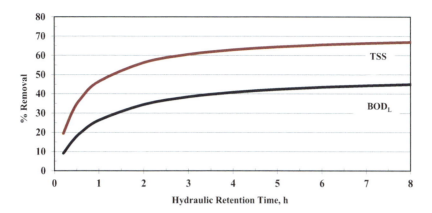

Figure 5.5 Removal of TSS and BOD_L in conventional primary sedimentation (from Metcalf and Eddy/AECOM (2014)).

B_0 is the maximum CH_4 producing capacity (0.25 kg CH_4/kg $BOD_{L,\,Removed}$) and MCF_{FP} is the methane correction factor for facultative pond (unitless).

The mass of methane produced is converted to the mass of carbon dioxide equivalents with Equation (5.11):

$$CO_{2,Equiv} = (CH_{4,\,Emissions,\,FP})(GWP) \quad (5.11)$$

$CO_{2,Equiv}$ is the carbon dioxide equivalent (kg $CO_{2,Equiv,FP}$/d) and GWP is the global warming potential for methane, 25 (IPCC, 2019).

The results can be compared with other treatment processes when the annual mass of carbon dioxide equivalents, $CO_{2,Equiv}$, is converted to mass per population equivalent:

$$CO_{2,Equiv}/pe\,yr = 365\left(\frac{CO_{2,Equiv}}{P}\right) \quad (5.12)$$

$CO_{2,Equiv}/pe\,yr$ = kg $CO_{2,Equiv}$ per population equivalent per year; P is the population connected to the facultative pond.

Finally, the estimated methane emissions can be compared to reported measured emissions in facultative ponds, usually expressed as mg CH_4/m^2 d, with Equation (5.13):

$$\lambda_{CH4,\,Emissions,\,FP} = CH_{4,\,Emissions,\,FP}\left(\frac{10^6}{A_{FP}}\right) \quad (5.13)$$

$\lambda_{CH4,\,Emissions,\,FP}$ is the CH_4 emissions at pond surface (mg CH_4/m^2 d); A_{FP} is the surface area of the facultative pond (m²); and 10^6 is the conversion from kg to mg.

The final design parameters are shown in Table 5.1. The values for B_0, MCF_{FP}, and EF_{FP} are directly from Tables 6.2 and 6.3 in Volume 5, Chapter 6, of the 2019 Revision of the IPCC Guidelines (IPCC, 2019).

Solving Equations (5.8), (5.9), (5.10), (5.7), (5.11), (5.12), and (5.13) yields the following:

(1) Influent wastewater BOD_L.

$$TOW_{WW} = 0.001(Q_{mean})(BOD_L) = 0.001(902\,m^3/d)(441\,mg/L) = 398\,kg/d$$

(2) Soluble wastewater BOD_L that will not be digested anaerobically.

$$S_{Sol} = 0.55\,TOW_{WW} = 0.55(398\,kg/d) = 219\,kg/d$$

Table 5.1 Methane emission parameters for facultative pond, Catacamas, Honduras.

Parameter	IPCC (2019)
B_0 (kg CH_4/kg BOD_L removed)	0.25
MCF_{FP}	0.20
EF_{FP}	0.05
GWP of CH_4	25
S_{FP} (kg BOD_L/d)	0.55 TOW_j^1

[1] Estimated from TSS removal in pond.

(3) Emission factor for methane production.

$$EF_{FP} = B_0 \cdot MCF_{FP} = (0.25 \text{ kg } CH_4/\text{kg } BOD_{L,Removed})(0.20) = 0.05 \text{ kg } CH_4/\text{kg } BOD_{L,Removed}$$

(4) Methane emissions in kg/d and kg/yr.

$$CH_{4,Emissions,FP} = [(TOW_{WW} - S_{Sol}) \cdot EF_{FP}] = [(398 \text{ kg/d} - 219 \text{ kg/d})(0.05)] = 8.95 \text{ kg } CH_4/\text{d}$$
$$= (8.95 \text{ kg } CH_4/\text{d})(365 \text{ d/yr}) = 3267 \text{ kg } CH_4/\text{yr}$$

(5) Carbon dioxide equivalent of methane emissions.

$$CO_{2,Equiv} = (CH_{4,Emissions,FP})(GWP) = (3267 \text{ kg } CH_4/\text{yr})(25) = 81\,670 \text{ kg } CO_{2,Equiv}/\text{yr}$$

(6) Carbon dioxide equivalent of methane emissions per population equivalent.

$$CO_{2,Equiv}/\text{pe yr} = \frac{CO_{2,Equiv}}{P} = \frac{81\,670 \text{ kg } CO_{2,Equiv}/\text{yr}}{3400 \text{ persons}} = 24.0 \text{ kg } CO_{2,Equiv}/\text{pe yr}$$

(7) Methane emissions flux at pond surface.

$$\lambda_{CH_4,Emissions,FP} = CH_{4,Emissions,FP}\left(\frac{10^6}{A_{FP}}\right) = 8.95 \text{ kg } CH_4/\text{d}\left(\frac{10^6 \text{ mg/kg}}{13\,800 \text{ m}^2}\right) = 649 \text{ mg } CH_4/\text{m}^2 \text{ d}$$

(b) Discussion

The value of $MCF_{FP} = 0.20$ (Table 5.1) from IPCC (2019) was reported without explanation of how it was selected; it essentially assumes that only 20% of the settled sludge BOD_L undergoes methanogenesis or that a large fraction of the methane produced is oxidized in the aerobic zone to carbon dioxide. A value derived from experience with unheated anaerobic digesters would be in the range of $MCF_{FP} \geq 0.50$ (ASCE, 1959). The following table presents the higher values obtained for $CO_{2,Equiv}$/pe yr, and methane flux, for $MCF_{FP} = 0.50$ and the original 0.20.

MCF_{FP}	EF_{FP} kg CH_4/kg $BOD_{L,Removed}$	kg $CO_{2,Equiv}$/pe yr	mg CH_4/m^2 d
0.20	0.050	24.0	649
0.50	0.125	64.1	1621

The annual per capita carbon dioxide equivalent emissions for the facultative pond and several wastewater treatment processes are presented in Table 5.2. As would be expected, septic tank emissions are higher than those of the facultative pond since both the sludge and water column are anaerobic.

Table 5.2 Annual per capita carbon dioxide equivalent emissions for several wastewater treatment processes.

Wastewater Process	kg $CO_{2,\,Equiv}$/pe yr
Facultative pond, Catacamas, Honduras (CH_4)	24.0–60.1
Septic tanks in the US[1] (CH_4)	96.5
Centralized WWTPs in the US[1]	
$\quad CH_4$	36.2
$\quad N_2O$	61.7
$\quad$ Electricity consumption in WWTP in the US[2]	45.7
$\quad$ Total	143.6

[1]USEPA (2021).
[2]Assuming median electricity consumption of 0.77 kWh/m³ for treatment and US average of 0.404 kg $CO_{2,\,Equiv}$/kWh (Energy Star, 2015).

Perhaps unexpectedly, aerobic centralized treatment plants, mostly activated sludge in the US, have high total emissions from methane, nitrous oxide, and electricity consumption:

- Methane emissions occur throughout the various treatment processes, including aerobic ones, as anaerobic pockets of soluble and solid organic matter form, such as in the settled sludge in sedimentation basins.
- Nitrous oxide is formed during the nitrification/denitrification processes in many plants (USEPA, 2021).
- Electricity consumed in wastewater treatment, with a median of 0.77 kWh/m³ for 1377 plants, with flowrates ranging from 757 to 280 090 m³/d, produced an estimated 45.7 kg $CO_{2,\,Equiv}$/pe yr (Energy Star, 2015).

In summary, the highest estimated per capita carbon dioxide equivalent emissions from methane in a well-operating facultative pond without excess sludge accumulation, assuming no oxidation to carbon dioxide by methanotrophs, would be $\approx$ 60% of the emissions from septic tanks (60.1/96.5) and $\approx$ 42% of the total $CO_{2,\,Equiv}$ emissions (60.1/143.6) from centralized wastewater treatment plants in the US.

Table 5.3 presents the methane flux estimates for the Catacamas pond and measured fluxes from facultative ponds in Mexico and Ecuador with excessive sludge accumulation. Excessive sludge accumulation greatly increased methane emissions, and the 2-m deep pond in Ecuador had sludge at the surface from the inlet to >1/3 the pond's length, with approximately 0.8-m sludge depth at mid-length (Ho et al., 2021). Unfortunately, there are few measured flux data in the literature for well-functioning ponds without excessive sludge accumulation.

Table 5.3 Methane flux (mg CH_4/m² d) reported for facultative ponds.

Wastewater Process	mg CH_4/m² d
Facultative pond, Catacamas, Honduras	649–1621
Facultative ponds, Mexico[1] (excess sludge in pond)	2976–4608
Facultative ponds, Ecuador[2] (excess sludge in pond)	
$\quad$ Inlet	13 000
$\quad$ Middle	7900
$\quad$ Outlet	4000

[1]Source: Paredes et al. (2015).
[2]Source: Ho et al. (2021).

Figure 5.6 Rising sludge from the anaerobic zone lifted with methane and carbon dioxide bubbles. This facultative pond had excessive sludge accumulation after more than 20 years of operation without sludge removal. In spite of the accumulation, the aerobic zone is still functioning well as seen by the bright green color. Unless the pond is desludged, the rising sludge will eventually cover the surface and the pond will turn anaerobic throughout its depth (Figure 5.7; Puno, Peru).

The minimization of methane production requires good design, construction, and operation and maintenance. When possible, deeper ponds (2 m) with longer detention times ($\geq$15 d) should be used. Designers should estimate sludge production and depths, and sludge depths should be measured annually. Sludge measurement and removal typically are not included in designs or operation manuals, and many ponds soon reach the conditions shown in Figures 5.6 and 5.7.

Sulfate reduction. When dissolved oxygen is depleted, sulfate reduction will occur before methanogenesis if significant concentrations of sulfate are present. Under these conditions, sulfate becomes the electron acceptor and the wastewater organic matter is oxidized according to unbalanced Equation (5.14):

$$C_{10}H_{19}O_3N + SO_4^{2-} + H^+ \rightarrow H_2S + HS^- + H_2O + CO_2 + HCO_3^- + NH_4^+ + C_5H_7O_2N$$

Organic matter   Bacterial cells   (5.14)

in wastewater

Sulfate reduction results in the formation of hydrogen sulfide, which diffuses to the aerobic zone, where it can be oxidized by *Acidothiobacillus*, and it may also diffuse to the atmosphere. If sulfate concentrations are relatively low in the raw wastewater, sulfate reduction will likely occur in the sewerage system leaving little or no sulfate in the influent of the facultative pond.

Anoxygenic photosynthesis and sulfide oxidation. The photosynthetic purple and green sulfur bacteria use light energy to fix carbon dioxide but, unlike algae that oxidize H_2O to O_2, they oxidize S^{2-} to SO_4^{2-} in two steps as shown in Equations (5.15a) and (5.15b) (Mara, 2003; Pepper *et al.*, 2015):

Theory and design of facultative ponds

Figure 5.7 A continuation of the process shown in Figure 5.6. Rising sludge from excessive accumulation is gradually covering a facultative pond. This pond operated for more than 15 years without desludging and is now anaerobic, with the maximum release of methane into the atmosphere (El Paraiso, Honduras).

$$21H_2S + 10CO_2 + 2NH_3 \xrightarrow{\text{Solar radiation}} 21S + 2H^+ + 16H_2O + \underset{\text{Bacterial cells}}{2C_5H_7O_2N} \quad (5.15a)$$

$$21S + 30CO_2 + 6NH_3 + 36H_2O \xrightarrow{\text{Solar radiation}} 3H_2O + \underset{\text{Sulfuric acid}}{21H_2SO_4} + \underset{\text{Bacterial cells}}{6C_5H_7O_2N} \quad (5.15b)$$

The purple and green sulfur bacteria live where both sulfide and light are present, in muds and stagnant waters, sulfur springs, and saline lakes in natural environments (Pepper *et al.*, 2015). In well-functioning facultative ponds, they are found at deeper depths below the algae; this is because they absorb light at longer wavelengths (750–900 nm) that penetrate deeper than the shorter wavelengths absorbed by algae (<700 nm) (Mara, 2003).

Organically overloaded ponds with high H_2S concentrations, which are toxic to algae, will have reddish colorations due to the dominance of the purple and green sulfur bacteria as shown in Figure 5.8. This effect also occurs in hydraulic dead spaces such as pond corners (Figure 5.8). Under these conditions, ponds function as anaerobic ponds with poorer treatment results.

Figure 5.8 Examples of reddish coloration in facultative ponds as a result of anoxygenic photosynthesis by the purple and green sulfur bacteria. The top left photo is the dead space in a corner of an otherwise well-functioning pond. The top right photo is a highly overloaded pond converting to an anaerobic one. Bottom photo is an overloaded pond (top left: Lagarto, Brazil; top right: Villanueva, Honduras; bottom: Mendoza, Argentina).

5.2 THEORY OF DESIGN OF FACULTATIVE PONDS

5.2.1 Maximum organic surface loading

The design of a facultative pond should be based on the daily oxygen mass balance of ultimate BOD and oxygen production by photosynthesis (Arceivala & Asolekar, 2007):

Mass of oxygen demand in influent wastewater ≤ Mass of oxygen produced by photosynthesis

kg BOD_L/d in influent wastewater ≤ kg O_2/d produced in facultative pond

Because a facultative pond is a natural solar collector, the kg O_2/d produced is distributed over the surface area of the pond. Thus, the production of O_2 must be expressed in terms of mass per day per unit area, usually hectares (ha). The area of the pond is then calculated as follows:

$$\text{Area of facultative pond} = \frac{\text{kg BOD}_L/\text{d in influent wastewater}}{\text{kg O}_2/\text{d ha produced}}(\text{SF})(\alpha_T) \qquad (5.16)$$

where SF is the safety factor and α_T is the temperature coefficient (decimal) for water temperature T (°C).

Historically, BOD_5 has been used instead of BOD_L without a safety factor, using empirical surface loading design values based on latitude or climate (Gloyna, 1971; USEPA, 1983), or empirical equations such as the widely used one by Mara (2003). This approach has worked since approximately 45% of the influent BOD_L would be typically removed by sedimentation, and thus, an influent BOD_L of 300 mg/L (BOD_5 = 200 mg/L) would be reduced to 165 mg/L in the water column. The hidden safety factor would then be 300/165 = 1.8.

The temperature coefficient was used in the design equations of Oswald and Gotaas (1957) to adjust for changes in algal growth rates due to temperature changes and should be used especially when the water temperature is below 20°C as discussed in Section 5.2.3.

The solar energy required to produce 1 kg of algal cells is 24 000 kJ (Rittmann & McCarty, 2001). Of the solar energy radiating on the surface of a facultative pond, only a small percentage is used by the algae as a result of their conversion efficiency; the conversion efficiency varies between species of algae, and the range has been reported from 2 to 7% (Arceivala et al., 1970). For design purposes, a value of 3% should be used (Rittmann & McCarty, 2001).

The photosynthesis equation (Equation (5.1)) is combined with the energy required to produce algal cells, and the efficiency of solar energy conversion, to give the following maximum oxygen production $\geq$ the maximum BOD_L surface loading (without SF and α_T). It should not be lost on designers that Equation (5.17) is based on photosynthetic oxygen production, in which BOD_L surface loadings must accommodate:

$$\lambda_{S,Max} = \frac{(\text{Solar radiation kJ/ha d})(\text{Fraction converted} = 0.03)(1.55\,\text{kg O}_2/\text{kg algae})}{24000\,\text{kJ/kg algae produced}} \qquad (5.17)$$

where $\lambda_{S,Max}$ is the maximum surface loading (kg BOD_L/ha d) $\leq$ maximum oxygen production (kg O_2/ha d).

Equation (5.17) reduces to Equation (5.18):

$$\lambda_{S,Max} = (1.937 \times 10^{-6}) \cdot (SR_i) \qquad (5.18)$$

where SR_i is the average daily solar radiation for design month, i (kJ/ha d).

Finally, Equation (5.19) is used to calculate the maximum oxygen production and hence the maximum surface BOD_L loading, based on water temperature for the design month with the lowest water temperature:

$$\lambda_{S,Tw} = \lambda_{S,Max}(\alpha_T) \qquad (5.19)$$

5.2.1.1 Sources of solar radiation data
5.2.1.1.1 CLIMWAT and CROPWAT
One excellent source of solar radiation data is CLIMWAT (http://www.fao.org/land-water/databases-and-software/climwat-for-cropwat/), published by the Water Development and Management Unit and the Climate Change and Bioenergy Unit of the Food and Agriculture Organization (FAO). CLIMWAT's climatic database must be used in combination with the program CROPWAT, which is also available from FAO.

CLIMWAT contains observed agroclimatic data of over 5000 stations worldwide and provides long-term data for the following climatic parameters:

- mean solar radiation (MJ/m² d);
- mean daily maximum temperature (°C);
- mean daily minimum temperature (°C);
- mean relative humidity (%);
- mean wind speed (km/d);
- mean sunshine (h/d);
- monthly rainfall (mm/mo);
- monthly effective rainfall (mm/mo); and
- reference evapotranspiration calculated with the Penman–Monteith method (mm/d).

CLIMWAT is especially useful since all data required for wastewater reuse (temperature, precipitation, and evapotranspiration) is downloaded with the solar radiation data.

Equation (5.18) can be converted to Equation (5.20) when CLIMWAT solar radiation data are expressed in units of MJ/m² d:

$$\lambda_{S,\text{Max}} = (1.937 \times 10^{-6}) \cdot (SR_{i,\text{MJ}}) \left(\frac{1000 \text{ kJ}}{1.0 \text{ MJ}}\right)\left(\frac{10\,000 \text{ m}^2}{\text{ha}}\right) = 19.37(SR_{i,\text{MJ}}) \quad (5.20)$$

where $SR_{i,\text{MJ}}$ is the solar radiation (MJ/m² d); 19.37 units $= \dfrac{\text{kg O}_2 \text{ m}^2}{\text{MJ ha}}$.

5.2.1.1.2 NASA POWER data access viewer

The NASA POWER Data Access Viewer (DAV) is a web mapping application (https://power.larc.nasa.gov/docs/tutorials/data-access-viewer/user-guide/) that contains geospatially enabled solar, meteorological, and cloud-related parameters formulated for assessing and designing renewable energy systems. The following parameters are useful for facultative pond designs when there are no CLIMWAT data for the design location:

- all-sky solar insolation on a horizontal surface (kWh/m² d);
- mean temperature at 2 m (°C).

CLIMWAT only has maximum and minimum air temperatures, and the mean temperature data at 2 m are especially useful to estimate water temperatures as discussed in Section 5.2.3.

Equation (5.18) can also be converted to Equation (5.21) with NASA POWER solar radiation data expressed in units of kWh/m² d:

$$\lambda_{S,\text{Max}} = (1.937 \times 10^{-6}) \cdot (SR_{i,\text{kWh}}) \left(\frac{3600 \text{ kJ}}{1.0 \text{ kWh}}\right)\left(\frac{10\,000 \text{ m}^2}{\text{ha}}\right) = 69.73(SR_{i,\text{kWh}}) \quad (5.21)$$

where $SR_{i,\text{kWh}}$ is the solar radiation (kWh/m² d); 69.73 units $= \dfrac{\text{kg O}_2 \text{ m}^2}{\text{kWh ha}}$.

5.2.1.2 Water temperature and algal growth
5.2.1.2.1 Design water temperature

Design water temperature is best determined by direct measurements of stabilization pond temperatures throughout the year from similar ponds near the design location. Nearby shallow lakes may also have water temperature data that could be used to estimate facultative pond temperatures. In the absence of field data, water temperatures have been estimated from air temperature data. For mean air temperatures between 15 and 35°C, the following equation has been proposed by Von Sperling (2007):

$$T_{\text{water}} = 12.7 + 0.54 T_{\text{air}} \quad (15°C \leq T_{\text{air}} \leq 35°C) \quad (5.22)$$

where T_{water} is the mean water temperature for mean air temperature T_{air} (°C).

Most of the data used to develop Equation (5.22) were from ponds in tropical or semi-tropical climates. For greater temperature ranges, Ali (2013) developed a simple linear regression model for data from an experimental pond in India that can be used with air temperatures from 8.5 to 38.8°C as shown in the following equation:

$$T_{water} = 0.86 T_{air} + 5.29 \quad (8.5°C \leq T_{air} \leq 38.8°C; (n = 1135; R^2 = 0.862) \tag{5.23}$$

The experimental pond had an area of 0.47 ha and a mean depth of 2.75 m. The water temperature was monitored 10 mm below the surface for 1135 consecutive days, with daily air temperatures obtained from the local meteorological station. The climate was dry semi-arid at 500 m elevation, with an annual rainfall of 723 mm, and mean daily air temperatures of 35°C in summer and 15°C in winter (Ali, 2013). While the quality and quantity of the data used to develop the model are as good as it gets, it is not known how applicable it is to vastly different climates (high elevations, high latitudes, high rainfall, etc.) for stabilization the pond design.

5.2.1.2.2 Temperature effects on algal growth

Water temperature has a significant effect on algal and bacterial growth rates, especially algal growth rates, which can be an order of magnitude lower than bacterial rates. Oswald and Gotaas (1957), for example, incorporated temperature coefficients for the growth of *Chlorella* species in their design equations for facultative ponds. The effect of temperature on algal growth needs to be considered especially when water temperatures fall below 20°C.

The minimum cell retention time for algae or bacteria to grow can be estimated using Equation (5.24) (Rittmann & McCarty, 2001):

$$\theta_{c,T}^{min} \approx \frac{1}{\mu_T - k_T} \approx \frac{1}{\mu_T} \tag{5.24}$$

where $\theta_{c,T}^{min}$ is the minimum algal cell residence time in the pond for growth (d); μ_T is the specific growth rate at temperature T (d^{-1}); and k_T is the cellular decay coefficient (d^{-1}). The cellular decay coefficient, k_T, is dropped because it is usually much lower than μ_T.

Specific growth rates and minimum cell retention times for bacteria and algae at 20°C are presented in Table 5.4. While different studies of algal growth cannot be directly compared because of additional growth factors such as the magnitude of solar radiation and nutrient concentrations, they can generally be compared to aerobic heterotrophic growth. From Table 5.4, the ratio of minimum algal to aerobic heterotrophic cell residence times is

$$\frac{\theta_{c,T,\ algae}^{min}}{\theta_{c,T,\ heterotrphic\ bacteria}^{min}} \approx \frac{5}{1} \text{ to } \frac{10}{1}$$

Algal growth rates, and the incorporation of an algal temperature coefficient for water temperature, thus govern the design of facultative ponds. As an example, Oswald and Gotaas (1957) presented the following temperature coefficients for the growth of *Chlorella* species in facultative ponds that were used in their design equations:

Water Temperature (°C)	Temperature Coefficient for *Chlorella* Growth (α_T)
10	0.49
15	0.87
20	1.00
25	0.91
30	0.82
35	0.69

Table 5.4 Specific growth rate and approximate minimum cell retention time for bacteria and algae at 20°C.

Microorganism	μ_T (d⁻¹)	$1/\mu_T \approx \theta_{c,T}^{min}$ (d)
Bacteria		
Aerobic heterotrophs	8.4–13.2	0.08–0.12
Sulfate reducers	0.29	3.45
Autotrophic sulfide oxidizers	1.4	0.71
Methanogens	0.30–0.50	2.00–3.33
Algae and Cyanobacteria		
Mixed cultures in lakes	1.00	1.00
Cultures of algal genera and species:		
Chlorophyta	0.58	1.72
Chrysophyta	0.50	2.00
Chryptophyta	0.35	2.86
Cyanophyta	0.27	3.70
Mixed culture in wastewater algal pond	0.99	1.01

Source: Developed from Buhr and Miller (1983), Nalley *et al.* (2018), and Rittmann and McCarty (2001).

Temperature coefficients for the algal genera most common in facultative ponds can be calculated by using existing data on temperature effects on algal growth. A detailed study by Nalley *et al.* (2018) presented growth rates for 26 algal species in five functional groups (genera). A plot of their data for the four genera common in facultative ponds is shown in Figure 5.9. Included is a curve from a different study for *Euglena gracilis*, a common algal species in facultative ponds.

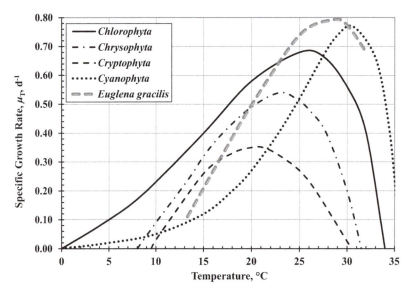

Figure 5.9 Growth curves for the algae and cyanobacteria genera common in facultative ponds as a function of temperature (Nalley *et al.*, 2018). Curve for *E. gracilis*, another common alga found in ponds, is by Buetow (1962).

All the algal genera and species in Figure 5.9 have their maximum growth rates between 20 and 30°C. Below the temperature of maximum growth, the specific growth rates decline more rapidly for a 10°C temperature decrease than would be expected for bacteria, which is typically around 50%. For *Chrysophyta*, for example, at 20°C, $\mu_{20} = 0.50$ d^{-1}, and at 10°C, $\mu_{10} = 0.10$ d^{-1}, a decrease of 80%.

The temperature coefficients for algal growth throughout the temperature range (5–35°C) can be estimated by dividing the growth rate at temperature T to the maximum growth rate for the genera or species of algae:

$$\alpha_T = \frac{\mu_T}{\mu_{\text{Max}}} \tag{5.25}$$

μ_T is the specific growth rate at temperature T (d^{-1}) and μ_{Max} is the maximum specific growth rate at temperature (T_{Max}, d^{-1}).

Figure 5.10 is developed using Equation (5.25) for the temperature range 5–30°C. *Chlorophyta*, *Chrysophyta*, and *Cryptophyta* all have temperature coefficients in the range of 0.85–1.0 at 20°C, but the coefficients decrease abruptly from a 10°C decrease, to the range of 0.10–0.35 at 10°C, which is a larger decrease than would be expected for bacteria. *Cyanophyta* and *E. gracilis* have coefficients near 1.0 at 30°C that drop to 0.35 and 0.50, respectively, at 20°C.

Chlorophyta, *Chrysophyta*, and *Cryptophyta* all have similar temperature coefficient curves down to 13°C in Figure 5.10, and any of the three could be used as a first approximation of temperature coefficients in facultative ponds. In Figure 5.11, *Chlorophyta* is selected for the design curve. The empirical equation developed by Mara (2003) is plotted for comparison and was developed as follows:

$$\lambda_{S,T} = 350(1.107 - 0.002T)^{T-25} \tag{5.26}$$

$\lambda_{S,T}$ is the surface loading rate at mean air temperature, T, in coldest month (kg BOD$_5$/ha d) and $\lambda_{S,\text{Max}}$ is 350 kg BOD$_5$/ha d (Mara, 2003).

Therefore, the ratio of surface loading at temperature T to the maximum at 350 kg BOD$_5$/ha d is as follows:

$$\frac{\lambda_{S,T}}{\lambda_{S,\text{Max}}} = \frac{350(1.107 - 0.002T)^{T-25}}{350} = (1.107 - 0.002T)^{T-25} \tag{5.27}$$

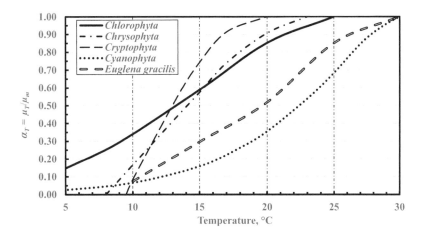

Figure 5.10 Temperature coefficients for the genera and species of algae and cyanobacteria in Figure 5.9. Plot for *E. gracilis* was calculated from the data reported by Torihara and Kishimoto (2015).

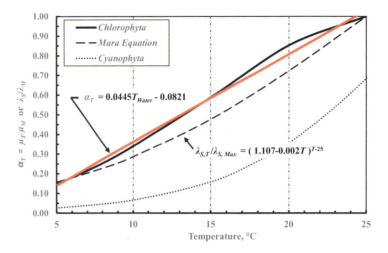

Figure 5.11 Design curve for temperature coefficient in facultative ponds using *Chlorophyta* as the baseline algal genera. The relative surface loading, $\lambda_{S,T}/\lambda_{S,Max}$, using the Mara equation (Mara, 2003) is shown for comparison.

The Mara equation uses the air temperature of the coldest month to estimate surface loading, with a maximum loading of 350 kg BOD_5/ha d at temperatures $\geq 25°C$. The plot of Equation (5.27) is similar to the temperature coefficient curve for *Chlorophyta*, but the curve is a function of air temperature and a maximum surface loading of 350 kg BOD_5/ha d.

A plot of temperature coefficients for high temperatures of 25–36°C is shown in Figure 5.12. The coefficients begin to decrease abruptly with temperatures > 31–32°C. Ponds in arid regions may have water temperatures > 32°C, and this should be considered in the design.

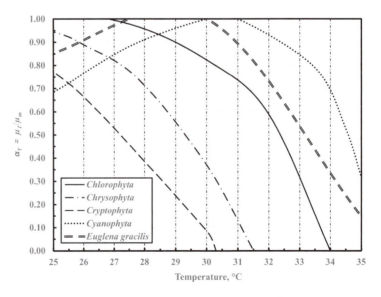

Figure 5.12 Temperature coefficients for high water temperatures (25–35°C) for the genera and species of algae and cyanobacteria in Figure 5.14. Plot for *E. gracilis* was calculated from data reported by Torihara and Kishimoto (2015).

Theory and design of facultative ponds

Table 5.5 Values of α_T as a function of water temperature for algal genus *Chlorophyta* (from Equation (5.28); Figures 5.16 and 5.17).

T_{water} (°C)	5	10	15	20	22.5	25	26	27	28	29	30	31	32	33
α_T	0.14	0.36	0.59	0.81	0.92	1.0	1.0	1.0	0.95	0.90	0.82	0.74	0.60	0.30

The values of α_T for *Chlorophyta* for the temperature ranges in Figures 5.11 and 5.12 are tabulated in Table 5.5. Equation (5.28) can also be used for temperatures of 5–25°C:

$$\alpha_T = 0.0445 T_{water} - 0.0821 \tag{5.28}$$

As discussed previously, T_{water} can be estimated down to 15°C with the von Sperling equation:

$$T_{water} = 12.7 + 0.54 T_{air} \quad (15°C \leq T_{air} \leq 35°C) \tag{5.29}$$

5.2.1.3 Case study: surface loading and facultative pond performance, Nagpur, India

Arceivala *et al.* (1970) presented the effects of increasing surface loadings on the depth of the aerobic zone for a facultative pond in Nagpur, India, which are shown in Figure 5.13. Measured solar radiation data for Nagpur will be used to verify the results from Figure 5.13 using Equation (5.18):

(1) No CLIMWAT data exist for Nagpur; thus, mean air temperature and solar radiation data from NASA POWER will be used for the coordinates of central Nagpur, 21.145 N, 79.085 E. These data are presented in Table 5.6.

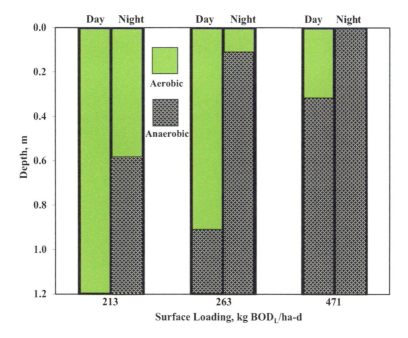

Figure 5.13 Diurnal depth of the aerobic zone as a function surface BOD_L loading for a facultative pond at Nagpur, India. (*Source*: Adapted from Arceivala *et al.* (1970).)

Table 5.6 Solar radiation[1], mean air temperature[1], calculated water temperatures, and maximum surface loadings, Nagpur, India (latitude 21.145 N; longitude 79.085 E).

1	2	3	4	5	6	7
Month	SR (kWh/m² d)	$\lambda_{S,Max}$ (kg BOD$_L$/ha d)	Mean Air Temperature at 2 m (°C)	Water Temperature (°C)	α_T	$\lambda_{S,T}$ (kg BOD$_L$/ha d)
J	4.62	322	20.46	23.7	0.92	296
F	5.50	384	23.79	25.5	1.00	384
M	6.22	434	28.83	28.3	0.95	412
A	6.77	472	33.48	30.8	0.74	349
M	6.59	460	36.30	32.3	0.60	276
J	4.95	345	31.91	29.9	0.82	283
J	3.89	271	27.13	27.4	1.00	271
A	3.72	259	26.15	26.8	1.00	259
S	4.40	307	26.04	26.8	1.00	307
O	5.13	358	24.24	25.8	1.00	358
N	4.80	335	21.62	24.4	1.00	335
D	4.49	313	19.81	23.4	1.00	313

[1]*Source:* Data from NASA POWER.

(2) Solar radiation is expressed in units of kWh/m² d; thus, Equation (5.29) is used to determine maximum surface loadings, $\lambda_{S,Max}$.

For January, the calculation is:

$$\lambda_{S,Max} = 69.73(SR_{i,kWh}) = \left(\frac{69.73 \text{ kg O}_2 \text{ m}^2}{\text{kWh ha}}\right)\left(\frac{4.62 \text{ kWh}}{\text{m}^2 \text{ d}}\right) = 322 \text{ kg BOD}_L/\text{ha d} \leq 322 \text{ kg O}_2/\text{ha d}$$

(3) The remaining monthly values of $\lambda_{S,Max}$ are listed in Table 5.6 (column 3).
(4) Water temperature is calculated from air temperature using Equation (5.22).
(5) The temperature coefficient, α_T, is estimated from Table 5.5 for the monthly water temperature.
(6) Monthly water temperatures are listed in column 6 in Table 5.6.
(7) The monthly surface loading adjusted for water temperature, $\lambda_{S,T}$, is calculated with Equation (5.19). For January (column 7 in Table 5.6):

$$\lambda_{S,T} = \lambda_{S,Max}(\alpha_T) = (322 \text{ kg BOD}_L/\text{ha d})(0.92) = 296 \text{ kg BOD}_L/\text{ha d}$$

(8) Table 5.6 and Figure 5.14 show the monthly results for $\lambda_{S,T}$.
(9) From Table 5.6 and Figure 5.14, the design surface loading is for April:

$$\lambda_{S,T, April} = 259 \text{ kg BOD}_L/\text{ha d} \leq 259 \text{ kg O}_2/\text{ha d}$$

The performance of the pond is illuminated by comparing actual surface loadings in Figure 5.13 with maximum surface loadings calculated from solar radiation data for Nagpur in Figure 5.14. The measured surface loading of 263 kg BOD$_L$/ha d in Figure 5.13, which maintained aerobic conditions at night, is close to the maximum of 259 kg BOD$_L$/ha d calculated in Table 5.6.

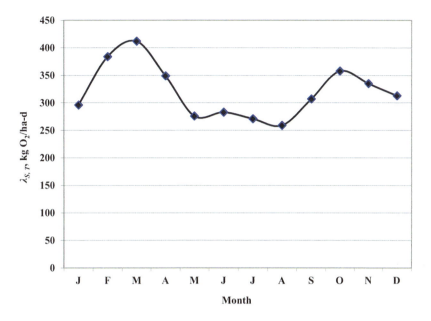

Figure 5.14 Water temperature adjusted BOD_L surface loadings, $\lambda_{S,T}$, for Nagpur, India. (*Source*: Data from Table 5.6.)

The depth of the aerobic zone is an important parameter for potential methane and hydrogen sulfide oxidation, and designs need to have sufficient safety factors to maintain aerobic conditions at night. An overloaded pond that is anoxic or anaerobic at night or day can still have good BOD_L removal but will release methane into the atmosphere and cause odor problems. Water temperature was considered in this example but did not affect the maximum surface loading for April ($\alpha_T = 1.0$ from Table 5.6).

5.2.1.4 Case study: organic overloading of facultative ponds in Honduras

The objective of a study in Honduras was to monitor 11 wastewater stabilization pond systems from different geographical areas, during the dry and wet seasons, to develop a database for improving design parameters, and operation and maintenance requirements (Oakley, 2005). It was also planned that the study would help develop effluent standards that would address the public health issues facing the municipalities and also promote effluent reuse in agriculture.

Nine of the selected systems were facultative–maturation pond configurations. All systems were monitored for 3 consecutive days during both the wet and dry seasons. The flowrates entering each pond system were continuously monitored over 3 days using an ISCO area–velocity flow meter; this allowed for the accurate determination of hydraulic retention time in each pond system monitored.

Table 5.7 presents the results for six facultative ponds for influent flowrates, calculated hydraulic retention times, mean influent BOD_L concentrations (calculated from BOD_5 data), and calculated organic surface loading rates. Basing the measured influent flowrate on the original design populations, the calculated per capita flow rates ranged 127–488 liters per person per day (Lppd), and five of six are much higher than the typical design assumption of 120–150 Lppd. The higher flows were likely due to increased connections due to the population growth, illegal connections from commercial and industrial sources, and inflow and infiltration into the sewer system.

As a result of the higher flow rates, the calculated hydraulic retention times were much shorter, and the organic surface loading rates were higher, than anticipated from the original designs. Only two

Table 5.7 Results of measured flow rates, hydraulic retention times (HRT), and organic surface loading rates for six monitored pond systems in Honduras. (Lppd = Liters/person/day)

Pond System	Mean Influent Flowrate (m³/d)	Original Design Population	Per Capita Flow Based on Design Population (Lppd)	Area of Facultative Pond (ha)	HRT (d)	Mean Influent BOD$_5$ (mg/L)	Mean Influent BOD$_L$ (mg/L)	Organic Surface Loading Rate (kg BOD$_L$ ha-d)	Mean Effluent BOD$_5$ (mg/L)
Catacamas Este	2610	5350	488	1.02	6.2	348	522	1332	131
Catacamas Oeste	924	3400	272	1.38	23.5	365	548	370	135
El Progreso	2932	23 000	127	2.83	20.9	71	107	110	28
Juticalpa	3510	11 422	307	1.23	6.1	177	266	621	110
Morocelí	218	705	309	0.12	7.0	220	330	615	60
Trinidad	1816	6108	297	0.98	7.8	76	114	211	42

facultative pond systems satisfied the recommendation that facultative ponds should have a minimum hydraulic retention time of 15 days (Von Sperling, 2007). The mean influent BOD$_L$ was found to vary greatly among systems, ranging 107–548 mg/L. As a result of the higher flowrates and influent BOD$_L$ values, the organic surface loading rates of three of the ponds far exceeded (615–1332 kg BOD$_L$/ha d) the estimated maximum loadings for the latitudes and climates of Honduras, which is estimated to be 275–350 kg BOD$_L$/ha d (Figure 5.15).

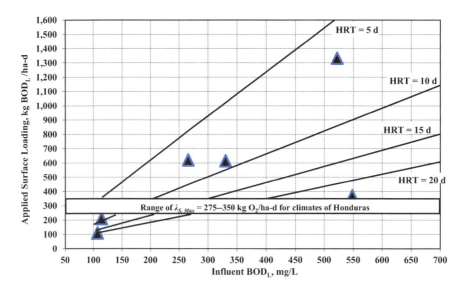

Figure 5.15 Influent BOD$_L$ concentrations and resultant applied BOD$_L$ surface loading rates for six facultative ponds in Honduras that were monitored intensively in winter and summer for 1 year. Only two ponds were within or under the maximum surface loading range of 275–350 kg BOD$_L$/ha d for the country's climatic conditions. The hydraulic retention timelines show that four ponds were far below the minimum 15 d recommended for a facultative pond (*Source*: Modified from Oakley (2005).)

Theory and design of facultative ponds

The overall BOD_5 removal, however, was what would be expected for normally loaded facultative ponds (Oakley, 2005), but with one caveat: the three ponds that were highly overloaded (615–1332 kg BOD_L/ha d) were operating as anaerobic ponds (Figure 5.15) with hydraulic retention times of 6.1–7.0 days. The pond at Trinidad (Figure 5.16), while not overloaded according to surface loading rate calculations, was also anaerobic due to overloading and sludge build-up in a baffled inlet, which should not be used in facultative ponds.

Figure 5.16 Top photo: The facultative pond at Trinidad, Honduras, was anaerobic throughout the monitoring project as a result of a short hydraulic retention time, sludge accumulation, and use of baffles that caused the inlet to be organically overloaded. Bottom photo: The facultative pond at El Progreso, Honduras, is an example of a well-designed and operated facultative pond, with brilliant green color from the algal growth. The pond is free of odors and has no floating scum or aquatic plant growth. Both ponds had similar BOD_5 removal.

A well-designed facultative pond, not organically overloaded and with a hydraulic retention time ≥15 d, should be aerobic in the upper depths 24 hours a day, have minimal odors, remove an important fraction of methane produced in the anaerobic zone, have ≥99% helminth egg reduction, and not need desludging for at least 10 years. The problems found in the Honduras study are common in many parts of the world, and remedies from the lessons learned need to be implemented by design engineers and the municipalities in charge of operation and maintenance, if integrated wastewater management is ever to be successful on a wide scale.

5.2.2 Wind effects in facultative ponds

The effect of reaeration by wind is not included in the design of facultative ponds because, as Oswald discussed long ago, the oxygen gain from reaeration is only a fraction of the gain from photosynthesis. For example, dissolved oxygen within a pond would have to have a deficit of 10 mg/L, a condition in which the pond would be anoxic or anaerobic, to obtain a gain of 18 kg O_2/ha d by reaeration, a minor gain compared to photosynthesis (Oswald, 1963).

Wind-induced mixing can generate circular flow patterns in ponds, however, contributing significantly to hydraulic short circuiting: The surface current of the pond flows in the direction of the wind, with the return flow in the upwind direction along the bottom. To minimize wind-induced short circuiting, the inlet–outlet axis of ponds should be perpendicular to prevailing wind directions (Figure 5.17) (USEPA, 1983). Wind fences can also be used for smaller ponds that are not oriented correctly to prevailing winds (Lloyd *et al.*, 2003a, 2003b).

5.2.3 Hydraulic considerations
5.2.3.1 Longitudinal dispersion

The hydraulic regime in facultative ponds, and in all waste stabilization ponds, is dispersed flow (Thirumurthi, 1969). The most efficient hydraulic regime is plug flow, in which designs should always attempt to approximate by minimizing as much as possible the effects of dispersion. Baffled channels are used to minimize dispersion effects in maturation ponds, but they cannot be used in facultative ponds because the daily surface BOD_L loading must be satisfied by photosynthetic oxygen production over the entire surface area of the pond. A baffled facultative pond will be organically overloaded at the inlet and will soon turn anaerobic.

Facultative ponds should be designed with the following guidelines (Figure 5.18):

- l/w ratios = 3/1, and no less than 2/1 if site conditions are limiting. ($l/w > 3/1$ causes organic surface overloading at the inlet end of the pond.)
- Multiple inlets and outlets, all connected with open channels for flow distribution control to each inlet/outlet.
- Adjustable sluice gates or flashboards for each inlet/outlet to maintain equal flowrate distribution (Figures 8.14 through 8.17).

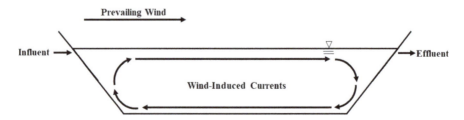

Figure 5.17 If the pond inlet and outlet are oriented in the direction of the prevailing wind, the surface current will flow with the wind and the return flow against the wind along the bottom. To avoid this effect, the inlet–outlet axis should be perpendicular to the prevailing wind direction.

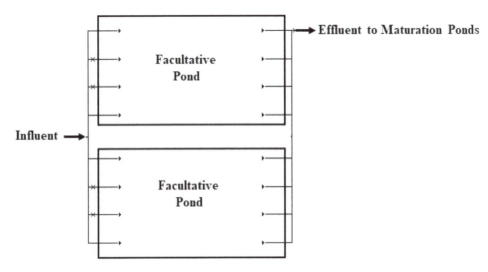

Figure 5.18 Facultative ponds should be designed to approximate plug flow, with at least two ponds in parallel (except for very small systems), with multiple inlets and outlets with open channels to control flow distribution. The *l/w* ratio should be 3/1 if not limited by site conditions, and no less than 2/1.

5.2.3.2 Thermal stratification and hydraulic short circuiting

Ponds become stratified as a result of temperature differences between the warmer upper zone (epilimnion), and the lower cooler zone (hypolimnion), which are separated by the thermocline (Figure 5.19). Thermal stratification can be diurnal, changing during day and night by heating and cooling of the epilimnion, and broken up by wind-induced mixing (Mara, 2003; Von Sperling, 2007). It can also be seasonal, driven by differences between influent temperature and pond temperature, especially at temperate and higher latitudes: In summer, the influent is cooler than the pond, and it sinks and flows along the bottom to the outlet structure; in winter, the influent is warmer than the pond and it skims along the surface layers to the outlet (USEPA, 1983).

Wind-induced mixing is important in controlling thermal stratification, and all pond designs should include proper orientation to prevailing wind directions, with inflow–outflow axes perpendicular to it. Pond designs should also not have irregular shapes that could prevent homogenous mixing of the entire pond (Von Sperling, 2007). Pond perimeters should be free of trees, vegetation, hills, buildings, or other obstacles that could block the normal path of winds.

Wind-powered slow mixers have been used with success in large facultative ponds to improve performance by eliminating large dead zones caused by thermal stratification and poor hydraulic

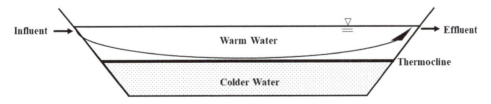

Figure 5.19 Thermal stratification in a facultative pond, where the thermocline forms a barrier to vertical mixing, can cause significant hydraulic short circuiting if it is not broken up by wind-induced mixing. (*Source*: Adapted from Geradi (2015).)

Figure 5.20 Wind-powered slow mixers in a facultative pond of the sanitation cooperative SAGAUPAC, Santa Cruz, Bolivia. If properly employed, the slow mixers greatly reduces vertical stratification and hydraulic short circuiting.

design (Figure 5.20). The mixers are strategically placed after dye studies are performed. After destratification, in addition to improved hydraulic performance, both dissolved oxygen produced by algae and the algae themselves are redistributed with depth.

5.2.3.3 Sludge accumulation effect on hydraulic short circuiting

The majority of sludge accumulation in facultative ponds occurs immediately downstream of inlet structures. If the sludge accumulation becomes excessive, it will divert flow direction to form preferential flow paths, reducing hydraulic retention times and creating dead zones (Figure 5.21). Ponds with a single inlet and outlet structures are especially prone to this problem, more so if they are not desludged on a regular basis based on routine sludge depth measurements (Alvarado *et al.*, 2012; Nelson *et al.*, 2004).

To control sludge accumulation effects for design, ponds should have multiple inlet and outlet structures, interconnected with open channels, with adjustable weirs to adjust flow distribution. For operation and maintenance, sludge accumulation should never be allowed to reach levels where it influences the flow patterns of the pond.

5.2.4 Pathogen reduction

Pathogen reduction should be the principal design objective for wastewater treatment if the effluent is to be reused in agriculture or aquaculture. The design approach with waste stabilization ponds is unique in that pathogen reduction occurs simultaneously with the removal of the conventional parameters of TSS and BOD_5, which are both sufficiently removed in facultative ponds that have been sized for the maximum surface loading at the design water temperature ($\lambda_{S,Tw}$).

Figure 5.21 Single inlets in facultative ponds such as this one cause sludge accumulation immediately downstream of the inlet. The sludge accumulation in turn diverts inflow, forming preferential flow paths and dead zones, reducing the hydraulic efficiency of the pond (Zaragoza, El Salvador).

5.2.4.1 Helminth egg reduction

Facultative ponds should be designed for maximum helminth egg reduction. For both restricted and unrestricted irrigation with treated wastewater, the WHO guideline is ≤1.0 egg/L (WHO, 2006). To meet this guideline, the following design parameters have been recommended for facultative ponds:

- Total hydraulic retention time of ponds in series (WHO, 1989): $t_{V/Q} \geq$ 8–10 d; and
- Pond overflow rate (OFR) ≤ 0.12 m³/m² d (Von Sperling, 2007)

$$\mathrm{OFR} = \frac{Q_m}{A_f} \leq 0.12\,\mathrm{m^3/m^2\,d} \tag{5.30}$$

where Q_m is the mean flowrate (m³/d) and A_f is the area of facultative pond (m²).

Equation (5.30) is identical to OFR equations used in the design of sedimentation basins. Here, it is assumed that the removal mechanism for helminth eggs is discrete settling associated with surface OFR independent of depth (Von Sperling, 2007). Studies with pilot-scale ponds in Brazil showed complete removal of helminth eggs at OFRs ranging 0.12–0.20 m³/m² d, with OFR ≤ 0.12 m³/m² d recommended for design (Von Sperling, 2007).

While the original WHO 1989 guidelines referred to ponds in series to meet the 8–10 d requirement, primary facultative ponds should always have $t_{V/Q} \geq$ 15 d and should be designed for 100% helminth egg reduction. Secondary facultative ponds following an anaerobic pond or UASB will likely have $t_{V/Q} < 10$ d if they are designed for BOD_5 removal. In this case, they should be designed for helminth egg reduction and not further BOD_5 removal; additional maturation ponds in series will also be needed for bacterial and viral pathogen reduction.

The following equation developed by Ayres *et al.* (1992) for helminth egg reduction in anaerobic, facultative, and maturation ponds has been recommended for design:

$$E_{95\% \text{ LCL}} = 100\left[1 - 0.41\exp(-0.49t_{V/Q} + 0.0085t_{V/Q}^2)\right] \tag{5.31}$$

where $E_{95\% \text{ LCL}}$ is the 95% lower confidence limit for percent helminth egg reduction and $t_{V/Q}$ is the hydraulic retention time, d, for $t_{V/Q} \leq 20$ d.

If Equation (5.31) is used for design, it is better to convert it to $\log_{10}$ reduction as shown in Equation (5.32). With Equation (5.32) $\log_{10}$ reductions are calculated for each pond in series and then added to calculate the total reduction in the pond system. Equation (5.32) is plotted in Figure 5.22.

$$\log_{10, 95\% \text{ LCL}} = -\log_{10}\left[1 - (1 - 0.41\exp(-0.49t_{V/Q} + 0.0085(t_{V/Q})^2))\right] \quad \text{for} \quad t_{V/Q} \leq 20 \text{ d} \tag{5.32}$$

where $\log_{10\ 95\% \text{ LCL}}$ is the $\log_{10}$ 95% lower confidence limit reduction.

Required $\log_{10}$ reduction will depend on raw wastewater helminth egg concentrations (Table 5.8), which in turn depend on the prevalence of infections in local populations. Helminth egg concentrations in raw wastewater typically range from <10 to 1000 eggs/L in cities in Africa, Asia, and Latin America (Oakley, 2005; Oakley & Mihelcic, 2019).

Equation (5.31) is a regression equation developed from data from a mixture of 23 ponds (anaerobic, facultative, and maturation) and cannot be used beyond the data limit of $t_{V/Q} = 20$ d. In addition, parameters such as depth, l/w ratio, temperature, wind effects, inlet/outlet structures, hydraulic short circuiting, sludge accumulation, operation, and maintenance conditions were not included in the model. Nevertheless, to date, there are few design equations to calculate helminth egg reduction, and Equations (5.30) and (5.32) can be used, but with caution and safety factors. For this reason, all facultative ponds should have a minimum hydraulic retention time of 15 d, which is equivalent to a 2.75 $\log_{10}$ reduction (Figure 5.22), with multiple inlets and outlets, and $l/w = 3/1$, to minimize hydraulic short circuiting. Downstream maturation ponds then serve as additional safety factors for further helminth egg reduction.

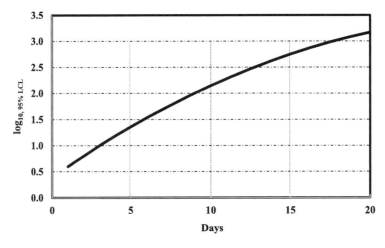

Figure 5.22 $\log_{10}$ reduction of helminth eggs at the 95% lower confidence limit using Equation (5.32) modified from Ayres *et al.* (1992).

Theory and design of facultative ponds

Table 5.8 Required log₁₀ helminth egg reduction for effluent concentrations < 1 egg/L.

Raw Wastewater (eggs/L)	Required Log₁₀ Reduction Effluent < 1 egg/L
1000	3.00
500	2.70
100	2.00
50	1.70
10	1.00

5.2.4.2 E. coli or fecal coliform reduction

The model of dispersion is used for *E. coli* or fecal coliform reduction in both facultative and maturation ponds. The Wehner and Wilhem equation (Wehner & Wilhelm, 1956), which was recommended by Thirumurthi (1969) for the design of facultative ponds, is used in the following form:

$$N_{\text{effluent}} = N_{\text{influent}} \frac{4a \cdot e^{1/2d}}{\left((1+a)^2 e^{a/2d} - (1-a)^2 e^{-a/2d}\right)} \tag{5.33}$$

where N_{effluent} is the concentration of *E. coli* or fecal coliforms in effluent (CFU/100 mL) and N_{influent} is the concentration of *E. coli* or fecal coliforms in influent (CFU/100 mL).

The term a is defined as follows:

$$a = \sqrt{(1 + 4k_{b,T} \cdot t_{V/Q} \cdot d)} \tag{5.34}$$

where $t_{V/Q}$ is the theoretical hydraulic retention time (d); $k_{b,T}$ is the mortality rate constant at temperature T (°C), d⁻¹; and d is the dispersion number.

Von Sperling (2005) developed the following mortality rate constant equation after evaluation of data from 186 facultative and maturation ponds around the world:

$$k_{b,20} = 0.549 \cdot h^{-1.456} \quad (R^2 = 0.845) \tag{5.35}$$

where h is the water depth in pond (m).

The temperature effect on $k_{b,T}$ is defined by Equation (5.35) (Von Sperling, 2007):

$$k_{b,T} = k_{b,20} \Theta^{(T-20)} = (0.549 \cdot h^{-1.456})(1.07)^{(T-20)} \tag{5.36}$$

The dispersion number, d, is traditionally defined by the following equation (Arceivala & Asolekar, 2007):

$$d = \frac{D}{U \times L} \tag{5.37}$$

where D is the longitudinal dispersion coefficient (m²/d); U is the mean flow velocity along the length of the pond (m/d); and L is the length of axial travel (m).

The values of D, U, and L can be determined only by conducting tracer tests on a full-scale pond or scaled model or using empirical equations to predict the probable dispersion model before a pond is constructed. As a point of reference, the following general ranges of dispersion numbers for different types of ponds have been reported by Arceivala and Asolekar (2007):

Waste Stabilization Pond Configuration	$d = \dfrac{D}{U \times L}$
Single rectangular ponds ($l/w \leq 4$)	1.0–4.0
Long, narrow ponds with baffles or multiple inlets	0.1–1.0
Overall dispersion number for two ponds in series	0.2–0.7
Overall dispersion number for three ponds in series	≤ 0.10

Von Sperling (2007) developed the following equation for estimation of the dispersion number for use in the Wehner and Wilhem equation:

$$d = \frac{1}{(l/w)_{\text{interior}}} \tag{5.38}$$

where $(l/w)_{\text{interior}}$ is the length-to-width ratio in the interior of pond.

For facultative ponds and ponds without interior baffles, $(l/w)_{\text{interior}} = (l/w)_{\text{exterior}}$.

Equation (5.38) is based on the results of two Monte Carlo simulations comparing four empirical equations available in the literature and is recommended for design (Von Sperling, 2007).

5.2.5 BOD$_5$ removal

Historically, the BOD rate constants for waste stabilization pond design and performance monitoring have been based on BOD$_5$ rather than BOD$_L$ (Gloyna, 1971; Mara, 2003; USEPA, 1983), continuing to the present day. (The use of BOD$_5$ to assess pond performance should not be confused with BOD$_L$ used to calculate maximum surface loadings to determine facultative pond areas.) The model of dispersion using the Wehner and Wilhem equation is also used for BOD$_5$ removal with the following equations (USEPA, 1983):

$$S_e = S_0 \frac{4a \cdot e^{1/2d}}{\left((1+a)^2 e^{a/2d} - (1-a)^2 e^{-a/2d}\right)} \tag{5.39}$$

where S_e is the concentration of BOD$_5$ in the effluent (mg/L) and S_o is the concentration of BOD$_5$ in the influent (mg/L)

$$a = \sqrt{(1 + 4k_{\text{BOD}_5,T} \cdot t_{V/Q} \cdot d)} \tag{5.40}$$

where $k_{\text{BOD}_5,T}$ is the first-order rate constant for BOD$_5$ removal at temperature T (°C), d^{-1}.

$$k_{\text{BOD}_5,T} = k_{\text{BOD}_5,20} \Theta^{(T-20)} = k_{\text{BOD}_5,20}(1.09)^{(T-20)} \tag{5.41}$$

$$k_{\text{BOD}_5,20} = 0.15\,\text{d}^{-1} \tag{5.42}$$

$$k_{\text{BOD}_5,T} = 0.15(1.09)^{(T-20)} \tag{5.43}$$

The terms $t_{V/Q}$ and d in Equation (5.39) are the same as in Equations (5.34) and (5.38).

Theory and design of facultative ponds

5.2.6 TSS removal

It can be assumed that the majority of raw wastewater TSS will be removed through sedimentation and, to a lesser extent, aerobic heterotrophic decomposition of finer solids that do not settle. Facultative ponds will still have effluent TSS concentrations ranging from 35 to >100 mg/L, but these solids are in the form of algae cells produced in the pond. The removal of raw wastewater TSS is used to estimate sludge production as discussed in the next section.

5.2.7 Sludge accumulation

Many, perhaps most, facultative ponds fail and are abandoned as a result of excess sludge accumulation. Sludge accumulation should be addressed and emphasized in the design, in the operation and maintenance manual, operator training, and the routine operation and maintenance of the system. All facultative pond designs should include the following:

(1) Projections of sludge accumulation, including effects of population growth, and so on.
(2) Estimates of the number of years of operation before desludging will be needed.
(3) Procedures for annual sludge accumulation field measurements.
(4) Procedures for sludge removal.
(5) Design for sludge treatment, storage, and final disposal or reuse (not recommended).

Points 1 and 2 are discussed in the following sections, and all are discussed in detail in Chapter 10.

5.2.7.1 Sludge accumulation reported in the literature

Sludge accumulation has traditionally been measured *in situ* at strategic locations within a pond, an estimated volume or mean depth calculated, and the results reported in terms of m³/capita yr, or mm/yr. Table 5.9 gives an example of accumulation rates reported for various parts of the world. Franci (1999) has noted that because the population connected to the treatment system can never be known with accuracy, accumulation rates based on sludge depth are preferable.

The ranges of accumulation rates vary widely in Table 5.9 and cannot be used with any confidence to project sludge accumulation from new designs.

5.2.7.2 Projection of sludge accumulation with flowrates and solids loadings

An alternative approach uses flowrate and solids loading estimates, parameters already required for the design of a pond system, and that should be routinely monitored when the pond is in operation, to estimate sludge accumulation by mass and volume.

Table 5.9 Reported sludge accumulation rates in facultative ponds.

Pond Location	Number of Ponds Studied	m³/capita yr	mm/yr	Years in Operation
Brazil[1]	2		12–28	
France[2]	19	0.04–0.148	19	12–24
India[3]	3	0.07–0.15		4–6.5
Mexico[4]	3	0.021–0.036	19–21	6–15
USA[3]	1		7–51	

[1]*Source:* Franci (1999).
[2]*Source:* Picot et al. (2005).
[3]*Source:* Arceivala et al. (1970).
[4]*Source:* Nelson et al. (2004).

The volume of in-pond digested sludge is estimated with Equation (5.44), which was originally developed to determine sludge volume–mass relationships in conventional wastewater treatment (Metcalf and Eddy/AECOM, 2014):

$$\dot{V}_{Sl,D\,yr} = \frac{\dot{M}_{Sl,D\,yr}}{\rho_{H_2O} \cdot S_{Sl} \cdot TS} \tag{5.44}$$

where $\dot{V}_{Sl,D\,yr}$ is the volume of digested sludge per year (sludge residence time >6 mo) (m³/yr); $\dot{M}_{Sl,D\,yr}$ is the mass of digested total solids per year (sludge residence time >6 mo) (kg/yr); ρ_{H_2O} is the density of water (1000 kg/m³); S_{Sl} is the specific gravity of sludge; and TS is the decimal fraction of total solids in sludge.

Equation (5.44) estimates the volume of well-digested sludge that has been in the pond for at least 6 months. After this time period, it is assumed the sludge is well digested, similar to the digestion process in unheated digesters and Imhoff tanks reported by Imhoff and Fair (1956) and is presented in Table 5.10.

The specific gravity of the sludge, S_{Sl}, and the decimal fraction of total solids, TS, in digested, consolidated sludge, can be estimated from values reported in the literature such as those in Table 5.11.

The mass of influent total solids that settle to the sludge layer is calculated with Equation (5.45).

$$\dot{M}_{Sl} = 0.001 \cdot Q_{mean} \cdot TSS_{Settleable} \tag{5.45}$$

where $\dot{M}_{Sl}$ is the mass of fresh sludge solids (kg/d); Q_{mean} is the mean influent flowrate (m³/d); $TSS_{Settleable}$ is the settleable suspended solids (mg/L) (assume 100% for safety factor); and 0.001 is the conversion factor (mg/L to kg/m³).

Table 5.10 Time required for 90% digestion in unheated digesters and Imhoff tanks

Sludge Temperature (°C)	Digestion Time (d)
10	80
15	58
20	42
25	33
30	28

Source: Imhoff and Fair (1956).

Table 5.11 Physical-chemical characteristics of digested sludge in facultative ponds.

Parameter	Range of Reported Values				
	Honduras[1]	Brazil[2]	Mexico[3]	India[4]	France[5]
Total solids decimal fraction (TS)	0.12–0.15	0.08–0.22	0.11–0.17	0.13–0.28	0.06–0.22
Volatile solids decimal fraction (VS)	0.24–0.31	0.36–0.42		0.17–0.31	0.21–0.59
Fixed solids decimal fraction (FS)	0.68–0.76	0.58–0.64		0.69–0.83	0.41–0.79
Specific gravity of sludge (S_{Sl})	1.049–1.076			1.11–1.165	

[1]*Source:* Oakley (2005).
[2]*Source:* Franci (1999).
[3]*Source:* Nelson et al. (2004).
[4]*Source:* Arceivala et al. (1970).
[5]*Source:* Picot et al. (2005).

Theory and design of facultative ponds

The annual mass of digested sludge produced in the pond is calculated with Equation (5.46).

$$\dot{M}_{Sl, D\, yr} = 365 \cdot \left[FS \cdot \dot{M}_{Sl} + (1 - VS_{Destroyed}) VS \cdot \dot{M}_{Sl} \right] \quad (5.46)$$

where $\dot{M}_{Sl, D\, yr}$ is the mass of digested sludge solids produced per year (kg/yr); FS is the decimal fraction of fixed solids; VS is the decimal fraction of volatile solids; and $VS_{Destroyed}$ is the decimal fraction of volatile solids destroyed (assume 0.50 for safety factor).

The values of FS and VS should be obtained from raw wastewater TSS analyses. In small installations, they can be assumed from typical values for raw primary sludge (Table 5.12). Although $VS_{Destroyed}$ can potentially be very high as a result of the time the sludge undergoes digestion, a value of $VS_{Destroyed} = 0.50$ should be assumed as a safety factor.

Finally, the time to fill 25% of the pond's volume with sludge, the recommended limit for pond operation without desludging, is estimated with Equation (5.47):

$$t_{25\%} = \frac{0.25 \cdot V_f}{\dot{V}_{Sl,\, D\, yr}} \quad (5.47)$$

where $t_{25\%}$ is the time to fill 25% of pond's volume (yr); V_f is the volume of facultative pond (m³).

5.2.7.3 Design example part 1: projection of sludge accumulation for TSS = 200 mg/L

A projection of sludge accumulation is required for a facultative pond with a design hydraulic retention time of 15 d. The projection will be made for two limiting conditions in digested and raw sludge taken from the range of reported data in Tables 5.11 and 5.12.

Condition 1: TS = 0.11 (Table 5.11, digested sludge); VS = 0.60 (Table 5.12, raw primary sludge);
Condition 2: TS = 0.15 (Table 5.11, digested sludge); VS = 0.85 (Table 5.12, raw primary sludge).

TSS = 200 mg/L for both conditions 1 and 2.
The following data apply:

- $Q_{mean} = 1000$ m³/d;
- volume of facultative pond: $V_f = 15,000$ m³;
- $VS_{Destroyed} = 0.50$; and
- $S_{Sl} = 1.08$ (assumed from the range of values in Table 5.11).

Condition 1:

(1) Determine the daily mass of settled sludge with Equation (5.45).

$$\dot{M}_{Sl} = 0.001 \cdot Q_{mean} \cdot TSS_{Settleable} = 0.001(1000\, m^3/d)(200\, mg/L) = 200\, kg/d$$

Table 5.12 Comparison of raw primary and digested facultative pond sludges.

Parameter	Raw Primary Sludge[1]	Digested Facultative Pond Sludge[2]
TS	0.01–0.06	0.11–0.17
VS	0.60–0.85	0.17–0.42
FS	0.15–0.40	0.58–0.76
S_{Sl}	1.02	1.05–1.16

[1]*Source:* Metcalf and Eddy/AECOM (2014).
[2]*Source:* Table 5.11.

(2) Determine the annual mass of digested sludge with Equation (5.46) for condition 1: TS = 0.11; VS = 0.60:

$$\dot{M}_{Sl,D\,yr} = 365 \cdot (FS \cdot \dot{M}_{Sl} + (1 - VS_{Destroyed})VS \cdot \dot{M}_{Sl})$$
$$= 365((0.40)(200\,kg/d) + (1 - 0.50)(0.60)(200\,kg/d)) = 51100\,kg/yr$$

(3) Determine the annual volume of digested sludge with Equation (5.44):

$$\dot{V}_{Sl,D\,yr} = \frac{\dot{M}_{Sl,D\,yr}}{\rho_{H_2O} \cdot S_{Sl} \cdot TS} = \frac{51100\,kg/yr}{(1000\,kg/m^3)(1.08)(0.11)} = 430\,m^3/yr$$

(4) Estimate the time to fill 25% of pond's volume with Equation (5.47):

$$t_{25\%} = \frac{0.25 \cdot V_f}{\dot{V}_{f-Sl}} = \frac{(0.25)(15000\,m^3)}{430\,m^3/yr} = 8.7\,yr$$

Condition 2:

(5) Calculate the annual mass of digested sludge with Equation (5.46) for condition 2 TS = 0.15; VS = 0.85:

$$\dot{M}_{Sl,D\,yr} = 365 \cdot (FS \cdot \dot{M}_{Sl} + (1 - VS_{Destroyed})VS \cdot \dot{M}_{Sl})$$
$$= 365((0.15)(200\,kg/d) + (1 - 0.50)(0.85)(200\,kg/d)) = 41975\,kg/yr$$

(6) Calculate the annual volume of digested sludge with Equation (5.44):

$$\dot{V}_{Sl,D\,yr} = \frac{\dot{M}_{Sl,D\,yr}}{\rho_{H_2O} \cdot S_{Sl} \cdot TS} = \frac{41975\,kg/yr}{(1000\,kg/m^3)(1.08)(0.15)} = 259\,m^3/yr$$

(7) Estimate the time to fill 25% of pond's volume:

$$t_{25\%} = \frac{0.25 \cdot V_f}{\dot{V}_{Sl,D\,yr}} = \frac{(0.25)(15000\,m^3)}{259\,m^3/yr} = 14.5\,yr$$

5.2.7.4 Design example part 2: projection of sludge accumulation for TSS = 350 mg/L

This example uses the same values as Part 1, but with a higher influent TSS of 350 mg/L.
Condition 1 (TS = 0.11; VS = 0.60; and TSS = 350 mg/L):

(1) Determine the daily mass of settled sludge with Equation (5.45):

$$\dot{M}_{Sl} = 0.001 \cdot Q_{mean} \cdot TSS_{Settleable} = 0.001(1000\,m^3/d)(350\,mg/L) = 350\,kg/d$$

(2) Determine the annual mass of digested sludge with Equation (5.46) for condition 1:

$$\dot{M}_{Sl,D\,yr} = 365 \cdot \left(FS \cdot \dot{M}_{Sl} + (1 - VS_{Destroyed})VS \cdot \dot{M}_{Sl}\right)$$
$$= 365((0.40)(350\,kg/d) + (1 - 0.50)(0.60)(350\,kg/d)) = 89425\,kg/yr$$

(3) Determine the annual volume of digested sludge with Equation (5.44):

$$\dot{V}_{Sl,D\,yr} = \frac{\dot{M}_{Sl,\,D\,yr}}{\rho_{H_2O} \cdot S_{Sl} \cdot TS} = \frac{89\,425\,kg/yr}{(1000\,kg/m^3)(1.08)(0.11)} = 753\,m^3/yr$$

(4) Estimate the time to fill 25% of pond's volume with Equation (5.47):

$$t_{25\%} = \frac{0.25 \cdot V_f}{\dot{V}_{f-Sl}} = \frac{(0.25)(15\,000\,m^3)}{753\,m^3/yr} = 5.0\,yr$$

Condition 2 (TS = 0.15; VS = 0.85; TSS = 350 mg/L):

(5) Calculate the annual mass of digested sludge with Equation (5.46) for condition 2:

$$\dot{M}_{Sl,\,D\,yr} = 365 \cdot \left[FS \cdot \dot{M}_{Sl} + (1 - VS_{Destroyed})VS \cdot \dot{M}_{Sl} \right]$$
$$= 365\left((0.15)(350\,kg/d) + (1 - 0.50)(0.85)(350\,kg/d) \right) = 73\,456\,kg/yr$$

(6) Calculate the annual volume of digested sludge with Equation (5.44):

$$\dot{V}_{Sl,D\,yr} = \frac{\dot{M}_{Sl,\,D\,yr}}{\rho_{H_2O} \cdot S_{Sl} \cdot TS} = \frac{73\,456\,kg/yr}{(1000\,kg/m^3)(1.08)(0.15)} = 453\,m^3/yr$$

(7) Estimate time to fill 25% of pond's volume:

$$t_{25\%} = \frac{0.25 \cdot V_f}{\dot{V}_{Sl,D\,yr}} = \frac{(0.25)(15\,000\,m^3)}{453\,m^3/yr} = 8.3\,yr$$

5.2.7.5 Discussion of design example results

The results of Parts 1 and 2 are presented in Table 5.13.

The following conclusions are drawn from this example:

(1) For a given influent TSS loading, lower TS and VS decimal fractions yield higher accumulation rates of in-pond digested sludge, and concomitant shorter time periods to reach $t_{25\%}$.
(2) Higher TSS loadings significantly increase accumulation rates.
(3) Even though the pond had a theoretical hydraulic retention time of 15 d, it could easily reach $t_{25\%}$ in less than 10 years if TS and VS are in the low ranges reported in Tables 5.11 and 5.12.

Yearly sludge accumulation projections for the data in Table 5.13 are plotted in Figure 5.23. Curves such as these, based on measured TSS concentrations, VS, and FS decimal fractions in the raw wastewater, should be developed and used for design. Results should also be included in pond design reports, with recommendations on in-pond verification monitoring of sludge accumulation as part of routine operation and maintenance.

Historically, pond designs have not focused on sludge accumulation, which is also unlikely to be mentioned in operation and maintenance manuals, if they exist. As a result, resource-limited municipalities are unprepared for the costs of a desludging operation. In a monitoring program in Honduras, for example, none of 10 pond systems studied had a sludge monitoring and removal plan, nor a planned budget to pay for sludge removal (Oakley, 2005). Several of these pond systems are now abandoned.

Table 5.13 Results for sludge accumulation projections.

Parameter	Part 1		Part 2	
	TSS = 200 mg/L		TSS = 350 mg/L	
	TS = 0.11; VS = 0.60	TS = 0.15; VS = 0.85	TS = 0.11; VS = 0.60	TS = 0.15; VS = 0.85
$\dot{M}_{Sl}$ (kg/d)	200	200	350	350
$\dot{M}_{Sl,\,D\,yr}$ (kg/yr)	51 100	41 975	89 425	73 456
$\dot{V}_{Sl,D\,yr}$ (m³/yr)	430	259	753	453
$t_{25\%}$ (yr)	8.7	14.5	5.0	8.3

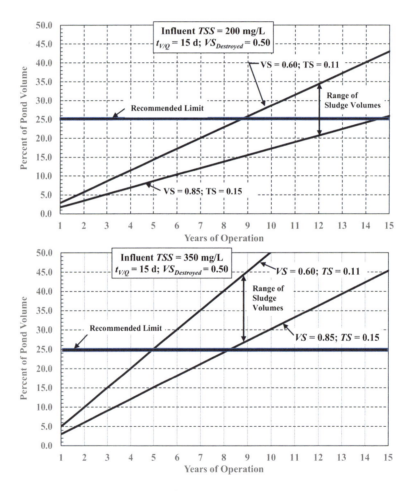

Figure 5.23 Sludge accumulation projections in a facultative pond: $t_{V/Q} = 15$ d, $VS_{Destroyed} = 0.50$, $S_{Sl} = 1.08$. Influent TSS concentrations are assumed to be 200 mg/L (top) and 350 mg/L (bottom). Potential sludge volumes range between the limits of VS = 0.60 with TS = 0.11, and VS = 0.85 with TS = 0.15. Higher values of both VS and TS result in less accumulation by volume. Greater influent TSS concentrations have a significant influence on accumulation rate as seen in the bottom graph.

Theory and design of facultative ponds

5.3 FACULTATIVE POND DESIGN PROCEDURE

(1) Determine the maximum BOD_L surface loading rate, which is also the maximum phototrophic oxygen production rate, for each month of the year using either Equation (5.20) or (5.21):
CLIMWAT Data: Solar radiation in units of MJ/m^2 d

$$\lambda_{S,\text{Max}} = (1.937 \times 10^{-6}) \cdot (SR_{i,\text{MJ}}) \left(\frac{1000\,\text{kJ}}{1.0\,\text{MJ}}\right) \left(\frac{10\,000\,\text{m}^2}{\text{ha}}\right) = 19.37(SR_{i,\text{MJ}}) \quad (5.20)$$

where $\lambda_{S,\text{Max}}$ is the maximum surface loading rate, kg BOD_L/ha d $\leq$ kg O_2/ha d; $SR_{i,\text{MJ}}$ is the solar radiation in units of MJ/m^2 d; 19.37 units $= \dfrac{\text{kg } O_2 \, \text{m}^2}{\text{MJ ha}}$.

NASA POWER Data: Solar radiation in units of kWh/m^2 d.

$$\lambda_{S,\text{Max}} = (1.937 \times 10^{-6}) \cdot (SR_{i,\text{kWh}}) \left(\frac{3600\,\text{kJ}}{1.0\,\text{kWh}}\right) \left(\frac{10\,000\,\text{m}^2}{\text{ha}}\right) = 69.73(SR_{i,\text{kWh}}) \quad (5.21)$$

where $SR_{i,\text{kWh}}$ is the solar radiation in units of kWh/m^2 d; 69.73 units $= \dfrac{\text{kg } O_2 \, \text{m}^2}{\text{kWh ha}}$.

(2) Determine monthly mean water temperatures.

Use existing water temperature data from nearby operating ponds or shallow lakes.

If water temperature data are not available, water temperatures from monthly air temperatures are estimated from the nearest meteorological station. If no close station exists, air temperatures at 2 m is used from NASA POWER Data in Equations (5.22) or (5.23), to estimate water temperatures within the respective air temperature ranges.

$$T_{\text{water}} = 12.7 + 0.54 T_{\text{air}} \quad (15°C \leq T_{\text{air}} \leq 35°C) \quad (5.22)$$

$$T_{\text{water}} = 0.86 T_{\text{air}} + 5.29 \quad (8.5°C \leq T_{\text{air}} \leq 38.8°C) \quad (5.23)$$

where T_{water} is the mean water temperature for mean air temperature, T_{air} (°C).

(3) Determine monthly water temperature coefficients from Equation (5.48), Table 5.5, or Figures 5.16 and 5.17.

$$\alpha_T = 0.0445 T_{\text{water}} - 0.0821 \quad (5°C \leq T_{\text{water}} \leq 25°C) \quad (5.48)$$

(4) Determine the maximum surface loading for each month for the calculated water temperature using Equation (5.19):

$$\lambda_{S,Tw} = \lambda_{S,\text{Max}}(\alpha_T) \quad (5.19)$$

where $\lambda_{S,Tw}$ is the maximum surface loading for water temperature (kg BOD_L/ha d $\leq$ kg O_2/ha d). The design month has the lowest value of $\lambda_{S,Tw}$.

(5) Determine the area of the facultative pond:

$$A_f = \frac{BOD_L,\,\text{kg/d}}{\lambda_{S,Tw}} = \frac{(0.001) \cdot (Q) \cdot (BOD_L)}{\lambda_{S,Tw}} \quad (5.49)$$

where A_f is the area of the facultative pond (ha); Q is the mean flowrate (m^3/d); BOD_L is the ultimate BOD of raw wastewater (mg/L) ($BOD_L = 1.5 BOD_5$).

(6) Determine the volume of the pond using the prismoid equation:

$$V_f = \frac{h}{6} \cdot [(l \cdot w) + (l - 2ih)(w - 2ih) + 4 \cdot (l - ih)(w - ih)] \quad (5.50)$$

where V_f is the pond volume (m³); h is the water depth (m); l is the length (m); w is the width (m); and i is the interior embankment slope, horizontal/vertical, typically 3/1.

(i) Select l/w of pond

For facultative ponds, $2 \leq l/w \leq 3$. $l/w = 3$ is preferred to limit dispersion effects.

(ii) Determine h, w, and l

$h = 1.5$–2.0 m for facultative ponds. $h = 2.0$ m is preferable for sludge accumulation safety factor.

With A_f in m²,
- $A_f = (l) \cdot (w)$,
- $l/w = x/1$ ($x = 2$–3 for facultative ponds, preferably 3), and
- $A_f = (xw) \cdot (w) = xw^2$.

$$w = \sqrt{\frac{A_f}{x}}$$

$l = x \cdot w$

(iii) Calculate V_f with the prismoid equation (Equation (5.50)).

(7) Calculate the theoretical hydraulic retention time.

$$t_{V/Q} = \frac{V_f}{Q} \tag{5.51}$$

$t_{V/Q}$ is the theoretical hydraulic retention time (d).

(8) Estimate helminth egg reduction

(i) Determine OFR and minimum hydraulic retention time (Equation (5.30)):

$$\text{OFR} = \frac{Q}{A_f} \leq 0.12 \, \text{m}^3/\text{m}^2 \, \text{d}; \; t_{V/Q} \leq 10 \, \text{d} \tag{5.30}$$

(ii) 95% lower confidence limit $\log_{10}$ reduction for $t_{V/Q} \leq 20$ d (Equation (5.32)):

$$\log_{10, 95\% \, \text{LCL}} = -\log_{10}\left[1 - (1 - 0.41\exp(-0.49 t_{V/Q} + 0.0085(t_{V/Q})^2))\right] \tag{5.32}$$

(9) Estimate *E. coli* or fecal coliform reduction using the Wehner and Wilhem dispersion equation (Equation (5.31)).

$$N_{\text{effluent}} = N_{\text{influent}} \frac{4a \cdot e^{1/2d}}{\left((1+a)^2 e^{a/2d} - (1-a)^2 e^{-a/2d}\right)} \tag{5.33}$$

where N_{effluent} is the concentration of *E. coli* or fecal coliforms in the effluent (CFU/100 mL); N_{influent} is the concentration of *E. coli* or fecal coliforms in the influent (CFU/100 mL); and d is the dispersion number.

$$d = \frac{1}{(l/w)_{\text{interior}}} \tag{5.38}$$

where $(l/w)_{\text{interior}}$ is the length-to-width ratio in interior of pond (same as the exterior for facultative ponds without baffles).

$$a = \sqrt{(1 + 4k_{b,T} \cdot t_{V/Q} \cdot d)} \tag{5.34}$$

where $t_{V/Q}$ is the theoretical hydraulic retention time (d); $k_{b,T}$ is the mortality rate constant at temperature T (d⁻¹) (Von Sperling, 2007):

Theory and design of facultative ponds

$$k_{b,20} = 0.542 \cdot h^{-1.259} \tag{5.52}$$

where h is the water depth in pond (m).
Temperature effect on $k_{b,T}$:

$$k_{b,T} = k_{b,20}\Theta^{(T-20)} = (0.549 \cdot h^{-1.259})(1.07)^{(T-20)} \tag{5.36}$$

(10) Determine BOD$_5$ removal in the facultative pond.
The model of dispersion for plug flow is assumed for BOD$_5$ removal using the Wehner and Wilhem equation (USEPA, 1983; Von Sperling, 2007):

$$S_e = S_0 \frac{4a \cdot e^{1/2d}}{\left((1+a)^2 e^{a/2d} - (1-a)^2 e^{-a/2d}\right)} \tag{5.39}$$

where S_e is the concentration of BOD$_5$ in the effluent (mg/L); S_o is the concentration of BOD$_5$ in the influent (mg/L); and d is the dispersion factor (unitless):

$$d = \frac{1}{(l/w)_{\text{interior}}} \tag{5.38}$$

where $(l/w)_{\text{interior}}$ = length/width ratio in the interior of pond (same as exterior for facultative ponds without baffles).

$$a = \sqrt{(1 + 4k_{\text{BOD}_5,T} \cdot t_{V/Q} \cdot d} \tag{5.40}$$

where $k_{\text{BOD}_5,T}$ is the first-order rate constant for BOD$_5$ removal at temperature T (°C) (d^{-1}):

$$k_{\text{BOD}_5,T} = k_{\text{BOD}_5,20}\Theta^{(T-20)} = k_{\text{BOD}_5,20}(1.09)^{(T-20)} \tag{5.41}$$

$$k_{\text{BOD}_5,20} = 0.15\,\text{d}^{-1} \tag{5.42}$$

$$k_{\text{BOD}_5,T} = 0.15(1.09)^{(T-20)} \tag{5.43}$$

(11) Estimate sludge accumulation rate and the time to fill 25% of the pond's volume:
   (i) Calculate the mass of influent total solids that settle to the sludge layer with Equation (5.45):

$$\dot{M}_{\text{Sl}} = 0.001 \cdot Q_{\text{mean}} \cdot TSS_{\text{Settleable}} \tag{5.45}$$

where $\dot{M}_{\text{Sl}}$ is the mass of fresh sludge solids (kg/d); Q_{mean} is the mean influent flowrate (m^3/d); $TSS_{\text{Settleable}}$ is the settleable suspended solids (mg/L) (assume 100% for safety factor); and 0.001 = conversion factor (mg/L to kg/m^3).
   (ii) Estimate the annual mass of digested sludge produced in the pond with Equation (5.44):

$$\dot{M}_{\text{Sl},D\,\text{yr}} = 365 \cdot \left(\text{FS} \cdot \dot{M}_{\text{Sl}} + (1 - \text{VS}_{\text{Destroyed}})\text{VS} \cdot \dot{M}_{\text{Sl}}\right) \tag{5.46}$$

where $\dot{M}_{\text{Sl},D\,\text{yr}}$ is the mass of digested sludge solids produced per year (kg/yr); FS is the decimal fraction of fixed solids; VS is the decimal fraction of volatile solids; and VS$_{\text{Destroyed}}$ is the decimal fraction of volatile solids destroyed (assume 0.50 for safety factor).
   (iii) Determine the volume of in-pond digested sludge using Equation (5.44)):

$$\dot{V}_{\text{Sl},D\,\text{yr}} = \frac{\dot{M}_{\text{Sl},D\,\text{yr}}}{\rho_{\text{H}_2\text{O}} \cdot S_{\text{Sl}} \cdot TS} \tag{5.44}$$

where $\dot{V}_{Sl,D\,yr}$ is the volume of digested sludge per year (sludge residence time > 6 mo) (m³/yr); $\dot{M}_{Sl,D\,yr}$ is the mass of digested total solids per year (sludge residence time > 6 mo) (kg/yr); ρ_{H_2O} is the density of water (1000 kg/m³); S_{Sl} is the specific gravity of sludge; and TS is the decimal fraction of total solids in sludge.

(iv) Determine the time to fill 25% of pond's volume (Equation (5.47)):

$$t_{25\%} = \frac{0.25 \cdot V_f}{\dot{V}_{f-Sl}} \qquad (5.47)$$

where S_{Sl} is the time to reach $0.25 V_f$ (yr).

5.4 DESIGN EXAMPLE: FACULTATIVE POND REDESIGN FOR AGRICULTURAL REUSE, COCHABAMBA, BOLIVIA

The city of Cochabamba waste stabilization pond system Albarrancho, built in 1986 and operated by SEMAPA (Servicio Municipal de Agua Potable y Alcantarillado), the municipal water and sewer provider, has been seriously overloaded for decades due to population growth and poor design and is presently abandoned. As a result, raw wastewater bypassing the system heavily pollutes the Río Rocha, which is used for agricultural irrigation downstream of the wastewater discharge (Coronado et al., 2001).

There is an urgent need to save the Albarrancho plant with a redesign using (i) a sustainable design flowrate and (ii) improved hydraulic design to minimize short circuiting, to enable effluent reuse in agriculture rather than direct discharge to the Rio Rocha losing the valorization potential. Implementation of the redesign will require excess flows to be diverted downstream, while treated effluent will be diverted directly to agricultural users. Data from the Albarrancho Wastewater Treatment Plant apply and are presented in Table 5.14 (SEMAPA, personal communication). A diagram of the waste stabilization pond facility is shown in Figure 5.24.

Design objectives:

- Redesign of the facultative and maturation ponds of the Albarrancho treatment facility to produce an effluent meeting the WHO guidelines for restricted wastewater use in agriculture.
- Redesign of the facultative ponds is presented in this section.
- Redesign of the maturation ponds is presented in Chapter 6.

Table 5.14 Design and operating data of the Albarrancho facultative ponds

Parameter	1986 (Design Values)	2012 (Measured Data)
Q_{mean} (m³/d)	34 733	70 243
Influent BOD_L (mg/L)	252	358
$\lambda_{S,T}$ (kg BOD_L/ha d)	400	997
$t_{V/Q}$ (d)	11.3	8.40
Total area (ha)	21.78	No change of physical parameters. Sludge accumulation causing hydraulic short circuiting has always been a problem.
Total volume (m³)	394 200	
Number of ponds in parallel	8	
Area/pond (ha)	2.7	
Length × width/pond (m)	164.3 × 164.3	
Volume/pond (m³)	49 275	
Depth (all ponds), h (m)	1.8	

Source: Data from SEMAPA, personnel communication.

Theory and design of facultative ponds

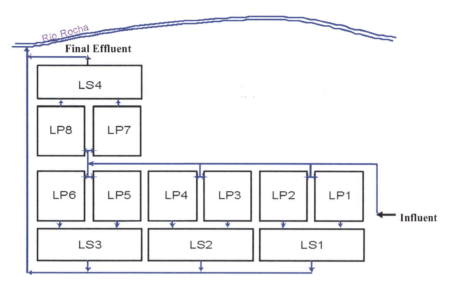

Figure 5.24 The Albarrancho waste stabilization pond facility consists of eight facultative ponds in parallel (LP, 1–8), followed by four maturation ponds (LS, 1–4), with each maturation pond receiving the effluent of two facultative ponds. The final discharge is to the Rio Rocha. (*Source*: Diagram courtesy of SEMAPA, Cochabamba, Bolivia.)

Following the design procedure from Section 5.3:

(a) Determine the maximum BOD_L surface loading rate for each month using either Equation (5.20) or (5.21):
  (1) CLIMWAT data exist for Cochabamba and are tabulated in column 2 in Table 5.14.
  (2) Solar radiation is expressed in units of MJ/m² d; thus, Equation (5.20) is used to determine maximum surface loadings, $\lambda_{S,Max}$.
  For January, the calculation is expressed as follows:

$$\lambda_{S,Max} = 19.37(SR_{i,MJ}) = \left(19.37 \frac{\text{kg}\,O_2\,\text{m}^2}{\text{MJ}\,\text{ha}}\right)(22.8\,\text{MJ/m}^2\,\text{d}) = 442\,\text{kg}\,BOD_L/\text{ha}\,\text{d}$$

  (3) The remaining monthly values of $\lambda_{S,Max}$ are tabulated in column 3 in Table 5.14.
(b) Determine monthly water temperatures.
  Mean monthly water temperature was measured in the facultative ponds, and the data are tabulated in column 4 in Table 5.15 and plotted in Figure 5.25.
(c) Determine monthly water temperature coefficients, α_T, from Table 5.5.
  The temperature coefficients, α_T, are estimated from Table 5.5 and listed in column 5 in Table 5.15.
(d) Determine the maximum surface loading for each month using Equation (5.19):
  (1) For January, the calculated value is:

$$\lambda_{S,T} = \lambda_{S,Max}(\alpha_T) = (442\,\text{kg}\,O_2/\text{ha}\,\text{d})(0.81) = 358\,\text{kg}\,BOD_L/\text{ha}\,\text{d}$$

  (2) From Table 5.15 and Figure 5.26, the design surface loading is presented for June:

$$\lambda_{S,T,\text{June}} = 217\,\text{kg}\,O_2/\text{ha}\,\text{d}$$

Table 5.15 Solar radiation[1], mean facultative pond water temperature[2], and maximum surface loadings for Albarrancho facultative ponds.

1 Month	2 SR (MJ/m² d)	3 $\lambda_{S, Max}$ (kg BOD$_L$/ha d)	4 Mean Water Temperature, T_{water} (°C)	5 α_T	6 $\lambda_{S, T}$ (kg BOD$_L$/ha d)
J	22.8	442	20.2	0.81	358
F	21.9	424	22.0	0.92	390
M	20.6	399	23.2	2.00	798
A	18.7	362	20.8	0.85	308
M	16.5	320	20.3	0.81	259
J	16.0	310	17.4	0.70	217
J	16.2	314	18.9	0.75	235
A	17.9	347	17.4	0.70	243
S	20.4	395	21.7	0.90	356
O	21.2	411	22.1	0.95	390
N	22.4	434	23.0	1.00	434
D	21.8	422	21.8	0.90	380

[1]*Source:* Data from CLIMWAT.
[2]*Source:* Data from SEMAPA.

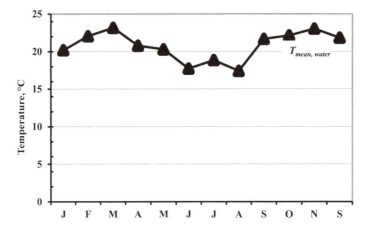

Figure 5.25 Mean monthly water temperatures in facultative ponds at Albarrancho waste stabilization pond facility, Cochabamba, Bolivia.

(e) Determine the area of facultative ponds required using Equation (5.49) for the original design flowrate and BOD$_L$ values:

$$A_f = \frac{(0.001) \cdot (Q) \cdot (BOD_L)}{\lambda_{S,Tw}} = \frac{(0.001)(34733 \, m^3/d)(252 \, mg/L)}{217 \, kg \, BOD_L/ha \, d} = \frac{8753 \, kg \, BOD_L/d}{217 \, kg \, BOD_L/ha \, d} = 40.3 \, ha$$

The required area is almost double that available based on the original design, which used a value of $\lambda_{S,T} = 400$ kg BOD$_L$/ha d. It is not known how this loading rate was selected, but

Theory and design of facultative ponds 147

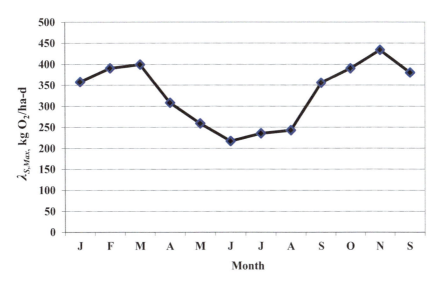

Figure 5.26 Monthly maximum surface loadings for water temperature, $\lambda_{S,T}$ calculated for Albarrancho waste stabilization pond facility.

the original design was overloaded from the start and became progressively worse with higher flowrates and BOD_L concentrations. In this situation, to save the existing system and valorize the effluent for agricultural reuse, the design will focus on the maximum flowrate and organic loading possible, along with a few key physical design modifications.

(1) Rearrange Equation (5.49) to solve for the maximum flowrate for the existing area of ponds.

$$Q = \frac{(A_f)(\lambda_{S,Tw})}{(0.001)(BOD_L)} = \frac{(21.78 \text{ ha})(217 \text{ kg } BOD_L/\text{ha d})}{(0.001 \text{ kg/m}^3/\text{mg/L})(358 \text{ mg } BOD_L/\text{L})} = 13{,}200 \text{ m}^3/\text{d}$$

(2) To improve hydraulic performance, the eight facultative ponds are combined into four facultative ponds as shown in Figure 5.27. Each pond has $l/w = 2/1$, $l = 328.6$ m, $w = 164.3$ m, and $A_f = 54\,000$ m² = 5.4 ha. The value of $l/w = 2/1$ cannot be increased to 3/1 because of site constraints, but is a significant improvement over the original design of $l/w = 1/1$.

(3) The depth of each pond will be increased to 2.0 m to increase the hydraulic retention time and as an additional safety factor for sludge accumulation.

(4) The original design had very poor hydraulic efficiency, with significant short circuiting from a single inlet to single outlet in a square shape, as visualized in Figure 5.28. The flow regime should be significantly improved with multiple inlets and outlets.

(f) Determine the volume of a pond with the new design using the prismoid equation (Equation (5.50)):

$$V_f = \frac{h}{6} \cdot \left[(l \cdot w) + (l - 2ih)(w - 2ih) + 4 \cdot (l - ih)(w - ih) \right]$$

(1) Select l/w of pond:

$l/w = 2/1$

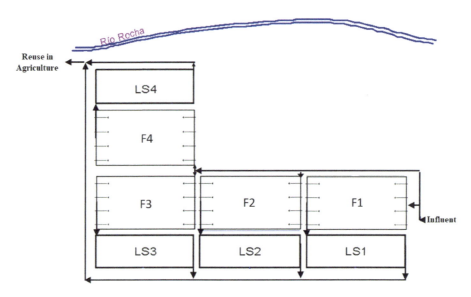

Figure 5.27 The redesign of facultative ponds combines two original ponds into one facultative pond with $l/w = 2/1$, for a total of four ponds with multiple inlets and outlets that discharge to each of the four maturation ponds. For each facultative pond $l = 328.6$ m, $w = 164.3$ m, and $A_f = 54\,000$ m^2 = 5.4 ha.

Figure 5.28 The inlet (top) and outlet (bottom) of facultative pond LP1 (Figure 5.24). The single inlet/outlet locations in a rectangular pond caused significant hydraulic short circuiting with large dead spaces in much of the volume, and sludge accumulation along the shoreline to the left of both the inlet and outlet in the direction of flow. Note the sludge accumulation to the left of the inlet (Albarrancho Wastewater Treatment Plant, Cochabamba, Bolivia).

(2) Determine h, w, and l:
   - $h = 2.0$ m,
   - $A_f = (l) \cdot (w)$,
   - $l/w = 2/1$, and
   - $A_f = (2w) \cdot (w) = 2w^2$.

$$w = \sqrt{\frac{A_f}{2}} = \sqrt{\frac{54{,}000\,\text{m}^2}{2}} = 164.3\,\text{m}$$

   - $l = 2 \cdot w = (2)(164.3) = 328.6$ m.

(3) Calculate V_f with the prismoid equation, Equation (5.50), for one pond.

$$\begin{aligned}V_f &= \frac{h}{6} \cdot \left[(l \cdot w) + (l - 2ih)(w - 2ih) + 4 \cdot (l - ih)(w - ih)\right] \\ &= \frac{2.0}{6}\left[((328.6)(164.3)) + (328.6 - 2(3)(2.0))(164.3 - 2(3)(2.0)) \right. \\ &\quad \left. + 4(328.6 - (3)(2.0))(164.3 - (3)(2.0))\right] \\ &= 102\,154\,\text{m}^3\end{aligned}$$

(g) Calculate the theoretical hydraulic retention time per pond with Equation (5.51).

$$t_{V/Q} = \frac{V_f}{Q}$$

$$Q_m/\text{pond} = \frac{13\,200\,\text{m}^3/\text{d}}{4} = 3300\,\text{m}^3/\text{d}$$

$$t_{V/Q} = \frac{V_f}{Q_m} = \frac{102\,154\,\text{m}^3}{3300\,\text{m}^3/\text{d}} = 31.0\,\text{d}$$

where $t_{V/Q}$ is almost three times greater than the original design and should greatly improve helminth egg and *E. coli* reduction and allow a much greater safety factor for sludge accumulation than was ever possible with the original design.

(h) Estimate helminth egg reduction:
(1) Calculate OFR and minimum hydraulic retention time.

$$\text{OFR} = \frac{Q_{\text{mean}}}{A_f} \leq 0.12\,\text{m}^3/\text{m}^2\,\text{d}; \text{and}\, t_{V/Q} \geq 10\,\text{d}$$

$$\text{OFR} = \frac{3300\,\text{m}^3/\text{d}}{54\,000\,\text{m}^2} = 0.032\,\text{m}^3/\text{m}^2\,\text{d} \ll 0.12\,\text{m}^3/\text{m}^2\,\text{d}; t_{V/Q} = 31.0\,\text{d} \gg 10.0\,\text{d}$$

(2) Calculate 95% lower confidence limit $\log_{10}$ reduction for $t_{V/Q} \leq 20$ d (Equation (5.32)):

$$\log_{10,95\%\,\text{LCL}} = -\log_{10}\left[1 - (1 - 0.41\exp(-0.49 t_{V/Q} + 0.0085(t_{V/Q})^2))\right]$$

$$\log_{10,95\%\,\text{LCL}} > -\log_{10}\left[1 - (1 - 0.41\exp(-0.49(20) + 0.0085(20)^2))\right] > 3.17$$

The facultative ponds should remove 100% of influent helminth eggs if they are operating properly without hydraulic short circuiting.

(i) Estimate *E. coli* reduction using the Wehner and Wilhem dispersed flow equation (Equation (5.33)):

$$N_{\text{effluent}} = N_{\text{influent}} \frac{4a \cdot e^{1/2d}}{\left((1+a)^2 e^{a/2d} - (1-a)^2 e^{-a/2d}\right)}$$

  (i) *E. coli* concentrations in raw wastewater.
  Using data from Coronado *et al.* (2001):

  $$N_{\text{influtent}} = 3.4 \times 10^8 \text{ MPN}/100\,\text{mL}$$

  (ii) Dispersion number:

  $$d = \frac{1}{(l/w)_{\text{interior}}} = \frac{1}{2/1} = 0.50$$

  (iii) Mortality rate constant at design water temperature, $k_{b,T}$ (d^{-1}) (Equation (5.36)):

  $$k_{b,T} = (0.542 \cdot h^{-1.259})(1.07)^{(T-20)}$$

  where h is the 2.0 m and T_{water} is the 17.4°C.

  $$k_{b,T} = (0.542 \cdot (2.0)^{-1.259})(1.07)^{(17.4-20)} = 0.19\,\text{d}^{-1}$$

  (iv) Determine a in Wehner–Wilhem Equation (Equation (5.34)):

  $$t_{V/Q} = 31.0\,\text{d}$$

  $$a = \sqrt{1 + 4k_{b,T} \cdot t_{V/Q} \cdot d} = \sqrt{1 + 4(0.19\,\text{d}^{-1})(31.0\,\text{d})(0.50)} = 3.6080$$

  (v) Determine effluent *E. coli* concentration and $\log_{10}$ reduction:

  $$N_{\text{effluent}} = N_{\text{influent}} \frac{4a \cdot e^{1/2d}}{\left((1+a)^2 e^{a/2d} - (1-a)^2 e^{-a/2d}\right)}$$

  $$= (3.40 \times 10^8) \left[\frac{4(3.608) \cdot e^{1/2(0.5)}}{\left((1+3.608)^2 e^{3.608/2(0.5)} - (1-3.608)^2 e^{-3.608/2(0.5)}\right)} \right] = 1.70 \times 10^7$$

  $$\log_{10,\text{Reduction}} = \log_{10}(3.40 \times 10^8) - \log_{10}(1.70 \times 10^7) = 1.28$$

The range and mean of $\log_{10}$ reductions for the 12 months of the year with different monthly water temperatures are as follows:

T_{water} (°C) (range)	Log$_{10}$ reduction Range	Mean
17.4–23.2	1.28–1.64	1.47

The mean 1.47 $\log_{10}$ reduction is within the range of values reported for facultative ponds in Brazil and the US for thermotolerant coliforms and *E. coli*, with 1.43 and 1.33 $\log_{10}$ reduction, respectively (Espinosa *et al.*, 2017).

Theory and design of facultative ponds

(j) Determine BOD_5 removal in using the Wehner and Wilhem dispersed flow equation (Equation (5.37)).

$$S_e = S_0 \frac{4a \cdot e^{1/2d}}{\left((1+a)^2 e^{a/2d} - (1-a)^2 e^{-a/2d}\right)}; \quad d = 0.50 \text{ determined previously}$$

(i) Influent BOD_5 concentration:

$$S_0 = BOD_{5,\text{Influent}} = \frac{BOD_L}{1.5} = \frac{358 \text{ mg/L}}{1.5} = 239 \text{ mg/L}$$

(ii) Rate constant for BOD_5 removal (Equation (5.43)) for $T_{\text{water}} = 17.4°C$:

$$k_{BOD_5,T} = 0.15(1.09)^{(T-20)} = 0.15(1.09)^{(17.4-20)} = 0.12 \text{ d}^{-1}$$

(iii) Determine a in the Wehner–Wilhem equation (Equation (5.40)) for $t_{V/Q} = 31.0$ d:

$$a = \sqrt{1 + 4k_{BOD_5,T} \cdot t_{V/Q} \cdot d} = \sqrt{(1 + 4(0.12 \text{ d}^{-1})(31.0 \text{ d})(0.50)} = 2.9040$$

(iv) Determine effluent BOD_5 concentration and % BOD_5 removal:

$$S_e = S_0 \frac{4a \cdot e^{1/2d}}{\left((1+a)^2 e^{a/2d} - (1-a)^2 e^{-a/2d}\right)}$$

$$= \frac{4(2.9040) \cdot e^{1/2(0.5)}}{\left((1+2.9040)^2 e^{2.9040/2(0.5)} - (1-2.9040)^2 e^{-2.9040/2(0.5)}\right)} = 27.2 \text{ mg/L}$$

$$\% \, BOD_5 \text{ Removed} = 100 \left[\frac{(239 \text{ mg/L} - 27.2 \text{ mg/L})}{239 \text{ mg/L}}\right] = 88.6\%$$

The BOD_5 removed and remaining is the soluble organic matter. The pond effluent will have a higher BOD_5 as a result of algal cells, which should be filtered out for the BOD analysis.

(k) Estimate the sludge accumulation rate and the time to fill 25% of the pond's volume.

(i) Calculate the mass of influent TSS that settle to the sludge layer with Equation (5.43): There are no reported concentrations so assume TSS = 350 mg/L (worse-case scenario):

$$\dot{M}_{Sl} = 0.001 \cdot Q_m \cdot TSS_{\text{Settleable}} = 0.001(3300 \text{ m}^3/\text{d})(350 \text{ mg/L}) = 1155 \text{ kg/d}$$

(ii) Estimate the annual mass of digested sludge produced in the pond with Equation (5.46) for the following worst-case scenario with VS = 0.60 and TS = 0.11 (yields highest digested volume):

VS	FS	TS	S_{Sl}
0.60	0.40	0.11	1.08

(iii) Mass of digested sludge solids per year (Equation (5.46)):

$$\dot{M}_{Sl,Dyr} = 365 \cdot \left[FS \cdot \dot{M}_{Sl} + (VS_{\text{Destroyed}})VS \cdot \dot{M}_{Sl}\right]$$

$$= 365 \text{ d/yr} \left(0.40(1155 \text{ kg/d}) + (0.50)(0.60)(1155 \text{ kg/d})\right) = 206\,955 \text{ kg/yr}$$

Table 5.16 Final results for redesign of facultative ponds Albarrancho wastewater treatment plant, Cochabamba, Bolivia.

Parameter	Single Facultative Pond	Total 4 Ponds in Parallel
Q_{mean} (m³/d)	3300	13,200
Influent BOD_L (mg/L)	358	
Influent BOD_5 (mg/L)	239	
Influent TSS (mg/L)	350	
Design water temperature T_{water} (°C)	17.4	
Design $\lambda_{S,T}$ (kg BOD_L/ha d)	217	
Area (ha)	5.40	21.6
Water depth, h (m)	2.0	
Volume (m³)	102,154	408,615
$t_{V/Q}$ (d)	31.0	31.0
Helminth egg reduction: OFR, m³/m² d (<0.12 m³/m² d)	0.032	
$\log_{10\,95\%,\,LCL}$	>3.17	
E. coli $\log_{10}$ reduction:		
Range	1.28–1.64	
Mean	1.47	
Sludge accumulation (m³/yr)	1742	5888
Frequency sludge removal (yr)	14.7	14.7
Effluent $BOD_{5,\,Soluble}$ (mg/L)	27.2	
Effluent quality for reuse		Objective of final design, including redesign of the maturation ponds, is to satisfy the WHO guidelines for restricted reuse in agriculture

(iv) Volume of in-pond digested sludge (Equation (5.44)):

$$\dot{V}_{Sl,D\,yr} = \frac{\dot{M}_{Sl,D\,yr}}{\rho_{H_2O} \cdot S_{Sl} \cdot TS} = \frac{206\,905\,\text{kg/yr}}{(1000\,\text{kg/m}^3)(1.08)(0.11)} = 1742\,\text{m}^3/\text{yr}$$

(v) Time to fill 25% of pond's volume (Equation (5.47)):

$$t_{25\%} = \frac{0.25 \cdot V_f}{\dot{V}_{f-Sl}} = \frac{(0.25)(102\,154\,\text{m}^3)}{1742\,\text{m}^3/\text{yr}} = 14.7\,\text{yr}$$

A summary of the final design is presented in Table 5.16. The maturation ponds will be redesigned in Chapter 6.

doi: 10.2166/9781789061536_0153

Chapter 6
Theory and design of maturation ponds

6.1 MATURATION PONDS AND PATHOGEN REDUCTION

The purpose of maturation ponds is to sufficiently remove or inactivate pathogens, with natural in-pond processes, to produce an effluent acceptable for wastewater reuse in agriculture and aquaculture. (Reduction as used in this chapter includes sufficient removal or inactivation of pathogens to protect public health to meet wastewater reuse guidelines.) Maturation ponds typically are designed to treat facultative pond effluents but have also been used with success for the treatment of UASB effluents (Dias *et al.*, 2014) and effluents from conventional treatment plants (Arceivala & Asolekar, 2007).

Maturation ponds are shallower than facultative ponds, with depths ranging from 0.4 to 1.0 m, and aerobic throughout depth without thermal stratification (Dias *et al.*, 2014; Maiga *et al.*, 2009; Mara, 2003). Depths less than 1.0 m require a liner to prevent the growth of emergent plants, which block solar insolation and promote mosquito breeding. Although maturation ponds have as great a diversity of algae as facultative ponds, the in-pond algal biomass is lower.

6.1.1 Factors affecting pathogen reduction

Table 6.1 lists the key factors influencing pathogen reduction in facultative, and especially, maturation ponds. While the design of facultative ponds is based on photosynthetic oxygen production and organic surface loading, with pathogen reduction a secondary consideration, maturation pond design focuses exclusively on pathogen reduction as influenced by sunlight (solar radiation), temperature, hydraulic retention time, and sedimentation.

6.1.1.1 Sunlight

Sunlight is considered to be the single most important primary factor for pathogen reduction in maturation ponds, especially for bacterial and viral pathogens (Davies-Colley, 2005; Verbyla *et al.*, 2017). The following sunlight-induced mechanisms occur (Table 6.1):

- Photobiological damage:
  UV-B in the 300–320 nm wavelength is absorbed by DNA causing direct damage, which can be repaired by enzymatic processes in bacteria.

© 2022 The Author. This is an Open Access book chapter distributed under the terms of the Creative Commons Attribution Licence (CC BY-NC-ND 4.0), which permits copying and redistribution for noncommercial purposes with no derivatives, provided the original work is properly cited (https://creativecommons.org/licenses/by-nc-nd/4.0/). This does not affect the rights licensed or assigned from any third party in this book. The chapter is from the book *Integrated Wastewater Management for Health and Valorization: A Design Manual for Resource Challenged Cities*, Stewart M. Oakley (Author)

Table 6.1 Factors influencing pathogen reduction in facultative and maturation Ponds.[1,2]

Factor	Mechanism	Affected Pathogens
Sunlight	Photobiological DNA damage by solar UV-B radiation	Bacteria, viruses, protozoa
Sunlight and dissolved oxygen	Photo-oxidative damage to (i) DNA or (ii) external structures (cell membranes)	Bacteria, possibly viruses
Sunlight, dissolved oxygen, high pH	Photo-oxidative damage to bacterial cell membranes at pH > 8.0	Bacteria, possibly viruses
Temperature	Affect rates of removal processes	Bacteria, viruses
Hydraulic retention time	Affects the extent of removal	Bacteria, viruses, protozoa, helminths
Sedimentation	Discrete settling of helminth eggs and larger protozoan cysts	Helminths, larger protozoa (e.g., *Entamoeba*, *Giardia*)
	Settling of aggregated particles with attached pathogens	Bacteria, viruses, protozoa, helminths
Predation	Ingestion by micro-fauna	Bacteria, protozoa, (possibly viruses)

[1]Adapted from Davies-Colley (2005) and Verbyla *et al.* (2017).
[2]Maturation pond design should address all factors and mechanisms minus predation.

- Photo-oxidative damage to internal structures:
  Short UV-B wavelengths are absorbed by endogenous photosensitizers (DNA and other cell constituents), which react with oxygen to form highly reactive photo-oxidizing species that damage genetic material within the cell.
- Photo-oxidative damage to external structures:
  UV and visible wavelengths are absorbed by exogenous photosensitizers, such as humic substances or natural organic matter, which react with oxygen to form highly reactive photo-oxidizing species that damage external structures such as the cell membrane or viral capsid proteins. The reaction is pH dependent (pH $\geq$ 8) for some bacteria such as *E. coli* (Davies-Colley, 2005).

6.1.1.2 Temperature
Temperature affects the rates of removal processes and has historically been used to estimate indicator bacteria die-off in pond systems, as well as BOD_5 removal. For the inactivation of pathogens, the temperature is considered a secondary factor that influences the rate of primary factors (Davies-Colley, 2005).

6.1.1.3 Hydraulic retention time
Hydraulic retention time controls the extent of the reduction of pathogens by primary factors. Since the theoretical hydraulic retention time is typically much less than the actual one as a result of hydraulic short-circuiting, maturation pond design should focus on plug flow with high length to width (l/w) ratios or baffled channels.

6.1.1.4 Sedimentation
Pathogen removal by sedimentation occurs through two processes: discrete settling and aggregated setting.
- Discrete settling:
  Discrete settling is a key removal mechanism for helminth eggs, which have settling velocities ranging from 9.4 to 188.9 m/d (Table 6.2).

Table 6.2 Helminth egg and protozoan (oo)cyst settling velocities in water at 20°C.[1]

Parasite	Egg or (oo) cyst Size (μm)	Settling Velocity $\nu_{settling}$ (m/d)	Safety Factor $SF = \dfrac{\nu_{settling}}{0.12 \text{ m/d}}$	Settling Time in Pond (d) $h = 1.0$ m	Settling Time in Pond (d) $h = 0.5$ m
Helminths					
Ascaris lumbricoides	60 × 45	15.6	130	0.06	0.03
Trichuris trichiura	50 × 22	36.7	306	0.03	0.01
Hookworms	60 × 40	9.4	78	0.11	0.05
Taenia saginata	40 × 30	17.3	144	0.06	0.03
Schistosoma mansoni	150 × 55	188.9	1574	0.005	0.0025
Protozoa					
Entamoeba histolytica	10–12	0.432	3.6[a]	2.3	1.16
Giardia lamblia	9.3 × 12.2	0.240	2.0[a]	2.1	1.04
Cryptosporidium parvum	4.5–5	0.048	0.4[b]	20.8	10.42

[1]After Stott (2003).
[a]Not likely removed by discrete settling.
[b]Not removed by discrete settling.

A hookworm egg, for example, with the lowest helminth settling velocity, could theoretically settle to the bottom of a 1.0 m deep quiescent maturation pond in 0.11 d:

$$T_{settling} = \frac{h}{\nu_{settling}} = \left(\frac{1.0 \text{ m}}{9.4 \text{ m/d}}\right) = 0.11 \text{ d}$$

A *Cryptosporidium parvum* oocyst, however, would take 20.8 d, to settle the same 1.0 m:

$$T_{settling} = \frac{h}{\nu_{settling}} = \left(\frac{1.0 \text{ m}}{0.048 \text{ m/d}}\right) = 20.8 \text{ d}$$

The design safety factor (SF) for helminth egg removal can be defined by Equation (6.1):

$$SF = \frac{\nu_{settling}}{0.12 \text{ m/d}} \tag{6.1}$$

$\nu_{settling}$ = settling velocity of egg, m/d and 0.12 m/d = maximum design overflow rate for helminth egg removal (0.12 m³/m² d) from Equation (5.28).

The safety factors for helminth eggs range from 78 to 1574. If protozoan (oo)cysts were to be removed by discrete settling with the same design overflow rate, the safety factors would range from 2.0 to 3.6 for *Entamoeba histolytica* and *Giardia lamblia*, values far too low for effective removal, and *C. parvum* would not be removed at all. While it is well known that helminth eggs can be removed in well-designed ponds by discrete settling, protozoan (oo)cysts will not likely be removed unless the pond is designed with an appropriate design overflow rate, similar to what has been done empirically with helminth eggs.

- Aggregated settling:

  All pathogens can be removed by aggregated settling by attaching to, or becoming embedded in, larger particles such as wastewater suspended solids, algae, and bacteria. Aggregated settling will be more significant in ponds containing higher concentrations of suspended solids.

Table 6.3 Survival times of select pathogens and indicators in water and wastewater.[1]

Organism	Survival Time in Water and Wastewater at 20–30°C (days)
Viruses	
Enteroviruses (polio-, echo-, coxsackievirus)	< 120 d; usually < 50 d
Bacteria	
Thermotolerant coliforms	< 60 d; usually < 30 d
Salmonella spp.	< 60 d; usually < 30 d
Shigella spp.	< 30 d; usually < 10 d
Vibrio cholerae 01[2]	$\approx$ 50 d @ T = 17.5°C
Protozoa	
Entamoeba histolytica cysts	< 30 d; usually < 15 d
Cryptosporidium oocysts	< 180 d; usually < 70 d
Helminths	
Ascaris eggs	Years
Tapeworm eggs	Many months

[1] Adapted from WHO (2006).
[2] Data from an $F/M_1/M_2$ waste stabilization pond system in Peru with a 4.26 $\log_{10}$ reduction of *Vibrio cholerae* 01 (Castro de Esparza *et al.*, 1992).

Helminth eggs are readily removed by discrete and aggregated settling in well-designed and maintained facultative-maturation pond systems, with the maturation pond adding an additional safety factor to ensure 100% removal. The smaller protozoan (oo)cysts will be largely removed by attachment with settleable particles. Pond systems can readily remove *Giardia* cysts and *Cryptosporidium* oocysts, but require longer retention times than many current designs. Grimason *et al.* (1993), for example, reporting on 11 waste stabilization pond systems in Kenya, concluded that the minimum hydraulic retention time required for the removal (non-detection) of both *Cryptosporidium* and *Giardia* was 37.3 d.

Pathogens removed by discrete or aggregated settling can retain viability from <10 to 120 days in the case of bacteria and viruses, <15 to 180 days for protozoa, and many months to years in the case of helminths (Table 6.3). These viable pathogens can be resuspended to the water column by wind-induced mixing, diurnal and seasonal overturning caused by temperature changes in the water column, and rising biogas bubbles from digesting sludge (Verbyla *et al.*, 2017). All of these processes occur mainly in facultative ponds rather than shallow, well-designed maturation ponds.

6.1.1.5 Predation

Predation is a natural removal mechanism for bacteria and protozoan (oo)cysts, and possibly viruses attached to larger particles, in water environments. Predators include nematodes, copepods, rotifers, and ciliated protozoa (Jasper *et al.*, 2013). While predation removes pathogens from the water column, it does not necessarily inactivate all of them after ingestion, and some bacterial pathogens remain activated after ingestion by ciliates and amoebas (Jasper *et al.*, 2013). The still viable pathogens, however, would be excreted in fecal pellets that could promote removal by discrete sedimentation (Davies-Colley, 2005).

6.1.2 Design strategies for pathogen reduction

In spite of the various complex factors interacting simultaneously in maturation ponds, waste stabilization ponds still need to be designed, and many others rehabilitated, by designers pressed to

Table 6.4 Design strategies related to pathogen reduction factors.

Design Strategy	Pathogen Reduction Factor

• Sunlight exposure
    ○ Avoid shading at low sun angles from baffles, berms, fences, vegetation, and so on.
    ○ E-W orientation for shading problems at low sun angles
• Depth
    ○ Range: 0.4–1.0 m
    ○ Lower depths give higher bacteria and virus reductions
    ○ Depths < 1.0 m require a pond liner to avoid emergent aquatic plants
	Sunlight inactivation

• Maximize hydraulic retention time and minimize dispersion
    ○ New designs based on hydraulic performance analyses of well-functioning ponds
    ○ Rehabilitate old ponds with hydraulic performance analyses
• Promote plug flow hydraulics
    ○ Avoid irregular pond shapes where design and performance data are non-existent
    ○ Unbaffled ponds, preferably in series: $(l/w)_{exterior} \geq 3/1$
    ○ Baffled ponds: $(l/w)_{interior} \geq 10/1$
    ○ Inlet structures
• Influent enters at or below the water surface
• Multiple inlets in large, unbaffled ponds
    ○ Outlet structures
• Multiple outlets in large, unbaffled ponds
• Weirs for control of surface effluent velocities
• Flashboards or similar structures to change water depth in unbaffled ponds
• Wind abatement
    ○ Direction of flow at right angles to prevailing winds
    ○ Wind fences and berms are used for strong prevailing winds, but should not shade ponds at low sun angles
	Hydraulic retention time (Promote plug flow)

• Low overflow rates, OFR ≪ 0.12 m/d, for helminth egg removal and promotion of aggregated settling of particles with attached pathogens • Rock filter for pathogen removal by aggregated settling	Sedimentation

solve problems of public health and the need for wastewater reuse. There are straightforward design strategies, continually being revised, that can significantly enhance pathogen reduction. Table 6.4 lists the strategies that address the key inactivation factors of sunlight, hydraulic retention time, and sedimentation. Briefly, maturation ponds should maximize solar insolation on the pond's surface, have shallow depths of 0.4–1.0 m, approximate plug flow with or without baffles, and have long hydraulic retention times.

6.1.2.1 Sunlight exposure
Low angle solar insolation on pond surfaces can be shaded by (i) high berms between or around ponds; (ii) adjacent ponds at different elevations; (iii) fences and vegetation; and (iv) in-pond baffles if the top edges are too high above water level. If natural shading effects cannot be eliminated, an East–West orientation of pond length would be preferred. In these circumstances, the designer should verify that low-angle sun shading will not significantly affect performance.

Solar radiation can vary greatly throughout the year as a result of latitude or rainy season, causing seasonal differences in bacterial and viral pathogen reductions. Under these conditions maturation pond design should ensure pathogen reduction is sufficient to meet reuse needs. Possible solutions include designs for the low solar radiation season or the use of stabilization reservoirs for further pathogen reduction.

6.1.2.2 Depth
Design maturation pond depths range from 0.4 to 1.0 m, with shallower depths providing higher levels of sunlight inactivation for bacteria and viruses, and possibly protozoa. At depths <1.0 m a liner must be used to prevent the growth of emergent aquatic plants. In unbaffled ponds depths can be changed, enabling operational control of the process (e.g., use of shallower depths in summer, yielding higher pathogen reduction for effluent reuse). In baffled ponds, the design depth cannot be significantly lowered as the baffles will shade the water surface at low sun angles.

6.1.2.3 Maximize theoretical hydraulic retention time and minimize dispersion
The hydraulic retention time controls the extent of pathogen reduction and is the main control the designer has over maturation pond processes. The theoretical retention time, $t_{V/Q}$, should be as long as site conditions allow to ensure adequate pathogen reductions, with an appropriate safety factor for low solar radiation periods. Unfortunately, often the tendency is to minimize $t_{V/Q}$ with poor site and technology selection (Figure 6.1).

The pond design should also minimize longitudinal dispersion to approximate plug flow to ensure sufficient pathogen reduction with a factor of safety. Numerous tracer studies have been performed on full-scale waste stabilization ponds and design insights can be learned from these studies to rehabilitate existing ponds, and to design new ones with improved hydraulic performance.

6.1.2.4 Longitudinal dispersion and mean hydraulic retention time
Historically, many pond systems were designed without consideration of hydraulic efficiency. The mean hydraulic retention time, which is the distance to the centroid of a residence time distribution developed from tracer studies, has often been found to be much less than the theoretical retention time, $t_{V/Q}$. Figure 6.2 shows hypothetical residence time distributions for different maturation pond hydraulic designs.

Much work has been done on minimizing dispersion in the design of disinfection contact chambers (MWH, 2005), and some of this knowledge is applicable to the design of maturation ponds. Figure 6.3 shows a typical residence time distribution curve for a well-designed, baffled chamber. True plug flow would appear as almost as a vertical line in the distribution (see Figure 6.2, Curve c), and in Figure 6.3 the left side does rise almost vertically at 6 minutes until it peaks at 15 minutes. The peak concentration is close to the calculated mean hydraulic retention time, $\bar{t}_{mean} = 16.5$ min, which is the time to the centroid of the distribution. The ratio of $\bar{t}_{mean}/t_{V/Q}(100) = 84.5\%$ with $d = 0.10$ also indicates the chamber has relatively low dispersion.

Historically, waste stabilization ponds have not been designed considering dispersion. Table 6.5 presents the results from various tracer studies of facultative and maturation ponds where the mean hydraulic retention time, $\bar{t}_m$, can be compared to the theoretical retention time, $t_{V/Q}$. The last column lists the mean retention as a percentage of the theoretical retention time. In the nine studies without baffles, the percentage of the mean retention time to the theoretical, $\bar{t}_m/t_{V/Q}$, ranged from 42.1 to 62.6% for maturation ponds and 21.4 to 49.0% for facultative ponds. In the one study where baffles were installed in the maturation pond in Colombia, giving $(l/w)_{interior} = 35/1$, $\bar{t}_m/t_{V/Q}$ only increased from 42.1 to 50.0% as a result of strong winds driving the hydraulic short-circuiting; the wind-induced short-circuiting was finally abated when wind fences were installed, yielding $\bar{t}_m/t_{V/Q} = 73.8\%$ (Lloyd et al., 2003b). Four ponds also had values of $t_{V/Q}$, ranging from 2.52 to 5.65 d, far too low for the adequate performance of both facultative and maturation ponds.

The results of these tracer studies and others indicate that many operating ponds are not well designed hydraulically, and as a result, a significant percentage of their volume, and $t_{V/Q}$, is lost to dead space. The resulting hydraulic short-circuiting in turn lowers pathogen reduction, causing many pond

Figure 6.1 A maturation pond following an anerobic reactor with river discharge. A pond system with agricultural reuse was originally proposed for this municipality, but the mayor chose this small area, the worst possible site, as the only available for a wastewater treatment plant. Groundwater is at river level, and a pump station had to be installed to pump to the elevated anaerobic reactors. The maturation pond design water depth is 1.5 m, with a hydraulic retention time of 2.8 d. After three years, the system is still not in operation (2021), and most likely will be abandoned. The plant was built to satisfy the requirement of wastewater treatment. Photo courtesy of CONASIN, SRL, Cusco, Peru.

systems to produce effluents with less than 2.0 $\log_{10}$ reduction of fecal coliforms or *E. coli* (Lloyd *et al.*, 2003a, 2003b). Simple tracer studies on existing ponds can, with careful analysis, play an important role in improving future designs, and in rehabilitating existing systems (Lloyd *et al.*, 2003a, 2003b).

Many older pond systems worldwide are failing, if they have not already, and, hopefully, they could be rehabilitated and redesigned, with improved hydraulics that minimizes longitudinal dispersion. The following sections present the equations used in a simple residence time distribution analysis, and a case study analysis of longitudinal dispersion and fecal coliform reduction in an operating maturation pond.

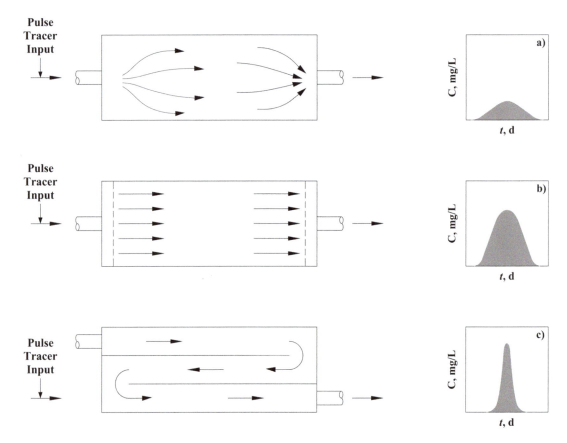

Figure 6.2 Longitudinal dispersion in different maturation pond designs as determined by residence time distributions of pulse tracer inputs (concentration versus time). Curve a): Inadequate inlet and outlet structures cause large dead spaces and hydraulic short-circuiting, resulting in a wide residence time curve with a low peak. Curve b): Multiple inlets and outlets with diffuser baffles distribute flow uniformly, causing the residence time curve to become narrower with a higher peak. Curve c): Baffles significantly increase *l/w*, causing the flow regime to approach plug flow, producing a residence time distribution with a still narrower width and higher peak. Redrawn by A. F. Orue Agramonte from MWH (2005).

6.1.2.5 Residence time distribution analysis to assess longitudinal dispersion

The following equations have been be used for dispersed flow analysis of maturation and facultative ponds with residence time distribution curves. The equations must be used with samples taken at uniform time intervals throughout the sampling period (Viessman & Hammer, 2005).

The mean hydraulic retention time, which is the distance to the centroid of the residence time distribution curve, is defined by Equation (6.2):

$$\bar{t}_m = \frac{\sum_i t_i C_i}{\sum_i C_i} \tag{6.2}$$

$\bar{t}_m$ = mean hydraulic retention time; distance to centroid of distribution, d; t_i = elapsed time of tracer sample, *i*, d; C_i = concentration of tracer at time, t_i, mg/L; μg/L, and so on.

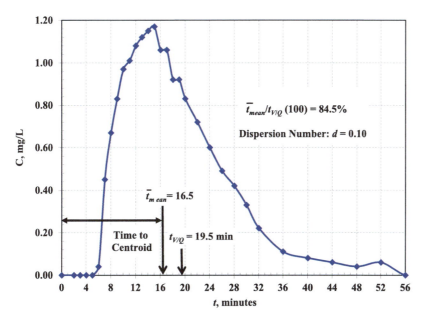

Figure 6.3 A residence time distribution of a pulse tracer input to a baffled disinfection contact chamber. The curve starts rising almost vertically at 6 minutes, and the peak at 15.0 minutes is very close to the mean hydraulic retention time (time to centroid) of 16.5 minutes, which is also close to the theoretical retention time, t_V/Q, of 19.5 minutes. The value of $t_{mean}/t_V/Q$ (100) = 84.5% is high, and should approach 90% for well-designed contact chambers. Developed from Viessman and Hammer (2005).

The variance of the distribution is defined by Equation (6.3):

$$\sigma^2 = \frac{\sum_i t_i^2 C_i}{\sum_i C_i} - \left(\bar{t}_m\right)^2 \quad (6.3)$$

σ^2 = variance of tracer data distribution, d².

The equation for the normalized variance, $\sigma^2/\left(\bar{t}_m\right)^2$, is solved by trial and error to determine the dispersion number, d, using Equation (6.4):

$$\frac{\sigma^2}{(\bar{t}_m)^2} = 2d - 2d^2(1 - e^{-1/d}) \quad (6.4)$$

$\sigma^2/(\bar{t}_m)^2$ = normalized variance, unitless and d = dispersion number; $d = D/(U \cdot L)$ from Equation (5.35).

The calculated dispersion number can then be used in the Wehner–Wilhem equation to estimate *E. coli* or fecal coliform reduction.

6.1.2.6 Limitations of residence time distribution studies

Unlike disinfection contact chambers, or primary sedimentation basins, the large areas and volumes of waste stabilization ponds cannot be easily controlled hydraulically, and flow patterns can change seasonally, monthly, or even daily, due to the following factors:

(1) Wide changes in flowrates due to infiltration/inflow, creating different patterns of dispersion and short-circuiting (Reynolds *et al.*, 1977).

Table 6.5 Reported theoretical and mean hydraulic retention times determined from tracer studies of facultative and maturation ponds in several countries.

Country	Type of Pond	$t_{V/Q}$ (d)	$\bar{t}_{mean}$ (d)	$\dfrac{\bar{t}_{mean}}{t_{V/Q}}$ (100%)
Colombia[1]	Maturation			
	Without baffles	2.52	1.06	42.1
	Baffles $(l/w)_{interior} = 35/1$	2.52	1.26	50.0
	Baffles and wind fences	2.52	1.86	73.8
Mexico[2]	Facultative	4.30	0.92	21.4
Peru[3]	Facultative:			
	Pond 1 without baffles	10.32	4.85	47.0
	Pond 3 without baffles	5.65	2.77	49.0
	Maturation:			
	Pond 1 without baffles with varying flowrates	15.26	6.98	45.7
		18.84	9.94	52.8
		13.13	5.86	44.6
	Pond 3 without baffles	3.23	2.02	62.5
USA[4]	Maturation			
	Without baffles	22.3	14.0	62.6
	Pond system:			
	1 facultative + 6 maturation in series without baffles	146.3	88.3	60.3

[1]Lloyd et al. (2003b).
[2]Lloyd et al. (2003a).
[3]Yánez (1993).
[4]Reynolds et al. (1977).

(2) Changing wind directions and speeds, inducing varying types of hydraulic short-circuiting that changes the volumes of dead spaces (Sweeney et al., 2003).
(3) Intentional changes in water depths in unbaffled maturation ponds can potentially change the effects of (1) and (2) above, for better or worse, in large ponds.
(4) Irregularly shaped ponds can form complex flow patterns and should be avoided if possible.
(5) In stratified facultative ponds, vertical mixing caused by (1) or (2) above can produce lower quality effluents.

In short, there are various residence time distributions occurring throughout the year that will never be assessed. Designers must take these uncertainties into consideration and develop hydraulic safety factors for improving and maintaining long-term performance.

6.1.2.7 Case study: residence time distribution analysis to assess fecal coliform reduction in a maturation pond, Corinne, Utah, USA

The Corinne, Utah waste stabilization pond system was built in 1970 and started operation in 1971 (Reynolds et al., 1977). The system is comprised of a facultative pond followed by six maturation ponds in series (Figure 6.4). The design flowrate was 265 m³/d, with a theoretical hydraulic retention time of 180 d for the system. A tracer study using rhodamine B dye was performed on maturation pond M2 in 1975 to determine the mean hydraulic retention time, $\bar{t}_m$, and the dispersion number, d. During the tracer study the influent flowrate averaged 275.8 m³/d.

Theory and design of maturation ponds

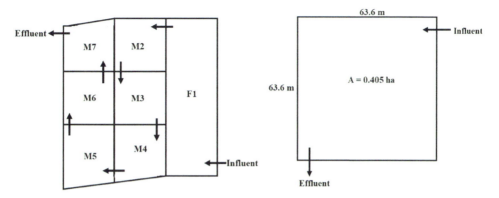

Figure 6.4 Left: The Corinne waste stabilization pond system, with one facultative pond (F1), followed by six maturation ponds in series (M2–M7). The depths of all ponds average 1.2 m. Right: maturation pond M1, with a theoretical hydraulic retention time, $t_{V/Q}$, of 22.3 d at the time of the monitoring study.

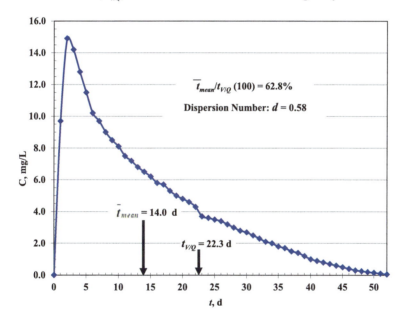

Figure 6.5 Residence time distribution for maturation pond M1 of the Corinne waste stabilization pond system. The skewness to the left is a sign of high longitudinal dispersion, also indicated by $d = 0.58$. The peak at 2 d is 12 d ahead of the mean hydraulic retention time of 14.0 d.

6.1.2.8 Determination of residence time distribution parameters

The calculations and results of the tracer study for pond M1 are shown in Table 6.5, and the residence time distribution curve in Figure 6.5. The results are:

(1) Mean hydraulic retention time (Equation (6.2)):

$$\bar{t}_m = \frac{\sum_i t_i C_i}{\sum_i C_i} = \frac{71{,}032.6 \text{ d mg/L}}{3165.1 \text{ mg/L}} = 13.96 \text{ d} = 14.0 \text{ d}$$

Table 6.6 Results of tracer analysis for maturation pond M2. Mean flowrate for time period = 275.8 m³/d.

1	2	3	4	5	6	7	8	9	10	11	12	13
t (d)	C (mg/L)	$\sum_i C_i$	$\sum_i t_i^2 C_i$	t (d)	C (mg/L)	$\sum_i C_i$	$\sum_i t_i^2 C_i$	$\bar{t}_m = \dfrac{\sum_i t_i C_i}{\sum_i C_i}$	$\sigma^2 = \dfrac{\sum_i t_i^2 C_i}{\sum_i C_i} - (\bar{t}_m)^2$	$\sigma^2/(\bar{t}_m)^2$	d	$2d - 2d^2(1-e^{-1/d})$
										$\sigma^2/(\bar{t}_m)^2 = 2d - 2d^2(1-e^{-1/d})$		
1	9.7	9.70	9.70	27	3.2	86.40	2332.80			0.61	1.00	0.74
2	14.9	29.80	59.60	28	3.0	84.00	2352.00				0.90	0.71
3	14.2	42.60	127.80	29	2.8	81.20	2354.80				0.80	0.69
4	12.8	51.20	204.80	30	2.7	81.00	2430.00				0.70	0.65
5	11.5	57.50	287.50	31	2.5	77.50	2402.50				0.60	0.62
6	10.2	61.20	367.20	32	2.3	73.60	2355.20				0.55	0.59
7	9.7	67.90	475.30	33	2.1	69.30	2286.90				0.57	0.60
8	9.0	72.00	576.00	34	2.0	68.00	2312.00				**0.58**	**0.61**
9	8.5	76.50	688.50	35	1.8	63.00	2205.00					
10	8.1	81.00	810.00	36	1.7	61.20	2203.20					
11	7.5	82.50	907.50	37	1.5	55.50	2053.50					
12	7.2	86.40	1036.80	38	1.4	53.20	2021.60					
13	6.8	88.40	1149.20	39	1.2	46.80	1825.20					
14	6.5	91.00	1274.00	40	1.0	40.00	1600.00					
15	6.2	93.00	1395.00	41	0.9	36.90	1512.90					
16	5.8	92.80	1484.80	42	0.8	33.60	1411.20					
17	5.7	96.90	1647.30	43	0.7	30.10	1294.30					
18	5.3	95.40	1717.20	44	0.6	26.40	1161.60					
19	5.0	95.00	1805.00	45	0.5	22.50	1012.50					
20	4.8	96.00	1920.00	46	0.4	18.40	846.40					
21	4.6	96.60	2028.60	47	0.3	14.10	662.70					
22	4.3	94.60	2081.20	48	0.25	12.00	576.00					
23	3.7	85.10	1957.30	49	0.20	9.80	480.20					
24	3.6	86.40	2073.60	50	0.15	7.50	375.00					
25	3.5	87.50	2187.50	51	0.10	5.10	260.10					
26	3.4	88.40	2298.40	52	0.05	2.60	135.20					
Sum of respective columns:		226.65				3165.10	71,032.60	13.96	118.39			

$C(2+6); \sum_i C_i(3+7); \sum_i t_i C_i(4+8)$

(2) Variance (Equation (6.3)):

$$\sigma^2 = \frac{\sum_i t_i^2 C_i}{\sum_i C_i} - (\bar{t}_m)^2 = \frac{71{,}032.6}{3165.1} - (13.96)^2 = 118.39\ \text{d}^2$$

(3) Dispersion number (Equation (6.7)):

$$\frac{\sigma^2}{(\bar{t}_m)^2} = 2d - 2d^2(1 - e^{-1/d})$$

$$\frac{\sigma^2}{(\bar{t}_m)^2} = \frac{118.39\ \text{d}^2}{(13.96\ \text{d})^2} = \frac{118.39\ \text{d}^2}{194.88\ \text{d}^2} = 0.61$$

$d = 0.58$ by trial and error (Table 6.6),

$$2d - 2d^2(1 - e^{-1/d}) = 2(0.58) - 2(0.58)^2(1 - e^{-1/0.58}) = 0.61$$

(4) Mean hydraulic retention time as a percentage of theoretical retention time:

$$\frac{\bar{t}_{\text{mean}}}{t_{V/Q}}(100\%) = \frac{14.0}{22.3}(100) = 62.8\%$$

The results show the M1 pond hydraulic characteristics, with $\bar{t}_m/t_{V/Q} = 62.8\%$, to be in the range of the ponds listed in Table 6.5. The highly skewed residence time distribution in Figure 6.5 is characteristic of high longitudinal dispersion. The peak concentration of dye occurs at 2 days, 12 days earlier than the mean hydraulic retention time at 14.0 d. The calculated dispersion number, $d = 0.58$, is characteristic of a large amount of dispersion (Viessman & Hammer, 2005).

6.1.2.9 Estimation of fecal coliform reduction using the Wehner and Wilhem equation

An estimation of fecal coliform reduction using the Wehner–Wilhem equation with the results of the residence time distribution analysis ($d = 0.58, \bar{t}_m = 14.0$ d) will be compared with the fecal coliform reduction measured during the Corinne monitoring study (Reynolds *et al.*, 1977). The following data apply for the period of July 15–August 14:

(i) *E. coli* concentrations in facultative pond effluent.
Using data from Reynolds *et al.* (1977) in Table 6.7:

$$N_{\text{influent}} = 5.16 \times 10^3\ \text{CFU}/100\ \text{mL}$$

(ii) Dispersion number:
$d = 0.58$ from residence time distribution analysis (Table 6.6).

Table 6.7 Measured Log$_{10}$ reduction of fecal coliforms in maturation pond M2.[1]

Sampling Period	Mean T_{water} (°C)	Pond Depth, h (m)	Solar Radiation (kJ/m²)	Fecal Coliforms[1] CFU/100 mL Influent	Effluent	Log$_{10}$ Reduction Fecal Coliforms
July–August	21.2	1.2	24,200–27,300	5.16E+03	1.45E+02	1.55

[1]Data from Reynolds *et al.* (1977).
[1]Geometric mean values from 31 days of continuous monitoring.

(iii) Mortality rate constant at design water temperature, $k_{b,T}$, d^{-1} using the von Sperling equation (Equation (5.34)):

$$k_{b,T} = (0.542 \cdot h^{-1.259})(1.07)^{(T-20)}$$

$h = 1.2$ m and $T_{water} = 21.2°C$.

$$k_{b,T} = (0.542 \cdot (1.2)^{-1.259})(1.07)^{(21.2-20)} = 0.47 \text{ d}^{-1}$$

(iv) Determine a in the Wehner–Wilhem equation (Equation (5.32)):

$t_{V/Q} = 14.0$ d

$$a = \sqrt{1 + 4k_{b,T} \cdot t_{V/Q} \cdot d} = \sqrt{1 + 4(0.47 \text{ d}^{-1})(14.0 \text{ d})(0.58)} = 4.0221$$

(v) Determine effluent *E. coli* concentration:

$$N_{effluent} = N_{influent} \frac{4a \cdot e^{1/2d}}{\left((1+a)^2 e^{a/2d} - (1-a)^2 e^{-a/2d}\right)}$$

$$= (5.16 \times 10^3)\left[\frac{4(4.0221) \cdot e^{1/2(0.58)}}{\left((1+4.0221)^2 e^{4.0221/2(0.58)} - (1-4.0221)^2 e^{-4.0221/2(0.58)}\right)}\right]$$

$$= 2.43 \times 10^2 \text{ CFU/100 mL}$$

(vi) Determine the $\log_{10}$ reduction of fecal coliforms:

$$\log_{10,\text{Reduction}} = \log_{10}(5.16 \times 10^3) - \log_{10}(2.43 \times 10^2) = 1.33$$

6.1.2.10 Comment on Corinne maturation pond case study

The 1.33 $\log_{10}$ reduction calculated with the von Sperling equation is close to the geometric mean value of 1.55 measured from 31 d of continuous monitoring (Table 6.7). At least for the period of time the pond was monitored, the residence time distribution analysis gave a close approximation of the pond's actual hydraulic performance in terms of fecal coliform reduction.

The environmental conditions in the Corinne pond during July–August, summer conditions in the northern hemisphere, are similar to those of the tropical and semi-tropical climates that von Sperling used to develop his equation from data for 186 facultative/maturation ponds around the world, most of which were in Brazil (von Sperling, 2005). The Corinne pond system, however, is at 41.5°N latitude, and during the northern hemisphere summer at this latitude, the solar insolation is higher than any month during the year in Brazil (Table 6.8). The von Sperling equation could not have been easily

Table 6.8 Solar radiation in Corinne, USA, and Belo Horizonte, Brazil.[1]

Pond System	Range of Solar Radiation (kJ/m² d)	
	June–August	December–February
Corinne, Utah, USA	24,200–27,300	6000–10,500
Belo Horizante, Brazil	14,900–17,500	18,300–20,600

[1]Data from CLIMWAT.

developed with a solar insolation term, but the equation may still be applicable at high north and south latitudes, during summer conditions, as long as the solar insolation and water temperatures are similar to those in the 186 pond systems from which the equation was developed (von Sperling, 2005).

6.1.2.11 Wind abatement
As discussed in Chapter 5, the direction of flow should be at right angles to prevailing winds. Baffled ponds should be oriented so the baffles are at right angles to prevailing winds. Wind fences and berms are used for strong prevailing winds, but should be designed so they do not shade ponds at low sun angles.

6.1.2.12 Overflow rate
The design overflow rate for maturation ponds should be ≤ 0.12 m/d based on the data and recommendations reported by von Sperling (2007) for helminth egg removal. Table 6.9 can be used as a guide for design in ponds with depths, h, ranging from 0.4 to 1.0 m. One well-designed unbaffled pond, with a maximum depth of 1.0 m and an area of 1000 m²/100 m³/d of influent flow, could meet the overflow requirement when operating at depths from 0.4 to 1.0 m, with $t_{V/Q}$ ranging from 3.3 to 9.0 d.

6.1.2.13 Rock filters
Rock filters have been in pond systems since the 1970s in the US for suspended solids removal in order to meet effluent discharge requirements (Middlebrooks, 1988; Middlebrooks *et al.*, 1982; Swanson & Williamson, 1980). The original rock filter designs used a bed of river rocks (5–20 cm) rising above the water surface, with a percentage voids of 40–42% (Figure 6.6), through which pond effluent was passed horizontally or vertically, allowing algae to settle, or rise, in the short distances in the void spaces and become attached to the rock surfaces (Swanson & Williamson, 1980). The rock filters with smaller diameter rock of 1.0–2.0 cm have also been used with success in Brazil, Jordan, and New Zealand (Dias *et al.*, 2014; Mara, 2003; Middlebrooks *et al.*, 2005).

Rock filters can also play an important role in pathogen removal by aggregated settling as evidenced in the 10-year study by Dias *et al.* (2014) in Brazil (Table 6.10). The $\log_{10}$ reduction of the geometric mean influent and effluent concentrations of *E. coli* was 0.59, a significant reduction. Rock filters could easily be implemented in existing systems and should be considered by designers. Rock filter design for pathogen removal is discussed in detail in Section 6.2.3.

6.2 DESIGN OF MATURATION PONDS
Unbaffled and baffled maturation ponds are both common worldwide, many with mixed performance results, and designers need to assess the advantages and disadvantages of a particular reuse design as listed in Table 6.11. A design of two to three ponds in series should meet, at minimum, the WHO guideline of a 4.0 $\log_{10}$ reduction of *E. coli* for restricted irrigation (WHO, 2006).

Table 6.9 Overflow rates ≤ 0.12 m/d as a function of depth and hydraulic retention time.

Q (m³/d)	h (m)	$t_{V/Q}$ (d)	V (m³)	A (m²)	OFR (m/d)
100	0.4	3.3	330	825	0.12
100	0.6	5	500	833	0.12
100	0.8	7	700	875	0.11
100	1.0	9	900	900	0.11

Figure 6.6 A rock filter installation in the city of Biggs, California. Top photo: filter construction with river rock, ranging in size from 8 to 15 cm, with void a volume equal to 40% of total volume. Bottom photo: Three parallel batteries of filters in operation. Rocks must be above water level to avoid algal growth on the surface.

6.2.1 Unbaffled ponds
6.2.1.1 Hydraulic retention time
Hydraulic retention times should be as long as possible, in a single pond or the total of ponds in series, to ensure the effluent meets the WHO guidelines for restricted and unrestricted irrigation ($\geq 4.0 \log_{10}$ reduction of *E coli*). As discussed previously (Table 6.5), various existing pond systems have been operating with insufficient retention times made worse by poor hydraulic design.

Table 6.10 *E. coli* reduction in a rock filter, Belo Horizonte, Brazil (Dias et al., 2014).

Parameter	Maturation Pond 3 Effluent	Rock Filter Effluent
TSS, mg/L	100	39
E. coli, CFU/100 mL	1.77E+03	4.50E+02
Log_{10} Reduction		0.59

Table 6.11 Application of unbaffled and baffled maturation ponds for pathogen reduction to satisfy WHO guidelines for restricted and unrestricted irrigation of crops.

Pond Type	Applications	Advantages/Disadvantages
Unbaffled	• First pond in series after anaerobic ponds or reactors; could also be classified as secondary facultative ponds. Designed so higher BOD and TSS loadings from anaerobic processes will be distributed through pond surface area at inlets, similar in concept to primary facultative ponds. • Pathogen reduction with adjustable depth capabilities. Pond depths can be adjusted to different seasons to maximize bacterial and viral pathogen removal. • Practical limits for design: $3/1 \leq l/w \leq 5/1$	• Construction costs without baffles may be much lower for larger ponds • Small ponds with $l/w \approx 5/1$, $t_{V/Q} \approx 6$ d, and $h = 0.6$ m, have achieved a 2.0 log_{10} reduction of *E. coli* at 25°C in Brazil (Dias et al., 2014) • Several ponds in series have obtained 4.0 log_{10} reduction of *E. coli* • Ponds can be designed to maintain design overflow rates independent of depth (Table 6.9) • Hydraulic short-circuiting can be a serious problem in large ponds as a result of poorly designed inlet/outlet structures, and wind effects
Baffled	• First pond in series after facultative ponds, designed for pathogen reduction • Second pond in series following unbaffled secondary facultative/maturation ponds, designed for pathogen reduction • Pathogen reduction based on dispersed flow hydraulics, with $(l/w)_{interior} \geq 16/1$ for highest log_{10} reductions	• Low dispersion numbers enable higher log_{10} pathogen reductions • One baffled pond can have better log_{10} reduction of *E. coli* than three unbaffled ponds in series with the same total areas and values of $t_{V/Q}$ (von Sperling, 2007) • Wind effects are minimized with baffles perpendicular to prevailing winds • Depths cannot be changed significantly as a result of baffle shading of water surface at low sun angles • Baffle configurations cannot be easily modified after start-up

6.2.1.2 Depths
Depths can range from 0.4 to 1.0 m, with shallower ponds giving higher pathogen reductions. Shallower depths, however, require liners to prevent the growth of emergent aquatic plants. Operating depths can be changed in unbaffled ponds, which can be an important operational control for pathogen reduction.

6.2.1.3 Length to width ratios
Unbaffled ponds with l/w ranging from 1/1 to 2/1 tend to form a circulation pattern in the bulk water mass as a result of the inlet jetting effect, which can significantly affect performance (Shilton & Sweeney, 2005). Ponds improve in hydraulic performance when $l/w \geq 3/1$, and can have excellent

170 Integrated Wastewater Management for Health and Valorization

Figure 6.7 Inadequately designed unbaffled maturation ponds with significant hydraulic short circuiting. In the top photo the single inlet, discharging above the water surface, creates a 'jetting effect' (Shilton and Sweeney, 2005), where the influent jets to the left sidewall, leaving half of the pond as a dead space ($l/w = 5/1$). The bottom photo shows a cascading influent jetting into the water surface, and drifting to the left-center, in a large, irregular pond. Note also the potential problem of shading of the water surface by trees in both ponds. (Top: Catacamas, Honduras; Bottom: Tela, Honduras).

pathogen reduction with $l/w \approx 5/1$, able to reach, for example, $\log_{10}$ reductions of *E. coli* > 4.0 for three ponds in series (Dias *et al.*, 2014). There are many poor designs, unfortunately, as shown in Figure 6.7. A better-designed pond is shown in Figure 6.8, but large ponds such as this pose a risk of hydraulic short-circuiting.

Figure 6.8 An unbaffled maturation pond with $l/w = 5.4/1$ (285 m × 53 m), with the inlet at the water surface. Short-circuiting is not apparent, but monitoring data are not available and performance cannot be assessed, a common problem. (Choloma, Honduras).

6.2.1.4 Inlet/outlet structures

Small ponds have operated well with a single inlet and outlet as reported in the extensive study of Dias *et al.* (2014). Larger ponds, however, are more problematic (Figure 6.7), and designs should use multiple inlets and outlets, or use smaller ponds built in parallel.

6.2.1.5 Case study: unbaffled maturation ponds in series, Belo Horizonte, Brazil

Dias *et al.* (2014) reported on the 10-year performance of 3 shallow maturation ponds in series, treating UASB reactor effluent, at the Center for Research and Training for Sanitation, Federal University of Minas Gerais-Copasa, Belo Horizonte, Brazil. A diagram of the pond system is shown in Figure 6.9.

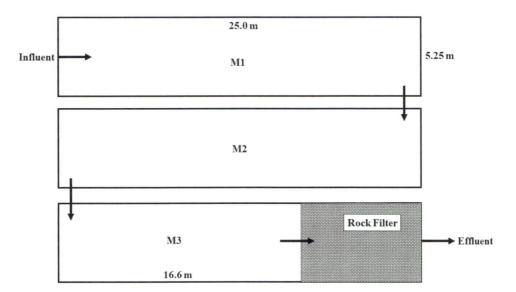

Figure 6.9 Flow diagram of maturation pond system with rock filter, Belo Horizonte, Brazil. The influent to the ponds is the effluent of a UASB reactor. Redrawn from Dias *et al.* (2014).

Table 6.12 Physical and operating characteristics of Belo Horizonte ponds, 2004–2013.[1]

Pond	*l* (m)	*w* (m)	*l/w*	*A* (m²)	Range of *h* (m)	Range of $t_{V/Q}$ (d)	T_{water} (°C)	Solar Radiation (kJ/m² *l*, m d)
M1	25.0	5.25	4.8	131.2	0.4–0.8	2–6	24.6–25.3	14,900–20,600
M2	25.0	5.25	4.8	131.2	0.4–0.8	2–6	24.6–25.3	14,900–20,600
M3	16.7	5.25	3.2	87.5	Not Reported	0.8–2.5	24.6–25.3	14,900–20,600

[1]Data from Dias *et al.* (2014). Solar radiation data from CLIMWAT.

Table 6.12 lists the physical and operating characteristics of the three ponds. Briefly, ponds M1 and M2 were operated with varying flowrates, at depths ranging from 0.4 to 0.8 m, with hydraulic retention times ranging from 2 to 6 d; and in pond M3, which was smaller in volume due to a rock filter (Figure 6.9), the hydraulic retention time varied between 0.8 and 2.5 d (Dias *et al.*, 2014).

Table 6.12 and Figure 6.10 also show solar radiation data for Belo Horizonte as reported by CLIMWAT. Solar radiation is an important parameter in pathogen inactivation in shallow ponds, and the data from Belo Horizonte can be used to compare pond performance at other locations with similar solar radiation and water temperatures.

The performance data for TSS and *E. coli* for the monitoring period are shown in Table 6.13. The following are key factors for designs under similar climatic conditions:

- The $\log_{10}$ reduction for *E. coli* of 4.41 easily meets the WHO guidelines for restricted irrigation with three shallow ponds in series with a maximum $t_{V/Q} = 14.5$ d.
- The 0.59 $\log_{10}$ reduction in the rock filter is significant, and aggregated settling of particles in rock filters in the final pond of a multiple pond system can be an important design strategy.
- The difference in $\log_{10}$ reduction between ponds M1 and M2 is assumed to be a result of intentionally varying flowrates and depths in each pond throughout the study period.

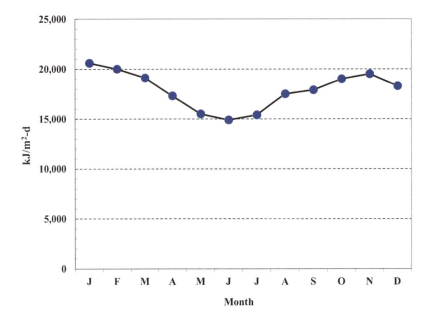

Figure 6.10 Monthly solar radiation at Belo Horizonte, Brazil. (Data from CLIMWAT).

Theory and design of maturation ponds

Table 6.13 Performance data of Belo Horizonte pond system, 2004–2013.[1]

	Raw Wastewater	Effluent UASB	M1	M2	M3	Rock Filter
TSS[1], mg/L	194	74	68	63	100	39
E. coli[2] MPN/100 mL	2.46E+08	4.61E+07	4.03E+05	2.47E+04	1.77E+03	4.50E+02
E. coli Log_{10} Reduction:		0.73	2.06	1.21	1.14	0.59

$$\sum_{M1}^{M3} Log_{10} \text{ Reduction} = 4.41$$

$$\sum_{M1}^{Rock\ Filter} Log_{10} \text{ Reduction} = 5.00$$

[1]Mean values for the monitoring period.
[2]Geometric mean values for the monitoring period.

- The total log_{10} reduction for the system of 5.73 almost meets the WHO guidelines value of 6.0 for unrestricted irrigation. If a facultative pond were used instead of a UASB, the unrestricted irrigation guidelines would have been met.
- Dias *et al.* (2014) reported that no helminth eggs were detected in the final effluent during the monitoring period. Since all values of $t_{V/Q}$ were less than 10 d, the calculation of overflow rate for helminth egg removal can be calculated by using the total area of the three ponds in series and the mean flowrate during the monitoring period:

$$A_{total} = (2)(131.2) + 87.5 = 350 \text{ m}^2; Q_{mean} = 33.0 \text{ m}^3/\text{d}$$

$$OFR = \frac{Q_{mean}}{A_{Total}} = \frac{33.0 \text{ m}^3/\text{d}}{350 \text{ m}^2} = 0.094 \text{ m/d} < 0.12 \text{ m/d}$$

- This case study clearly demonstrates that well-designed and operated maturation ponds can consistently meet >4.0 log_{10} reduction of *E. coli*, and non-detection of helminth eggs, for hydraulic retention times $\leq$ 14.5 d, with the water temperatures and solar radiation ranges at this site.

A plot of log_{10} reduction of *E. coli* as a function of water depth, calculated with the von Sperling Equation for ponds M1 and M2, is shown in Figure 6.11 for three hydraulic retention times used in the Belo Horizonte study. A sample calculation is shown is shown below.

Design Condition in Pond M1:

- $T_{water} = 24.6°C$
- $t_{V/Q} = 6$ d
- $h = 0.6$ m

(i) *E. coli* concentrations in M1 influent = UASB effluent.
$N_{influent} = 4.61 \times 10^7$ MPN/100 mL

(ii) Dispersion number:

$$l/w = \frac{25.0 \text{ m}}{5.25 \text{ m}} = 4.8$$

$$d = \frac{1}{l/w} = \frac{1}{4.8} = 0.21$$

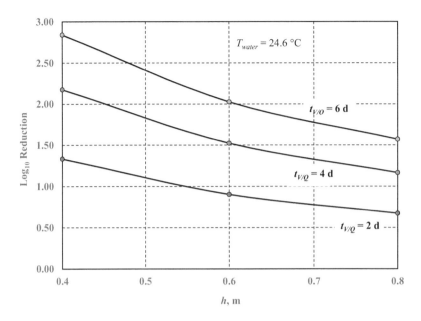

Figure 6.11 Log₁₀ reduction of *E. coli* as a function of depth and hydraulic retention time, $t_{V/Q}$, for maturation ponds M1 and M2, at water temperature $T_{water} = 24.6°C$, using the von Sperling (von Sperling, 2007) and Wehner and Wilhem equations.

(iii) Mortality rate constant at design water temperature, $k_{b,T}$, d⁻¹ using the von Sperling equation (Equation (5.34)):

$$k_{b,T} = (0.542 \cdot h^{-1.259})(1.07)^{(T-20)}$$

$h = 0.6$ m and $T_{water} = 24.6°C$.

$$k_{b,T} = (0.542 \cdot (0.6)^{-1.259})(1.07)^{(24.6-20)} = 0.21 \text{ d}^{-1}$$

(iv) Determine *a* in the Wehner–Wilhem equation (Equation (5.32)):

$t_{V/Q} = 6.0$ d

$$a = \sqrt{1 + 4k_{b,T} \cdot t_{V/Q} \cdot d} = \sqrt{1 + 4(0.21 \text{ d}^{-1})(6.0 \text{ d})(0.21)} = 2.8351$$

(v) Determine effluent *E. coli* concentration:

$$N_{effluent} = N_{influent} \frac{4a \cdot e^{1/2d}}{\left((1+a)^2 e^{a/2d} - (1-a)^2 e^{-a/2d}\right)}$$

$$= (4.61 \times 10^7) \left[\frac{4(2.8351) \cdot e^{1/2(0.21)}}{\left((1+2.8351)^2 e^{2.8351/2(0.21)} - (1-2.8351)^2 e^{-2.8351/2(0.21)}\right)} \right]$$

$$= 4.35 \times 10^5 \text{ MPN/100 mL}$$

(vi) Determine the log₁₀ reduction of *E. coli*:

$$\log_{10,Reduction} = \log_{10}(4.61 \times 10^7) - \log_{10}(4.35 \times 10^5) = 2.03$$

Theory and design of maturation ponds

Comment. The calculated $\log_{10}$ reduction is very close to the actual geometric mean of measured values during the 10-year monitoring study for pond M1, which gives confidence in using the von Sperling and Wehner and Wilhem equations for similar environmental conditions of solar radiation and water temperature.

6.2.2 Baffled ponds
Baffles greatly improve the hydraulic performance of ponds if properly designed and spaced. Baffles promote plug flow, and reduce the jetting and wind-driven effects occurring in unbaffled ponds. Baffles also cause a spiral flow at the end of each channel, which is induced as flow passes around the end of the baffle, that enhances vertical mixing and prevents stratification (Middlebrooks et al., 1982).

6.2.2.1 Depths
Baffled ponds require careful design. Depths cannot be significantly changed during operation or shading of the water surface by baffles at low sun angles will occur, which could significantly affect pathogen reduction depending on the length of time of shading.

6.2.2.2 Length to width ratios
The external l/w ratio should be $\geq 2/1$ if site conditions permit. Higher external l/w ratios yield higher internal l/w ratios with fewer baffles. Internal l/w ratios should be $\geq 25/1$ as discussed below.

Figure 6.12 shows examples of transverse and longitudinal baffling and the design equations for both configurations are presented below.

6.2.2.3 Transverse baffle equations: baffles parallel to width
The interior length to width ratio, $(l/w)_{interior}$, for transverse baffling is calculated with Equation (6.5) (von Sperling, 2007).

$$(l/w)_{interior, trans} = \left(\frac{1}{(l/w)_{exterior}}\right)(n+1)^2 \tag{6.5}$$

$(l/w)_{interior, trans}$ = interior length to width ratio with transverse baffles; $(l/w)_{exterior}$ = exterior length to width ratio; n = number of baffles; and $n + 1$ = number of channels.

The total length of interior channels is calculated with Equation (6.6).

$$l_{channel, trans} = (n+1)(w) \tag{6.6}$$

$l_{channel, trans}$ = total length of the interior channel with transverse baffles, m and w = exterior width of pond, m.

Equation (6.7) is used to calculate the width of interior channels.

$$w_{channel, trans} = \left(\frac{1}{(n+1)}\right)(l) \tag{6.7}$$

$w_{channel, trans}$ = width of transverse channels, m.

6.2.2.4 Longitudinal baffle equations: baffles parallel to length
The interior length to width ratio, $(l/w)_{interior}$, for longitudinal baffling is calculated with Equation (6.8) (von Sperling, 2007)

$$(l/w)_{interior, long} = \left[(l/w)_{exterior}\right] \cdot (n+1)^2 \tag{6.8}$$

$(l/w)_{interior, long}$ = interior length to width ratio with longitudinal baffles.
Equation (6.9) is used to calculate the total length of interior channels

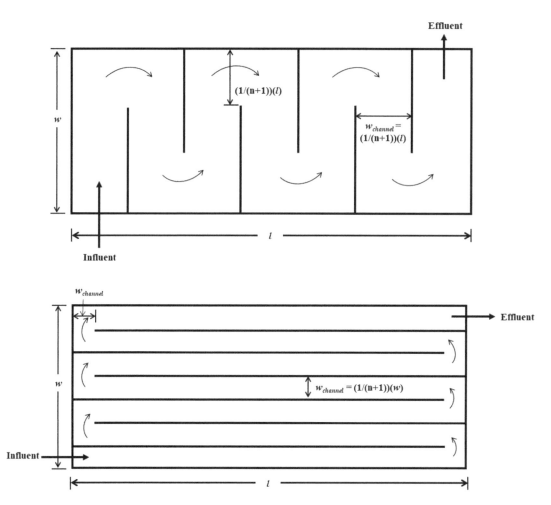

Figure 6.12 The two baffle arrangements in ponds are transverse baffles parallel to the width (top), and longitudinal baffles parallel to the length (bottom). Transverse baffling requires more baffles, and hence channels, than longitudinal baffling to obtain the same interior length to width ratio, $(l/w)_{interior}$. $(l/w)_{exterior} = 2.5/1$ for both configurations, but $(l/w)_{interior} = 19.6$ for the transverse baffled pond, and 122.5 for the longitudinal pond. The transverse baffled pond would need 16 baffles to have $(l/w)_{interior} \approx 115$. A suggested design value is $(l/w)_{interior} \geq 25/1$. The distance from the end of the baffle to the wall is equal to the channel width for both arrangements, enabling the cross-sectional area of flow to remain as close to a constant as possible, thus preventing large velocity changes (Middlebrooks et al. 1982).

$$l_{channel,long} = (n+1)(l) \tag{6.9}$$

$l_{channel,long}$ = total length of the interior channel with longitudinal baffles, m.

The width of interior channels is calculated with Equation (6.10).

$$w_{channel,long} = \left(\frac{1}{(n+1)}\right)(w) \tag{6.10}$$

$w_{channel,long}$ = width of longitudinal channels, m.

6.2.2.5 Design example: comparison of transverse and longitudinal baffled ponds

In Figure 6.12 the external length/width ratio, $(l/w)_{exterior}$, is equal to 2.5/1 for both pond configurations. Determine the interior length/width ratio, $(l/w)_{interior}$, and the dispersion number, d, for both configurations.

6.2.2.5.1 Transverse baffled pond
Equation (6.5) is used with $(l/w)_{exterior} = 2.5$ and $n = 6$

$$(l/w)_{interior,trans} = \left[\frac{1}{(l/w)_{exterior}}\right](n+1)^2 = \left[\frac{1}{2.5}\right](6+1)^2 = 19.6$$

Equation (5.36) is used to determine the dispersion number, d

$$d = \frac{1}{(l/w)_{interior,trans}} = \frac{1}{19.6} = 0.051$$

6.2.2.5.2 Longitudinal baffled pond
Equation (6.8) is used with $(l/w)_{exterior} = 2.5$ and $n = 6$

$$(l/w)_{interior,long} = \left[(l/w)_{exterior}\right]\cdot(n+1)^2 = [2.5]\cdot(6+1)^2 = 122.5$$

Using Equation (5.36) to determine the dispersion number, d.

$$d = \frac{1}{(l/w)_{interior,long}} = \frac{1}{122.5} = 0.008$$

Comment. For the same values of $(l/w)_{exterior}$ and n, the longitudinal baffled pond has a much higher $(l/w)_{interior}$, and a much lower dispersion number, d, and should have a significantly better performance for pathogen reduction. The width of the channels, however, could be too narrow at the inlet, causing excessive BOD_L surface loadings, and this would need to be checked during design. The cost of construction would also be greater as a result of the longer length of interior baffles. The design depth for both ponds is critical as it cannot be significantly changed because of baffle shading at low sun angles.

6.2.2.6 Design strategies for baffled and unbaffled ponds
Currently, there is a paucity of clear design recommendations for $(l/w)_{interior}$ values for both baffled and unbaffled ponds. What should be the minimum $(l/w)_{interior}$ to have a 1.5–2.0 $\log_{10}$ reduction of *E. coli*? What is the maximum $(l/w)_{interior}$ that is practical to construct, for both baffled and unbaffled ponds, and at what water depths?

As an example, Figure 6.13 shows plots of $\log_{10}$ reduction of *E. coli* versus $t_{V/Q}$ for two water depths (1.0 and 0.6 m), and four values of $(l/w)_{interior}$ (3, 5, 25, and 50), using the von Sperling and Wehner and Wilhem equations at 20°C. The calculated data used in the plot are tabulated in Table 6.14. It is assumed ponds with $l/w = 3$ and 5 would be unbaffled, and those with $l/w = 25$ and 50, baffled.

For $h = 1.0$ m in Figure 6.13, there are minor differences in $\log_{10}$ reduction between $l/w = 25$ and 50, and also between $l/w = 3$ and 5. In these cases, the lower value of $l/w = 25$ would be preferable for baffled ponds to lower construction costs. For $h = 0.6$ m, there are slightly greater differences between $l/w = 25$ and 50, and between $l/w = 3$ and 5, especially at 15 d retention times; the increases for the baffled ponds, however, may not be sufficient to warrant the added costs for construction. A design value of $l/w = 25$, with $t_{V/Q} \geq 10$ d, which would give a predicted $\log_{10}$ reduction of at least 2.0, seems an appropriate recommendation for design for baffled ponds.

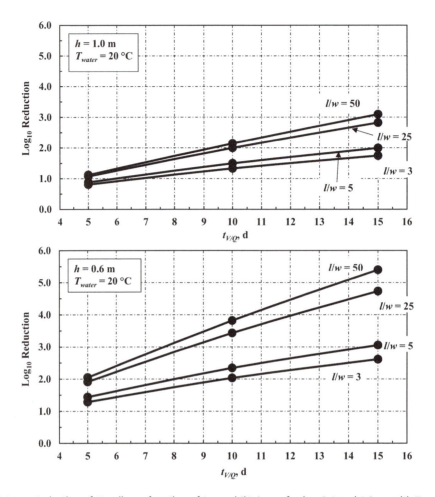

Figure 6.13 Log_{10} Reduction of *E. coli* as a function of $t_{V/Q}$ and $(l/w)_{interior}$ for $h = 0.6$ and 1.0 m, with $T_{water} = 20°C$.

Table 6.15 presents the differences between log_{10} reduction for the baffled ($l/w = 25$) and unbaffled ($l/w = 3$) ponds, for each depth, at 10 and 15 d hydraulic retention times. The advantage of baffles increases with increasing retention times. The advantage of unbaffled ponds, however, is the ability to change water depths for varying environmental conditions.

6.2.2.7 Case study: E. coli reduction in transverse baffled maturation pond, Helidon, Australia

Stratton *et al.* (2015) presented a detailed analysis of the performance of the Helidon baffled maturation pond shown in Figure 6.14. A summary of their results relevant to *E. coli* reduction is summarized in Table 6.16.

6.2.2.7.1 Determination of Helidon pond performance using design equations

The transverse baffle, von Sperling, and Wehner and Wilhem equations will be used to calculate *E. coli* reduction for the Helidon maturation pond, and the results compared with the measured data in Table 6.16.

Table 6.14 Log$_{10}$ reduction of *E. coli* at 20°C as a function of h, $t_{V/Q}$, and $(l/w)_{interior}$, using the von Sperling and Wehner and Wilhem equations for $h = 0.6$ and 1.0 m.

h (m)	$t_{V/Q}$ (d)	T (°C)	$k_{b,20}$ (d^{-1})	$(l/w)_{interior}$	d	a	$N_{influent}$	$N_{effluent}$	Log$_{10}$ Reduction
1.0	5	20	0.54	3	0.33	2.1479	1.00E+06	1.55E+05	0.81
1.0	5	20	0.54	5	0.20	1.7799	1.00E+06	1.31E+05	0.88
1.0	5	20	0.54	25	0.04	1.1973	1.00E+06	8.42E+04	1.07
1.0	5	20	0.54	50	0.02	1.1031	1.00E+06	7.58E+04	1.12
1.0	10	20	0.54	3	0.33	2.8682	1.00E+06	4.65E+04	1.33
1.0	10	20	0.54	5	0.20	2.3100	1.00E+06	3.19E+04	1.50
1.0	10	20	0.54	25	0.04	1.3665	1.00E+06	1.00E+04	2.00
1.0	10	20	0.54	50	0.02	1.1973	1.00E+06	7.14E+03	2.15
1.0	15	20	0.54	3	0.33	3.4409	1.00E+06	1.79E+04	1.75
1.0	15	20	0.54	5	0.20	2.7393	1.00E+06	1.01E+04	1.99
1.0	15	20	0.54	25	0.04	1.5168	1.00E+06	1.50E+03	2.82
1.0	15	20	0.54	50	0.02	1.2847	1.00E+06	7.99E+02	3.10
0.6	5	20	1.03	3	0.33	2.8061	1.00E+06	5.16E+04	1.29
0.6	5	20	1.03	5	0.20	2.2637	1.00E+06	3.61E+04	1.44
0.6	5	20	1.03	25	0.04	1.3509	1.00E+06	1.22E+04	1.91
0.6	5	20	1.03	50	0.02	1.1885	1.00E+06	8.92E+03	2.05
0.6	10	20	1.03	3	0.33	3.8403	1.00E+06	9.25E+03	2.03
0.6	10	20	1.03	5	0.20	3.0412	1.00E+06	4.53E+03	2.34
0.6	10	20	1.03	25	0.04	1.6278	1.00E+06	3.68E+02	3.43
0.6	10	20	1.03	50	0.02	1.3509	1.00E+06	1.52E+02	3.82
0.6	15	20	1.03	3	0.33	4.6500	1.00E+06	2.44E+03	2.61
0.6	15	20	1.03	5	0.20	3.6570	1.00E+06	8.79E+02	3.06
0.6	15	20	1.03	25	0.04	1.8640	1.00E+06	1.85E+01	4.73
0.6	15	20	1.03	50	0.02	1.4958	1.00E+06	3.98E+00	5.40

Table 6.15 Difference in Log$_{10}$ reduction of *E. coli* for $(l/w)_{interior} = 3$ and 25 ($T = 20$°C).

h (m)	$t_{V/Q}$ (d)	$(l/w)_{interior}$	Log$_{10}$ Reduction	Δ Log$_{10}$ Reduction
1.0	10	3	1.33	0.67
1.0	10	25	2.00	
1.0	15	3	1.75	1.07
1.0	15	25	2.82	
0.6	10	3	2.03	1.40
0.6	10	25	3.43	
0.6	15	3	2.61	2.12
0.6	15	25	4.73	

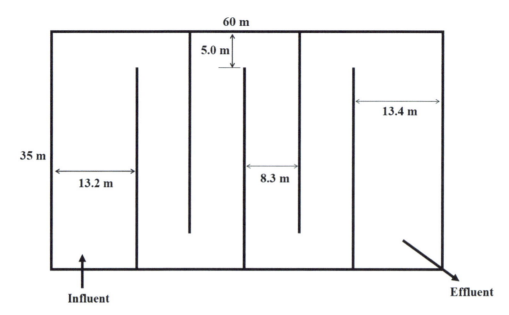

Figure 6.14 The Helidon, Australia transverse baffled maturation pond. The average flowrate was reported as 60 m³/d, with $t_{V/Q}$ ranging from 12 to 20 d. The peak tracer concentration in the pond effluent, determined from simulated tracer studies using the MIKE 3 software package, was 14.0 d (Stratton *et al.* 2015). Note the varying channel widths, and the end of baffle openings (5.0 m) that are less than the width of the channels. Drawing developed from Stratton *et al.* (2015).

Design conditions in Helidon maturation pond:

- $T_{water} = 27.0\,°C$
- $t_{V/Q} = 12$ d
- $h = 1.0$ m
- Geometric mean *E. coli* influent concentration (Stratton *et al.*, 2015): 1.70×10^5 CFU/100 mL

(i) Transverse baffled pond (Figure 6.14).
Equation (6.5) is used with $n = 5$.

$$(l/w)_{interior,trans} = \left(\frac{1}{(l/w)_{exterior}}\right)(n+1)^2 = \left(\frac{1}{60/35}\right)(5+1)^2 = 21$$

Table 6.16 Results of Helidon maturation pond performance study.[1]

Parameter	Value
Average flowrate, Q, m³/d	60
Range of $t_{V/Q}$, d	12–20
Time of peak tracer concentration in effluent, determined from simulated tracer studies, d	14.0
h, m	1.0
T_{water}, °C	27.0
Geometric mean measured $\log_{10}$ reduction, *E. coli*	2.0

[1]Data from Stratton *et al.* (2015).

Equation (5.36) is used to determine the dispersion number, d.

$$d = \frac{1}{(l/w)_{\text{interior,trans}}} = \frac{1}{21} = 0.048$$

(ii) *E. coli* concentrations in pond influent.
$N_{\text{influent}} = 1.70 \times 10^5$ CFU/100 mL.

(iii) Mortality rate constant at design water temperature, $k_{b,T}$, d^{-1} using the von Sperling equation (Equation (5.34)):

$$k_{b,T} = (0.542 \cdot h^{-1.259})(1.07)^{(T-20)}$$

$h = 1.0$ m and $T_{\text{water}} = 27.0\,°C$

$$k_{b,T} = (0.542 \cdot (1.0)^{-1.259})(1.07)^{(27.0-20)} = 0.87 \text{ d}^{-1}$$

(iv) Determine a in the Wehner–Wilhem equation (Equation (5.32)):

$t_{V/Q} = 12.0$ d

$$a = \sqrt{(1 + 4k_{b,T} \cdot t_{V/Q} \cdot d)} = \sqrt{(1 + 4(0.87)(12.0)(0.048))} = 1.7290$$

(v) Determine effluent *E. coli* concentration:

$$N_{\text{effluent}} = N_{\text{influent}} \frac{4a \cdot e^{1/2d}}{\left((1+a)^2 e^{a/2d} - (1-a)^2 e^{-a/2d}\right)}$$

$$= (1.70 \times 10^5)\left(\frac{4(1.7290) \cdot e^{1/2(0.048)}}{\left((1+1.7290)^2 e^{1.7290/2(0.048)} - (1-1.7290)^2 e^{-1.7290/2(0.048)}\right)}\right)$$

$$= 74.8 \text{ CFU/100 mL}$$

(vi) Determine the $\log_{10}$ reduction of *E. coli*:

$$\log_{10,\text{Reduction}} = \log_{10}(1.70 \times 10^5) - \log_{10}(7.48 \times 10^1) = 3.36$$

Commentary. The calculated $\log_{10}$ reduction is 1.36 log units higher than that measured in the monitoring study. The difference could be due to:

- Lower values of solar radiation at Helidon than those at ponds used to develop the von Sperling equation.
- Poor in-pond hydraulics with hydraulic short-circuiting.
- High concentrations of TSS in the facultative pond effluent, or high turbidity in the maturation pond, which inhibited solar radiation penetration.

Figure 6.15 shows the monthly solar radiation for Belo Horizonte and Helidon. For 3 months, May, June and July, the solar radiation at Helidon is significantly lower than in other months. The monitoring data reported by Stratton *et al.* (2015), however, were from January through March, when the solar radiation at Helidon was similar to that at Belo Horizonte.

Hydraulic short-circuiting likely played a significant role as a result of varying channel widths, and inadequate spacing between the end of baffles and the wall, as shown in Figure 6.14. To minimize short-circuiting, the cross-sectional area throughout a channeled pond should be kept constant, promoting a constant horizontal velocity (Middlebrooks *et al.*, 1982). Unfortunately, a residence time distribution curve and dispersion number were not presented in the report by Stratton *et al.* (2015), and the calculation performed with the Wehner and Wilhem equation would be for an improved design of the existing pond.

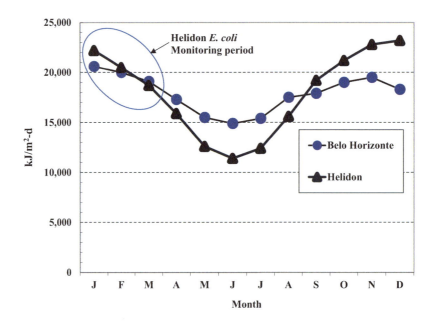

Figure 6.15 Incident solar radiation on a flat surface for Belo Horizonte, Brazil, and Helidon, Australia. Data from CLIMWAT.

No data were presented by Stratton *et al.* (2015) on TSS concentrations, and no conclusions can be drawn about this possible effect.

This case study shows the continuous difficulties encountered in designing maturation ponds, and waste stabilization ponds in general, in attempting to combine theory with data from existing systems. In the case of the Helidon pond, as an example, a redesign of the baffles with continued monitoring could offer valuable information for improving designs.

6.2.3 Rock filters for pathogen reduction
6.2.3.1 Design of rock filters

While rock filters have traditionally been designed for additional TSS removal in pond effluents, their use should also be considered for additional pathogen reduction as demonstrated in the results of Dias *et al.* (2014) for the pond system in Belo Horizonte (Section 6.2.1). Table 6.17 summarizes the design criteria for the two rock sizes that have been used with success for TSS removal.

A rock filter is designed with the following steps:

(1) Select the design hydraulic loading rate, HLR_{filter}, and determine the volume of submerged rock with Equation (6.11).

$$V_{submerged} = \frac{Q_{mean}}{HLR_{filter}} \quad (6.11)$$

$V_{submerged}$ = volume of submerged rock, m³; Q_{mean} = mean flowrate, m³/d and HLR_{filter} = hydraulic loading rate to filter, 0.10–0.30 m³/m³ d.

(2) Determine the top area of the filter with Equation (6.12).

$$A_{filter} = \frac{V_{submerged}}{h} \quad (6.12)$$

A_{filter} = top area of the filter, m² and h = water depth of the filter, m.

Table 6.17 Design criteria for rock filters.[1]

Parameter	Rock Size	
	1–2 cm	3–20 cm
Porosity, %	≈40	40–42
Hydraulic loading rate, HLR$_{filter}$, m³/m³ d	0.08–0.20	0.10–0.30
Height above the water surface, m	≥0.2	≥0.2
TSS removal, %	60–90	60–90
Mean effluent TSS, mg/L	6–12	4–12
Operational life without cleaning, yr	>15	>15

[1]Developed from Middlebrooks et al. (2005) and Swanson and Williamson (1980).

(3) Determine the total volume of rock required using Equations (6.13) and (6.14).

$$V_{total} = V_{submerged} + V_{dry} \tag{6.13}$$

V_{total} = total volume of rock required, m and V_{dry}, = volume of rock above the waterline, m³

$$V_{dry} = A_{filter} h_2 \tag{6.14}$$

h_2 = height of the rock above the waterline, m.

(4) Determine the total volume of voids with Equation (6.15).

$$V_{voids} = V_{submerged}(P_{rocks}) \tag{6.15}$$

V_{voids} = volume of voids below the waterline, m³ and P_{rocks} = porosity of rocks expressed as a decimal ($P_{rocks} \approx 0.40$).

(5) Determine the hydraulic retention time in the filter with Equation (6.16)

$$t_{V/Q,filter} = \frac{V_{voids}}{Q_{mean}} \tag{6.16}$$

$t_{V/Q,filter}$ = hydraulic retention time in the filter, d.

(1) Estimate the number of years to fill 50% of the voids in the filter.
 (a) Determine the mass of influent TSS that settle in the void space with Equation (5.43).

$$\dot{M}_{Sl} = 0.001 \cdot Q_{mean} \cdot TSS_{Settleable} \tag{5.43}$$

$\dot{M}_{Sl}$ = mass of fresh sludge solids, kg/d and $TSS_{settleable}$ = settleable algal suspended solids, mg/L (≈30–50 mg/L)

 (b) The annual mass of digested sludge produced in the filter is calculated with Equation (5.44).

$$\dot{M}_{Sl, D-yr} = 365 \cdot \left[FS \cdot \dot{M}_{Sl} + (1 - VS_{Destroyed}) VS \cdot \dot{M}_{Sl} \right] \tag{5.44}$$

$\dot{M}_{Sl, D-yr}$ = mass of digested sludge solids produced per year, kg/yr; FS = decimal fraction of fixed solids ($FS \approx 0.15$ for algae); VS = decimal fraction of volatile solids ($VS \approx 0.85$ for algae) and $VS_{Destroyed}$ = decimal fraction of volatile solids destroyed (assume 0.40 for safety factor).

 (c) Determine the volume of the digested sludge produced in the filter per year with Equation (5.42).

$$\dot{V}_{Sl,D-yr} = \frac{\dot{M}_{Sl, D-yr}}{\rho_{H_2O} \cdot S_{Sl} \cdot TS} \tag{5.42}$$

$\dot{V}_{Sl,D-yr}$ = volume of digested sludge per year in rock filter, m³/yr; $\dot{M}_{Sl,D-yr}$ = mass of digested total solids per year, kg/yr; ρ_{H_2O} = density of water, 1000 kg/m³; S_{Sl} = specific gravity of sludge and TS = decimal fraction of total solids in wet sludge (0.05 for wet sludges; >0.05–0.15 for sludges that have been air dried).

(d) Finally, the time to fill 50% of the filter's volume with sludge, the recommended limit for filter operation, is estimated with a modified Equation (5.45):

$$t_{50\%} = \frac{0.50 \cdot V_{voids}}{\dot{V}_{Sl,D-yr}} \quad (5.45)$$

$t_{50\%}$ = time to fill 50% of the filter's volume, yr and V_{voids} = volume of voids in the filter, m³.

The following section presents a case study of the design and operating characteristics of a rock filter design for algal TSS removal in a waste stabilization pond system.

6.2.3.2 Case study: rock filter design, City of Biggs, California waste stabilization pond system

Rock filters were designed for the City of Biggs waste stabilization pond system in 1999 in order to meet more stringent TSS effluent regulations of the State Regional Water Quality Board. The pond system originally consisted of two aerated ponds in series, followed by a ballast pond designed to provide a continuous flowrate to the disinfection chamber (Figure 6.16). The ponds were later operated without aeration with success, and a rock filter system was constructed to (i) meet the discharge requirements

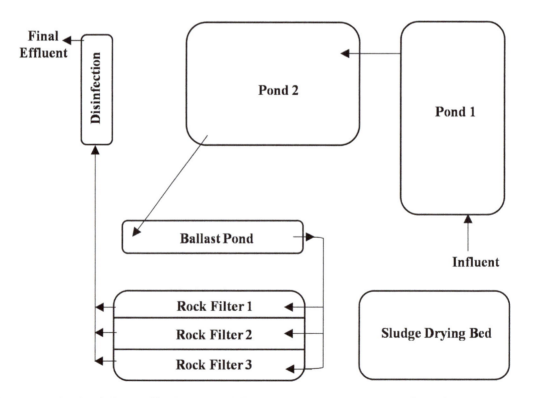

Figure 6.16 The city of Biggs, California waste stabilization pond system. The battery of rock filters was designed for improved TSS removal to meet surface water discharge requirements.

and (ii) to lower chlorine demand from high TSS concentrations in the disinfection chamber. The rock filters consisted of three batteries in parallel so one battery could be taken out of service for maintenance. The filters receive the effluent from the ballast pond, and the hydraulic loading can be varied for each filter.

The following design criteria were used to design the rock filter system:

- Design mean flowrate: $Q_{mean} = 1136$ m³/d.
- Hydraulic loading rate to rock filter: $HLR_{filter} = 0.15$ m³/m³ d.
- Water depth, $h = 2.0$ m; height of rock above water surface, $h_2 = 0.2$ m.
- Use of river rock ranging in size from 7 to 15 cm, with measured porosity, $P_{rocks} = 42\%$.
- The fraction of volatile solids digested anaerobically within the filter is assumed to be 0.40, which is the minimum value reported for various species of algae by Foree and McCarty (1970).

(1) The volume of submerged rock is found using Equation (6.11)

$$V_{submerged} = \frac{Q_{mean}}{HLR_{filter}} = \frac{1136 \text{ m}^3/\text{d}}{0.15 \text{ m}^3/\text{m}^3 \text{ d}} = 7575.3 \text{ m}^3$$

(2) The top area of the filter is determined with Equation (6.12)

$$A_{filter} = \frac{V_{submerged}}{h} = \frac{7575.3 \text{ m}^3}{2.0 \text{ m}} = 3786.7 \text{ m}^2 \Rightarrow \text{use } 4000 \text{ m}^2$$

Final design: $V_{submerged} = (2.0 \text{ m})(4000 \text{ m}^2) = 8000 \text{ m}^3$

(3) Equations (6.13) and (6.14) are used to determine the total volume of rock required

$$V_{total} = V_{submerged} + V_{dry} = 8000 \text{ m}^3 + A_{filter}h_2$$
$$= 8000 \text{ m}^3 + (4000 \text{ m}^2)(0.2 \text{ m}) = 8800 \text{ m}^3$$

(4) The volume of voids in submerged rocks is determined with Equation (6.15)

$$V_{voids} = V_{submerged}(P_{rocks}) = (8000 \text{ m}^3)(0.42) = 3360 \text{ m}^3$$

(5) The hydraulic retention time in the filter is calculated with Equation (6.16).

$$t_{V/Q,filter} = \frac{V_{voids}}{Q_{mean}} = \frac{3360 \text{ m}^3}{1136 \text{ m}^3/\text{d}} = 2.96 \text{ d}$$

Settling velocities of algae have been reported to vary from 0.02 to 0.3 m/d (Swanson & Williamson, 1980) and the value of $t_{V/Q,filter}$ should allow complete settling with a safety factor.

Figure 6.17 Rock Filter 1 under construction. The three filters in parallel were designed for a water depth of 2.0 m, with 0.2 m of filter above the water surface, with a design $HLR_{filter} = 0.15$ m³/m³ d. See also Figure 6.6.

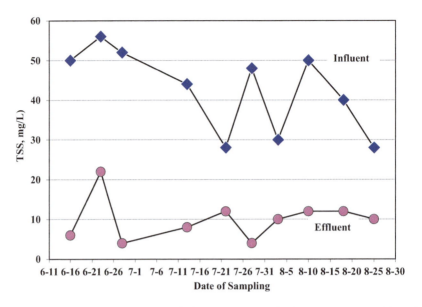

Figure 6.18 Suspended solids removal in Filter 1 during summer months for HLR$_{filter}$ = 0.15 m³/m³ d. TSS removal was approximately 30 mg/L, with effluent TSS averaging 10 mg/L. Graph developed from unpublished monitoring data, City of Biggs and California State University, Chico.

Figure 6.17 shows one of the rock filters under construction (see also Figure 6.6 for filters in operation).

Figure 6.18 presents influent/effluent TSS monitoring data for Filter 1 during summer months when algae concentrations are highest. The final effluent TSS concentration averaged 10 mg/L and the Biggs pond system has been able to meet its TSS permit requirements to the present day; unfortunately, *E. coli* reduction through the filter was never monitored. The treatment system has been unable to meet the new ammonia discharge limits, however, and a new proposal has been developed to avoid surface water discharge and use the final effluent for land application, with crop irrigation, on 60.7 ha (Bennett Engineering, n.d.), which would be an excellent improvement and a move away from the linear paradigm dominant in California. In this case, it would be important to monitor *E. coli* reduction through the filter.

There has been concern about the useful life of rock filters but many filters have operated for years without problems. It is important to estimate the volume of sludge accumulation and to plan for the cleaning of the filter when it becomes necessary.

(6) Estimate the number of years to fill 50% of the voids in the filter.
  (a) Determine the mass of influent TSS that settle in the void space with Equation (5.43).

$$\dot{M}_{Sl} = 0.001 \cdot Q_{mean} \cdot TSS_{Settleable} = 0.001(1136 \text{ m}^3/d)(30 \text{ mg/L}) = 34.1 \text{ kg/d}$$

  (b) The annual mass of digested sludge produced in the filter is calculated with Equation (5.44), with $VS_{Destroyed} = 0.40$ (Foree & McCarty, 1970).

$$\dot{M}_{Sl, D-yr} = 365 \cdot (0.15 \cdot (34.1 \text{ kg/d}) + (1 - 0.40)(0.85)(34.1 \text{ kg/d}) = 8215 \text{ kg/yr}$$

  (c) The volume of digested sludge produced in the filter per year is calculated with Equation (5.42) assuming the decimal fraction of digested solids equals 0.05.

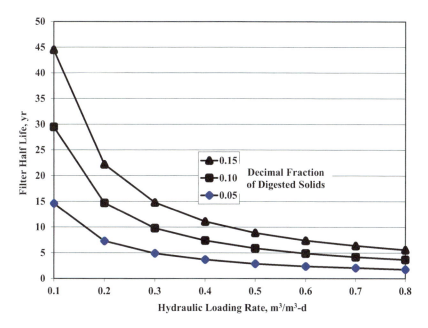

Figure 6.19 Rock filter half-life as a function of hydraulic loading rate and digested solids decimal fraction. TSS Removal = 30 mg/L; porosity = 0.42.

$$\dot{V}_{Sl,D-yr} = \frac{\dot{M}_{Sl,D-yr}}{\rho_{H_2O} \cdot S_{Sl} \cdot TS} = \frac{8215 \text{ kg/yr}}{(1000 \text{ kg/m}^3)(1.007)(0.05)} = 163 \text{ m}^3/\text{yr}$$

(d) Finally, the time to fill 50% of the filter's volume with sludge, the recommended limit for filter operation, is estimated with the modified Equation (5.45):

$$t_{50\%} = \frac{0.50 \cdot V_{voids}}{\dot{V}_{Sl,D-yr}} = \frac{(0.50)(3360 \text{ m}^3)}{163 \text{ m}^3/\text{yr}} = 10.3 \text{ yr}$$

Comment. A value of $t_{50\%}$ equal to 10.3 years is low for a design hydraulic loading rate of only 0.15 m³/m³ d. Figure 6.19 shows a plot of filter half-life versus hydraulic loading rate for three values of the decimal fraction of digested solids: 0.05, 0.10, and 0.15. The solids fraction of 0.05 will occur if digestion is slow, and the filter is not taken out of service for drying and cleaning. A solids fraction of 0.10 and higher can occur if the filter solids are well digested, either anaerobically or aerobically, and the fraction of volatile solids digested is higher than 0.40; Foree and McCarty (1970) reported that volatile solids destruction can range from 40 to 80% for various algal cultures; in this case, the solids fraction would be similar to that found in facultative ponds.

A better approach is to drain and dry the filters every few years. The filter at Veneta was drained and dried during dry weather months until the slime layer on the rocks completely dried, forming a thin layer of dried organic matter; this thin layer does not rehydrate, and is washed out when the filter is put back in service (Swanson & Williamson, 1980). Figure 6.20 shows an example of this with air-dried algae from the Biggs pond effluent. The rock filters at Biggs were built in parallel so one could be taken out of service and dried, as was done at the rock filter in Veneta, Oregon. Thus, the useful life of a rock filter can easily be more than 20 years if draining and drying of the rock media are carried out on a periodic basis.

Figure 6.20 Top photo: A beaker of settled algae from the Biggs pond effluent. Middle photo: Air-dried algae in a crucible that was filled with algae from the top beaker. Bottom photo: The air-dried algae does not rehydrate when placed in a beaker of water.

6.3 MATURATION POND DESIGN PROCEDURE

(1) Select the design hydraulic retention time and calculate the volume of the pond with Equation (6.17).

$$V_m = Q \cdot t_{V/Q} \tag{6.17}$$

V_m = volume of maturation pond, m³ and $t_{V/Q} \geq 10$ d to ensure helminth egg removal, especially in the case where the maturation pond follows an anaerobic reactor or pond instead of a facultative pond.

(2) Select the depth of the pond and calculate the area with Equation (6.18).

$$A_m = \frac{V_m}{h_m} \tag{6.18}$$

A_m = area of pond, m²; h_m = depth of pond, m; and h_m = 0.4–1.0 m.

(3) Select the exterior l/w ratio and determine the length and width of the pond.

$(l/w)_{exterior} = x/1 \quad (x = 1 - 3 \text{ depending on site conditions})$

$w = \sqrt{\dfrac{A_m}{x}}; \quad l = xw$

For example: $(l/w)_{exterior} = 3/1$ and $w = \sqrt{A_m/3}; l = 3w$.

(4) Select the interior embankment slope, i, and calculate the pond volume using the prismoid equation.
Select i:
$1/1 \leq i \leq 3/1$ depending on pond depth and site conditions.

$V_{m,cal} = \dfrac{h}{6} \cdot \left[(l \cdot w) + (l - 2ih)(w - 2ih) + 4 \cdot (l - iw)(w - ih) \right]$

If $V_{m,cal} < V_m$, re-dimension until $V_{m,cal} \geq V_m$.

(5) Design of interior baffles.
  (i) Select the number of baffles to be used.
  (ii) Transverse baffle equations: baffles parallel to width.
    The interior length to width ratio, $(l/w)_{interior}$, is calculated with Equation (6.5).

$(l/w)_{interior, trans} = \left(\dfrac{1}{(l/w)_{exterior}} \right)(n+1)^2 \qquad (6.5)$

$(l/w)_{interior, trans}$ = interior length to width ratio with transverse baffles; n = number of baffles; and $n + 1$ = number of channels.
The total length of interior channels is calculated with Equation (6.6).

$l_{channel, trans} = (n+1)(w) \qquad (6.6)$

$l_{channel, trans}$ = total length of the interior channel with transverse baffles, m and w = exterior width of pond, m.
Equation (6.7) is used to calculate the width of interior channels

$w_{channel, trans} = \left(\dfrac{1}{(n+1)} \right)(l) \qquad (6.7)$

$w_{channel, trans}$ = width of transverse channels, m.
  (iii) Longitudinal baffle equations: baffles parallel to the length.
    The interior length to width ratio, $(l/w)_{interior}$, is calculated with Equation (6.8).

$(l/w)_{interior, long} = \left[(l/w)_{exterior} \right] \cdot (n+1)^2 \qquad (6.8)$

$(l/w)_{interior, long}$ = interior length to width ratio with longitudinal baffles.
Equation (6.9) is used to calculate the total length of interior channels.

$l_{channel, long} = (n+1)(l) \qquad (6.9)$

$l_{channel, long}$ = total length of the interior channel with longitudinal baffles, m.
The width of interior channels is calculated with Equation (6.10).

$w_{channel, long} = \left(\dfrac{1}{(n+1)} \right)(w) \qquad (6.10)$

$w_{channel, long}$ = width of longitudinal channels, m.

(6) Determine the overflow rate and estimate helminth egg reduction with Equations (5.28) and (5.30).

$$\mathrm{OFR} = \frac{Q}{A_m} \leq 0.12 \text{ m}^3/\text{m}^2 \text{ d}; t_{V/Q} \geq 10 \text{ d} \tag{5.28}$$

$$\log_{10\ 95\%\ \mathrm{LCL}} = -\log_{10}\left[1 - (1 - 0.41\exp(-0.49 t_{V/Q} + 0.0085(t_{V/Q})^2))\right]$$
for ≤ 20 d $\tag{5.30}$

The calculation of the overflow rate, OFR, is especially important if the maturation pond follows an anerobic reactor or pond.

(7) Calculate the reduction of fecal coliforms or *E. coli* using the von Sperling and Wehner and Wilhelm equations.

$$N_{t,m} = N_{0,m} \frac{4a \cdot e^{1/2d}}{\left((1+a)^2 e^{a/2d} - (1-a)^2 e^{-a/2d}\right)}$$

$N_{t,m}$ = concentration of fecal coliforms or *E. coli* in the effluent, CFU/100 mL; $N_{0,m}$ = concentration of fecal coliforms or *E. coli* in the influent, CFU/100 mL; and d = dispersion number.

$$d = \frac{1}{(l/w)_{\text{interior}}}$$

$$a = \sqrt{(1 + 4 k_{b,T} \cdot t_{V/Q} \cdot d)}$$

$t_{V/Q}$ = theoretical hydraulic retention time, d and $k_{b,T}$ = mortality rate constant at temperature T, d^{-1}.

$$k_{b,20} = 0.542 \cdot h^{-1.259}$$

h = depth of the pond, m.

$$k_{b,T} = k_{b,20} \Theta^{(T-20)} = (0.542 \cdot h^{-1.259})(1.07)^{(T-20)}$$

6.4 MATURATION POND DESIGN EXAMPLE: REDESIGN FOR EFFLUENT REUSE, COCHABAMBA, BOLIVIA

This is a continuation of the facultative pond redesign in Section 5.4. A summary of the redesign concept will be repeated here with emphasis on the maturation ponds. In Chapter 7 the design of the wastewater reuse in agriculture systems will be covered.

The City of Cochabamba waste stabilization pond system Albarrancho, built in 1986 and operated by SEMAPA (Servicio Municipal de Agua Potable y Alcantarillado), the municipal water and sewer provider have been seriously overloaded for decades due to population growth and poor design, and is presently abandoned. As a result, raw wastewater bypassing the system heavily pollutes the Río Rocha, which is used for agricultural irrigation downstream of the wastewater discharge (Coronado et al., 2001).

There is an urgent need to save the Albarrancho plant with a redesign using (i) a sustainable design flowrate and (ii) improved hydraulic design to minimize short-circuiting, to enable effluent reuse in agriculture rather than direct discharge to the Rio Rocha losing the valorization potential of the wastewater. The implementation of the redesign will require excess flows to be diverted downstream, while treated effluent will be diverted directly to agricultural users. Data from the Albarrancho

Theory and design of maturation ponds

Table 6.18 Design data for the Albarrancho maturation ponds.

Parameter	1986[1] Design Values	2021 Design Values[2]
Q_{mean}, m³/d	34,733	3300 (per pond) 13,200 (total)
$t_{V/Q}$, d	18	To be determined
Total area, ha	13.65	No change
Total volume (approximate), m³	205,200	To be determined
Number of ponds in parallel	4	No change
Area/pond, ha	3.4	No change
Length × width/pond, m	325 × 105	No change
Depth (all ponds), h, m	1.5	To be determined

[1]Data from SEMAPA, personnel communication.
[2]From Section 5.4.

Wastewater Treatment Plant that apply to the maturation ponds are presented in Table 6.18 (SEMAPA, personal communication). A diagram of the waste stabilization pond facility with the facultative pond redesign from Section 5.4 is shown in Figure 6.21.

Design objectives:

- Redesign of the maturation ponds of the Albarrancho treatment facility to produce an effluent meeting the WHO guidelines for restricted wastewater use in agriculture.

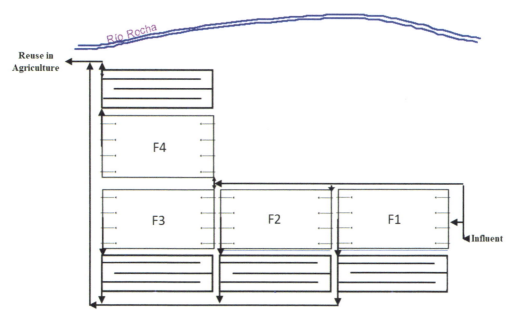

Figure 6.21 The redesign of facultative ponds combines two original ponds into one facultative pond with $l/w = 2/1$, for a total of 4 ponds with multiple inlets and outlets that discharge to each of the 4 baffled maturation ponds. For each facultative pond $l = 328.6$ m, $w = 164.3$ m, and $A_f = 54,000$ m² $= 5.4$ ha. The design flowrate is 3300 m³/d/pond or 13,200 m³/d total (Section 5.4).

- The facultative ponds were redesigned in Section 5.4.
- The maturation ponds will be redesigned in this section.
- The wastewater reuse in agriculture design is in Chapter 7

Following the design procedure from Section 6.3:

(1) Select the design hydraulic retention time and calculate the volume of each pond with Equation (6.17).
The ponds already exist and this step is not necessary.

(2) Select the depth of the pond and calculate the area with Equation (6.18).
The depth of the existing ponds: $h = 1.5$ m. Use $h = 1.0$ m to start.
Area of exiting ponds: $A_m = 3.4$ ha.

(3) Select the exterior l/w ratio and determine the length and width of the pond.
Data from existing ponds:

$l = 325$ m; $w = 105$ m;

$(l/w)_{exterior} = 325/105 = 3.10$

(4) Select the interior embankment slope, i, and calculate the pond volume using the prismoid equation, and then calculate the hydraulic retention time.

$i = 2/1$ (existing ponds)

$$V_{m,cal} = \frac{h}{6} \cdot \left[(l \cdot w) + (l - 2ih)(w - 2ih) + 4 \cdot (l - iw)(w - ih) \right]$$

$$= \frac{1.0}{6} \cdot \left[(325 \cdot 105) + (325 - 2(2)(1.0))(105 - 2(2)(1.0)) + 4 \cdot (325 - 2(105))(105 - 2(1.0)) \right]$$

$$= 33\,267 \text{ m}^3$$

$$t_{V/Q,m} = \frac{V_m}{Q_m} = \frac{33\,267 \text{ m}^3}{3300 \text{ m}^3/\text{d}} = 10.1 \text{d}$$

The hydraulic retention time meets the minimum desired value of 10 d.

(5) Design of interior baffles.
Longitudinal baffles will be used with $n = 4$.
The interior length to width ratio, $(l/w)_{interior}$, is calculated with Equation (6.8).

$$(l/w)_{interior,long} = \left[(l/w)_{exterior} \right] \cdot (n + 1)^2 = (3.10) \cdot (4 + 1)^2 = 77.5/1$$

Equation (6.9) is used to calculate the total length of interior channels.

$$l_{channel,long} = (n + 1)(l) = (5 + 1)(325) = 1950 \text{ m}$$

The width of interior channels is calculated with Equation (6.10).

$$w_{channel,long} = \left(\frac{1}{(n+1)} \right)(w) = \left(\frac{1}{(4+1)} \right)(105 \text{ m}) = 21 \text{ m}$$

(6) Estimate the overflow rate and helminth egg reduction with Equations (5.28) and (5.30).

$$\text{OFR} = \frac{Q}{A_m} \leq 0.12 \text{ m}^3/\text{m}^2 \text{ d}; \, t_{V/Q} \geq 10 \text{ d}$$

$$\text{OFR} = \frac{Q}{A_m} = \frac{3300 \text{ m}^3/\text{d}}{34{,}000 \text{ m}^2} = 0.097 \text{ m}^3/\text{m}^2 \text{ d}$$

Table 6.19 Solar radiation[1] and mean maturation pond water temperature (pond LS1),[2] Albarrancho Maturation Ponds, Cochabamba, Bolivia.

Month	Solar Radiation (kJ/m² × d)	Maturation Pond LS1 T_{water} (°C)
J	22,800	18.7
F	21,900	24.5
M	20,600	24.4
A	18,700	22.0
M	16,500	20.5
J	16,000	18.6
J	16,200	19.1
A	17,900	20.4
S	20,400	23.1
O	21,200	23.4
N	22 400	24.6
D	21,800	23.3

[1]Data from CLIMWAT.
[2]Data from SEMAPA.

The OFR meets the minimum requirement for helminth egg reduction and adds an important safety factor for agricultural reuse. The calculation of the overflow rate is especially important if a maturation pond follows an anerobic reactor or pond.

Helminth egg reduction:

$$\log_{10 \text{ 95\% LCL}} = -\log_{10}\left[1-(1-0.41\exp(-0.49t_{V/Q}+0.0085(t_{V/Q})^2))\right]$$
$$= -\log_{10}\left[1-(1-0.41\exp(-0.49(10.1)+0.0085(10.1)^2))\right] = 2.16$$

(7) Calculate the reduction *E. coli* using the von Sperling and Wehner and Wilhelm equations. Solar radiation data and pond water temperatures for Cochabamba are shown in Table 6.19.

The range of solar radiation data for Cochabamba is slightly higher than that at Belo Horizonte, therefore the von Sperling equation should be applicable to calculate the mortality rate constant:

Pond System	Range of Solar Radiation (kJ/m² d)
Belo Horizonte	14,900–20,600
Cochabamba	16,000–22,800

(i) *E. coli* concentrations in influent wastewater (facultative pond effluent)
$N_{\text{influent}} = 1.70 \times 10^7$ MPN/100 mL (Section 5.4)

(ii) Dispersion number:

$$d = \frac{1}{(l/w)_{\text{interior}}} = \frac{1}{77.5} = 0.013$$

(iii) Mortality rate constant at design water temperature, $k_{b,T}$, d⁻¹ (Equation (5.34)):

$$k_{b,T} = (0.542 \cdot h^{-1.259})(1.07)^{(T-20)}$$

$h = 1.0$ m.
$T_{\text{water}} = 18.6°C$.

$$k_{b,T} = (0.542 \cdot (1.0)^{-1.259})(1.07)^{(18.6-20)} = 0.49 \text{ d}^{-1}$$

(iv) Determine a in the Wehner–Wilhem equation (Equation (5.32)):
$t_{V/Q} = 10.1 \text{ d}$

$$a = \sqrt{1 + 4k_{b,T} \cdot t_{V/Q} \cdot d} = \sqrt{(1 + 4(0.49 \text{ d}^{-1})(10.1 \text{ d})(0.013)} = 1.1212$$

(v) Determine effluent *E. coli* concentration and $\log_{10}$ reduction:

$$N_{\text{effluent}} = N_{\text{influent}} \frac{4a \cdot e^{1/2d}}{\left((1+a)^2 e^{a/2d} - (1-a)^2 e^{-a/2d}\right)}$$

$$= (1.70 \times 10^7) \left[\frac{4(1.1212) \cdot e^{1/2(0.013)}}{\left((1+1.1212)^2 e^{1.1212/2(0.013)} - (1-1.1212)^2 e^{-1.1212/2(0.013)}\right)} \right]$$

$$= 1.55 \times 10^5 \text{ CFU/100 mL}$$

$$\log_{10,\text{Reduction}} = \log_{10}(1.70 \times 10^7) - \log_{10}(1.55 \times 10^5) = 2.04$$

The range and mean of $\log_{10}$ reductions for the 12 months of the year with different monthly water temperatures are shown below:

T_{water} (°C)	Log_{10} Reduction	
Range	Range	Mean
18.6–24.6	2.04–2.98	2.54

The mean 2.54 $\log_{10}$ reduction is higher than the values reported for maturation ponds in Brazil by Dias *et al.* (2014).

A summary of the redesign results is presented in Table 6.20, with a summary of the design results for both the facultative and maturation ponds in Table 6.21. The wastewater reuse design for the Cochabamba waste stabilization pond system is presented in Chapter 8.

Table 6.20. Final results for redesign of maturation ponds Albarrancho wastewater treatment plant, Cochabamba, Bolivia.

Parameter	Single Maturation Pond	Total 4 Ponds in Parallel
Q_{mean}, m³/d	3300	13,200
Design Water Temperature T_{water}, °C	17.4	
Area, ha	5.40	21.6
Water depth, h, m	1.0	
Volume, m³	33,267	133,068
$t_{V/Q}$, d	10.1	
Helminth egg reduction:	0.032	
OFR, m³/m² d (<0.12 m³/m² d)		
$\log_{10,95\%, \text{LCL}}$	>3.17	
E. coli $\log_{10}$ reduction:		
Range	1.28–1.64	
Mean	1.47	
Effluent quality for reuse	Objective of the final design is to satisfy the WHO guidelines for restricted reuse in agriculture.	

Table 6.21 Final results for facultative/maturation pond system, Cochabamba, Bolivia.

Parameter	Facultative (Per Pond; 4 Ponds in Parallel)	Maturation (Per Pond; 4 Ponds in Parallel)	Total
Q_{mean}, m³/d	3300	3300	13,200
Influent BOD_5, mg/L	200		
Influent BOD_Lm, g/L	300		
Influent TSS, mg/L	200		
Temperature T_{water}, °C	17.4–23.2	18.6–24.6	
Design $\lambda_{S,T}$, kg BOD_L/ha d	217		
Area, ha			
• per pond	5.4	3.4	35.2
• total	21.6	13.6	
Water depth, m	2.0	1.0	
Volume, m³			
• per pond	102,154	33,267	
• total	408,616	133,068	
$t_{V/Q}$, d	31.0	10.1	41.1
Sludge accumulation, m³/yr			
• per pond	1742		
• total	6968		
Frequency sludge removal, yr	10–15		
Helminth egg removal $\log_{10,95\%,LCL}$	3.17	2.16	5.57
E. coli mean $\log_{10}$ reduction	1.47	2.54	4.01
Effluent $BOD_{5,Soluble}$, mg/L	27.2		
Effluent quality for reuse		Meets WHO guidelines for restricted reuse in agriculture	

doi: 10.2166/9781789061536_0197

Chapter 7

Wastewater reuse in agriculture: guidelines for pathogen reduction and physicochemical water quality

7.1 THE SAFE USE OF WASTEWATER FOR REUSE IN AGRICULTURE

It is estimated that, on a global scale, 29.3 million ha of cropland is irrigated with wastewater in countries where less than 75% of all wastewater is treated (Thebo *et al.*, 2017), which is not to say it is adequately treated for pathogen reduction to protect public health. Irrigation with wastewater, whether adequately treated or not, is commonly practiced by farmers who know the value of the nutrients in the wastewater but are unaware of serious public health risks (Figure 7.1). The design of wastewater treatment and reuse systems must ensure that treated wastewater is safe to use.

Section 7.2 covers the WHO guidelines for pathogen reduction in wastewater treatment for reuse in agriculture, focusing on key design issues and including various case studies. Section 7.3 addresses the physical–chemical water quality parameters that should be assessed to ensure the final effluent treated for pathogen reduction also meets water quality objectives for crop production. Various case studies of successes and failures are also discussed.

7.2 PATHOGEN REDUCTION

7.2.1 WHO guidelines for wastewater use in agriculture
7.2.1.1 Development of the WHO guidelines
In 1989, the World Health Organization published its health-based guidelines for wastewater use in agriculture and aquaculture (WHO, 1989). This new health-based approach focused on measurable risks based on epidemiological studies of exposed groups and emphasized sustainable wastewater treatment processes such as waste stabilization ponds (WSPs). This health-based approach, which is to this day the only approach capable of succeeding in resource-limited areas of the world, replaced the expensive and technically inappropriate treatment technology emphasis (primary and secondary treatment, advanced processes, nitrogen and phosphorus removal, and chemical disinfection) presented in the 1973 WHO publication, Reuse of Effluents: Methods of Wastewater Treatment and Health Safeguards (WHO, 1973).

Table 7.1 summarizes the 1989 WHO microbiological guidelines, which were, and still are, considered the most important aspect of wastewater reuse in agriculture. Helminth infections were

© 2022 The Author. This is an Open Access book chapter distributed under the terms of the Creative Commons Attribution Licence (CC BY-NC-ND 4.0), which permits copying and redistribution for noncommercial purposes with no derivatives, provided the original work is properly cited (https://creativecommons.org/licenses/by-nc-nd/4.0/). This does not affect the rights licensed or assigned from any third party in this book. The chapter is from the book *Integrated Wastewater Management for Health and Valorization: A Design Manual for Resource Challenged Cities*, Stewart M. Oakley (Author).

Figure 7.1 A farmer irrigating his fields with raw wastewater directly from the trunk sewer in a small city. In the left photo, the suction pipe for the pump is placed directly in the trunk sewer, intentionally broken by the farmer. The right photo shows the irrigated fields, with produce later harvested and sent to city markets. While this case seems extreme, it is not uncommon, as farmers know the nutrient value of wastewater, but unfortunately are not aware of the serious public health risks. Design of wastewater treatment for reuse in agriculture should address the pathogen reduction as the first priority. (*Source*: Photos from Guastatoya, Guatemala.)

Table 7.1 WHO (1989) microbiological guidelines for wastewater use in agriculture.

Category	Reuse Conditions	Exposed Group	Intestinal Helminths[1] (Arithmetic Mean Number of Eggs per Liter)	Fecal Coliforms (Geometric Mean Number per 100 mL)	Wastewater Treatment Expected to Achieve the Required Microbiological Guideline
A: unrestricted irrigation	Irrigation of crops likely to be eaten uncooked, sports fields, public parks	Workers Consumers Public	≤ 1	≤ 1000	A series of stabilization ponds designed to achieve the microbiological quality indicated, or equivalent treatment
B: restricted irrigation	Irrigation of cereal crops, industrial crops, fodder crops, pasture, and trees[2]	Workers	≤ 1	No standard recommended	Retention in stabilization ponds for 8–10 days or equivalent helminth and fecal coliform removal

Source: From WHO (1989).
[1]*Ascaris* and *Trichuris* species and hookworms.
[2]In the case of fruit trees, irrigation should cease 2 weeks before fruit is picked, and no fruits should be picked off the ground. Sprinkler irrigation should not be used.

considered the main health risk, and a reduction to less than 1.0 helminth egg/L was recommended for both unrestricted and restricted irrigation. (Restricted irrigation is the use of wastewater to grow crops not eaten raw by humans, and other nonedible crops; unrestricted irrigation is the use of treated wastewater to grow crops that are normally eaten raw (WHO, 2006).) For unrestricted irrigation, including crops eaten raw, sport fields, and public parks, ≤1000 fecal coliforms/100 mL (geometric mean) was also recommended. In addition, the sustainable wastewater treatment technologies expected to achieve these recommendations, WSP systems, were also included in the guidelines as presented in Table 7.1 (WHO, 1989).

For almost three decades, the 1989 guidelines have been fully or partially applied in countries around the world. They are used as a reference for the National Water Authority (Autoridad Nacional de Agua) in Peru to evaluate and approve wastewater reuse authorizations throughout the country (Moscoso, 2016).

In the decades following the 1989 WHO Guidelines, the interest in wastewater reuse in agriculture increased, driven by water scarcity, food insecurity, and lack of availability of fertilizer nutrients, combined with concerns about public health and environmental effects. As a result, WHO updated the guidelines in 2006 to incorporate newer scientific evidence on pathogens, chemicals, risk assessment, and epidemiological studies, as well as changes in population characteristics and sanitation practices (WHO, 2006). The 2006 guidelines were published as the Guidelines for the Safe Use of Wastewater, Excreta, and Greywater in four volumes: (1) policy and regulatory issues; (2) wastewater use in agriculture; (3) wastewater and excreta use in aquaculture; and (4) excreta and greywater use in agriculture.

The 2006 guidelines, for the first time, do not emphasize effluent quality standards. Instead, they offer the flexibility to select a range of options throughout the sanitation chain to achieve health protection objectives. This change recognizes that high levels of treatment are not always feasible or cost-effective, or necessary, and that the use of treated or partially treated wastewater is common in many countries. This approach is the most appropriate for resource-poor cities and peri-urban areas worldwide, where sophisticated wastewater treatment with disinfection will not be economically feasible, thus allowing the development of regulatory and management systems commensurate with the local socio-economic and environmental realities.

The WHO guidelines incorporate the multiple barrier approach as shown in Figure 7.2, where each barrier along the water reuse chain plays a key role in the protection of public health. Barriers are incorporated at the following places:

(1) At the point of wastewater generation: sustainable treatment plants designed for pathogen reduction and effluent reuse;
(2) On the agricultural land: safe irrigations practices, worker protection, crop restrictions;
(3) In food markets: hygienic food handling and washing of harvested crops with potable water; and
(4) In homes of consumers: disinfection, peeling, and cooking of produce.

Various combinations of these barriers should be sufficient to achieve the appropriate level of acceptable risk. One of the simplest measures is crop restriction based on irrigation water quality: for example, the use of partially treated wastewater on nonfood crops such as cotton, or in irrigation of tree crops. As an example, during the cholera outbreak in 1993 in the city of Santiago, Chile, there was a 90% reduction in cholera cases, which were originally caused by the consumption of wastewater-contaminated vegetables, as a result of the cessation of irrigation of these crops (Moscoso, 2016). Another potential barrier would be the cessation of irrigation 1 or 2 weeks before harvest, allowing time for natural mortality of bacterial and viral pathogens on crop surfaces.

Rather than focus solely on the quality of the wastewater used for agricultural irrigation, with emphasis on treatment processes, the WHO guidelines recommend setting realistic health goals using

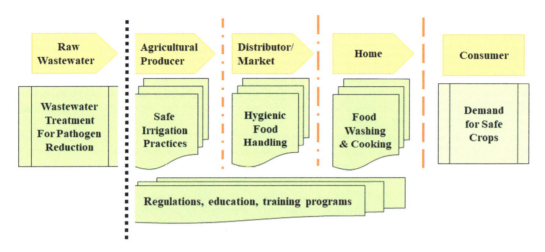

Figure 7.2 Multiple barriers for wastewater reuse in agriculture, from wastewater discharge to consumer. In this framework, the chain of wastewater treatment, agricultural production, market distribution, and home preparation comprise the multiple barriers for pathogen reduction and public health protection. In high-income countries, wastewater treatment is typically the only barrier relied upon for pathogen reduction. (*Source*: Adapted from Ilic et al. (2009)).

the multiple barrier concept, continuously evaluating and managing the risks, from the generation of wastewater to the consumption of products irrigated with the wastewater. This allows the development of a regulatory and monitoring system appropriate to the socioeconomic realities of the country or locality.

The 2006 WHO guidelines are also designed as an aid to developing national and international approaches and to offer a framework for making national and local decisions regarding the identification and management of health risks associated with the use of wastewater in agriculture and aquaculture. It is recognized that country or local changes in policy, and investments in improvements, infrastructure works, operational measures, and behavioral modifications, involve multiple actors and will take time (Moscoso, 2016).

7.2.1.2 Health risks and pathogen reductions for wastewater reuse

Table 7.2 summarizes the epidemiological studies of infectious disease transmission related to wastewater reuse in agriculture as reported by WHO (2006). In areas where wastewater is used in agriculture with inadequate treatment and with inadequate protection of field workers, the greatest health risks for all exposed groups are intestinal helminth infections. Given the high global prevalence of soil-transmitted helminth infections as shown in Figure 7.3, wastewater treatment for reuse should focus first, above all else, on helminth egg reduction.

The health risks from bacteria and viruses, which are presented in Table 7.2, have been documented for diarrheal disease when thermotolerant coliform (TTC) concentrations exceed 10^4/100 mL. *Salmonella* infections have been documented in children exposed to raw wastewater, and norovirus seroresponse has been detected in adults exposed to partially treated wastewater. The hepatitis A outbreak in the US has been associated with green onions imported from Mexico, and it is likely the onions were irrigated or washed with partially treated or raw wastewater.

The evidence for protozoan infections from wastewater is lacking (Table 7.2) although parasitic protozoa have been found on wastewater-irrigated vegetables. Five outbreaks of cyclosporiasis in the US and Canada have been linked to raspberries imported from Guatemala, and it is again likely the fruit was irrigated or washed with partially treated or raw wastewater.

Table 7.2 Health risks associated with the use of wastewater for agricultural irrigation.

Exposed Group	Health Risks		
	Helminth Infections	Bacterial/Viral Infections	Protozoal Infections
Farm workers and their families	Significant risk of *Ascaris* infection from contact with untreated or inadequately treated wastewater or sludge; risk remains, especially for children, with effluent concentrations <1.0 helminth egg/L; also increased risk of hookworm infections from contact with untreated or inadequately treated wastewater.	Increased risk of diarrheal disease in young children with wastewater contact if water quality exceeds 10^4 TTC/100 mL; elevated risk of *Salmonella* infections in children exposed to untreated wastewater; elevated serorespose to norovirus in adults exposed to partially treated wastewater.	Increased risk of amoebiasis observed with contact with untreated wastewater.
Nearby communities (risks from aerosols from sprinkler irrigation)	Significant risk of *Ascaris* infection for both adults and children with heavy contact with untreated or inadequately treated wastewater.	Sprinkler irrigation with poor water quality (10^6–10^8 TTC/100 mL) and high aerosol exposure associated with increased rates of infection.	No data on transmission of protozoal diseases from sprinkler irrigation.
Consumers	Significant risk of *Ascaris* infection for both adults and children from crops irrigated with untreated or inadequately treated wastewater.	Cholera, typhoid, and shigellosis outbreaks reported from use of untreated wastewater on consumed crops; increase in nonspecific diarrheal disease when water quality exceeds 10^4 TTC²/100 mL. The largest outbreak of hepatitis A associated with fresh produce in the history of the US occurred in 2003, with more than 700 cases in four states, all linked to green onions imported from Mexico (Acheson & Fiore, 2004). It is likely the onions were irrigated or washed with wastewater-contaminated water.	Five outbreaks of cyclosporiasis from 1995 to 2000 in the US and Canada, caused by the protozoan pathogen *Cyclospora cayetanensis*, have been linked to raspberries imported from Guatemala (Bern et al., 1999; Ho et al., 2002). Another outbreak in the US linked to snow peas imported from Guatemala occurred in 2004 (CDC, 2004). It is likely the raspberries and snow peas were irrigated or washed with wastewater-contaminated water. In other studies, evidence of parasitic protozoa has been found on wastewater-irrigated vegetable surfaces, but with no direct evidence of disease transmission.

Source: Modified from WHO (2006).
TTC, thermotolerant coliforms, previously called fecal coliforms.

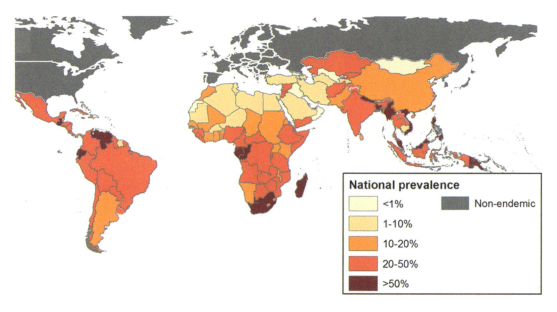

Figure 7.3 Global prevalence of soil-transmitted helminth infections with *Ascaris lumbricoides*, *Trichuris trichiura*, and Hookworms in 2010. An estimated 1.45 billion people were infected with at least one species of helminth, resulting in 5.18 million DALYs, which is the population metric of life years lost due to morbidity and mortality (WHO, 2006), equivalent to 3.57×10^{-3} DALYs/ person yr (Pullan *et al.*, 2014). (*Source*: Pullan *et al.* (2014)).

As a result of a detailed analysis of the newer epidemiological data, WHO developed a revised set of health-based targets for the 2006 guidelines for two scenarios, restricted and unrestricted irrigation, which are presented in Table 7.3. Unrestricted irrigation is the irrigation of crops that are eaten raw, with the principal exposed group at risk being the consumers of these crops; restricted irrigation is the consumption of crops not eaten raw, with fieldworkers being the main group at risk. The acceptable health-based target selected for bacterial and viral diseases was set at $\leq 10^{-6}$ disability-adjusted life

Table 7.3 Health-based targets for wastewater use in agriculture.

Exposure Scenario	Exposed Group	DALY per person per year	Log₁₀ Pathogen Reduction Needed[1]	Helminth Egg Concentration[2] eggs/L
Unrestricted irrigation (crops eaten raw)	Consumers	$\leq 10^{-6}$		
Leaf crops (lettuce)			6	≤ 1
Root crops (onions)			7	≤ 1
Restricted irrigation (crops not eaten raw)	Fieldworkers	$\leq 10^{-6}$		
Labor intensive			4	≤ 1
Highly mechanized			3	≤ 1

Source: WHO (2006).
[1]Rotavirus reduction achieved by a combination of wastewater treatment and additional health protection measures; rotavirus was found to have a higher risk than bacterial and protozoan infections.
[2]Helminth eggs should be removed by treatment methods, such as sedimentation in WSPs; when children younger than 15 are exposed, additional health measures should be used (e.g., protective equipment such as gloves and boots, periodic examinations for helminth infections).

years (DALYs)/person yr, which is to be equivalent to the acceptable risk used by WHO for drinking water (WHO, 2006). Quantitative microbial risk assessments (QMRAs) of rotavirus were performed for several scenarios to determine the required $\log_{10}$ reductions necessary to meet the DALY target; rotavirus was selected because it was found to have a higher risk than bacterial or protozoan infections, and thus, rotavirus reduction to meet $\leq 10^{-6}$ DALYs/person yr will provide sufficient protection against bacterial and protozoan infections (WHO, 2006). For unrestricted irrigation, leaf crops above the soil surface require a 6.0 $\log_{10}$ reduction of rotavirus; however, root crops because of contact with the soil require a 7.0 $\log_{10}$ reduction. For labor-intensive restricted irrigation, a 4.0 $\log_{10}$ reduction is required to protect fieldworkers, while highly mechanized irrigation, with less manual labor, requires only a 3.0 $\log_{10}$ reduction

The health-based targets for helminth egg infections were based on epidemiological studies showing that excess helminth infections, for both farmers and consumers, could not be measured when wastewater with ≤ 1 helminth egg/L was used for irrigation (WHO, 2006).

Table 7.4 lists pathogen reductions achievable by different control measures that can be used to meet the health-based targets for unrestricted and restricted irrigation in Table 7.3 using the multiple barrier approach. There are various combinations of control measures that can be used to meet the 6 or 7 $\log_{10}$ reduction for viral/bacterial pathogens and the ≤ 1.0 helminth egg/L requirements, without relying solely on wastewater treatment as is done in the EU and the US.

Table 7.4 Achievable pathogen reductions by various public health protection measures.

Control Measure	Log$_{10}$ Pathogen Reduction[1]	Comments
Wastewater treatment	1–6	Depends on combination of technologies; wastewater treatment, particularly disinfection, is typically the sole control measure for pathogen reduction in high-income countries
Facultative/maturation ponds in series[2]	≥ 4	If well designed to local climates, with sufficient hydraulic retention time and minimal hydraulic short circuiting
Wastewater storage reservoir[2]	≥ 5	Operated as a batch reactor during the irrigation season at water temperatures 10–30°C (Liran et al., 1994)
Pathogen die-off	0.5–2 (per day)	Occurs between last irrigation and consumption; log$_{10}$ reduction depends on climate, time, crop type, and so on.
Produce washing	1	Washing fruits and vegetables with clean water
Produce disinfection	2	Washing fruits and vegetables with disinfectant solution; rinsing with clean water
Produce peeling	2	Fruits, vegetables, root crops
Produce cooking	6–7	Temperatures at or near 100°C until food is cooked
Spray irrigation control[3]	1	Drift control and buffer zones
Localized drip irrigation[4] Low growing crops	2	Root crops and crops such as lettuce that have contact with the soil
Localized drip irrigation[4] High growing crops	4	Crops that have minimal or no contact with soil

Source: Modified from WHO (2006).
[1]For rotavirus, which is estimated to have a higher risk from exposure than bacterial and protozoan pathogens.
[2]Fecal coliform or E. coli reduction rather than rotavirus.
[3]Spray irrigation has a higher risk from exposure to aerosols than furrow or flood irrigation. It also may have limited application in resource-limited areas as a result of pumping requirements and sprinkler maintenance.
[4]Drip irrigation will never be used in resource-limited areas as a result of high cost and strict water quality requirements to prevent emitter clogging.

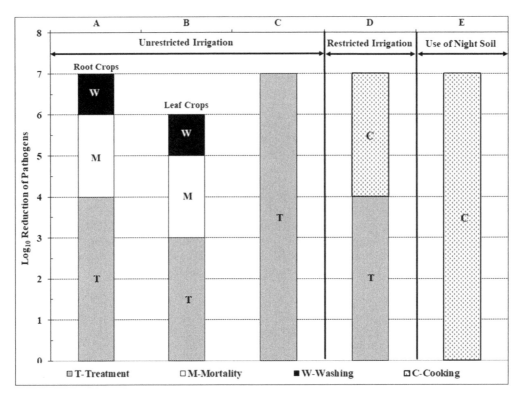

Figure 7.4 Examples of different combinations of public health protection measures (Table 7.4) for pathogen reduction to achieve the health-based target of ≤10^{-6} DALY/ person yr. Options A and B represent combinations best suited for resource-limited areas where treatment in waste stabilization ponds, for example, can be combined with other measures from Table 7.3, in this case combining natural mortality and washing, to provide crops that can be eaten raw; root crops such as onions require a 7.0 log$_{10}$ reduction because of contact with the soil, while leaf crops such as lettuce that do not come in contact with the soil require a 6.0 log$_{10}$ reduction. In Option B, treatment only yields a 3 log$_{10}$ reduction, with the additional reduction required to allow farmers to enter the fields obtained by natural die-off. Option C would be typical of high-income areas, such as the EU and the US, where wastewater treatment is the sole public health protection measure used for pathogen reduction. Option D for restricted irrigation has sufficient treatment to protect field workers and to irrigate all crops except those eaten uncooked; it is assumed that edible crops would be cooked in this option, so cooking is the additional control measure. Option E is an example of the night soil system traditionally used in China, where food is cooked and water is boiled, in areas where night soil was used as fertilizer; there would be high risk to agricultural workers and consumers in this option, which is rare but still used in parts of East Asia and Africa. All options can pose potential risks to agricultural workers and their families that must be mitigated with the use of proper treatment, operation and maintenance, clothing, safety equipment, and hygienic measures.

To protect field workers in labor-intensive agriculture, a 4.0 log$_{10}$ reduction is required. This can be satisfied solely by treatment or a combination of other measures, such as treatment and natural mortality effected by leaving crops in the field for a defined period after irrigation before harvesting (Table 7.4).

Figure 7.4 shows examples of different combinations of public health protection measures from Table 7.4 for both unrestricted and restricted irrigation for pathogen reduction (bacterial, viral, and protozoan) to achieve the health-based target of ≤10^{-6} DALY/person yr. Helminth egg reduction must be met by a treatment technology such as WSPs.

To ensure pathogen reduction targets are being met, the WHO guidelines recommend verification monitoring of (i) wastewater treatment and (ii) public health measures implemented in the reuse system.

7.2.1.3 Verification monitoring of wastewater treatment

Table 7.5 lists the options shown in Figure 7.4, with verification monitoring using *Escherichia coli* as the indicator for bacterial/viral/protozoan pathogens in the final effluent.

From Table 7.5 for different levels of wastewater treatment:

(1) Unrestricted irrigation:
Option A: Root crops require a 4.0 $\log_{10}$ reduction; thus, the final effluent should have $\leq 10^3$ *E. coli*/100 mL, assuming raw wastewater has a concentration of 10^7 *E. coli*/100 mL.
Option B: Leaf crops require a 3.0 $\log_{10}$ reduction, with a required effluent $\leq 10^4$ *E. coli*/100 mL.
Option C: Root crops require a 7.0 $\log_{10}$ reduction, and leaf crops require a 6.0 $\log_{10}$ reduction, with required effluent concentration being $\leq 10^0$ *E. coli*/100 mL and $\leq 10^1$ *E. coli*/100 mL, respectively. This option requires a sophisticated wastewater treatment plant, which would routinely monitor effluent concentrations.

(2) Restricted irrigation:
Option D: Requires a 4.0 $\log_{10}$ reduction, and thus, the final effluent should have $\leq 10^3$ *E. coli*/100 mL.

Table 7.5 Verification monitoring of wastewater treatment for the different levels of treatment in options A–E in Figure 7.4.

Type of Irrigation	Option From Figure 7.4	Required Log$_{10}$ Reduction by Treatment (Figure 7.4)	Verification Monitoring of Final Effluent[1] *E. coli*/100 mL	Comments	Helminths Verification Monitoring or TT[2]
Unrestricted	A	4	$\leq 10^3$	Root crops	≤ 1 egg/L or TT
	B	3	$\leq 10^4$	Leaf crops	
	C	7	$\leq 10^0$	Root crops	
		6	$\leq 10^1$	Leaf crops	
Restricted	D	4	$\leq 10^3$	Labor-intensive agriculture protective of adults[3]	≤ 1 egg/L or TT
Night soil	E	Few public health measures	N/A	Food crops are traditionally cooked; all agricultural workers, children, and general public are at risk. Still exists in parts of East Asia and Africa. High risk of infections with *Ascaris* and other helminths (Figure 7.4), and with bacterial, viral, and protozoan pathogens	N/A

Source: Adapted from WHO (2006).
[1]Assuming a raw wastewater concentration of 10^7 *E. coli*/100 mL.
[2]Treatment technology such as WSPs with the total overflow rate, OFR <0.12 m^3/m^2 d and mean hydraulic retention time, $t_{w/o}$>10 d.
[3]If children younger than 15 are exposed, additional health measures should be used (e.g., protective equipment such as gloves and boots, periodic examinations for helminth infections).

(3) Night soil:
Option E: The night soil option is meant as an example not to be emulated and will be discussed in Section 7.2.3 as a case study.

Helminth eggs are not likely to be monitored, and reduction should be verified by the treatment technology used, such as overflow rates and hydraulic retention times in WSP systems (Table 7.5).

Under most circumstances, in small- to medium-sized cities, effluent monitoring likely will never take place unless performed by a research team from the capital city or another country. It is thus imperative that designers ensure the treatment plant will meet the required $\log_{10}$ reductions with a safety factor, designing a WSP system, for example, for a 5.0 $\log_{10}$ reduction for a restricted irrigation application requiring a 4.0 $\log_{10}$ reduction.

7.2.1.4 Verification monitoring of health protection measures

The various health protection measures presented in Table 7.4 also need to be monitored to ensure the multiple barriers of protection are in place and working (WHO, 2006). Some can be monitored by visual inspection, including

- Types of crops grown in wastewater-treated irrigation areas;
- Types of wastewater application methods used;
- Use of protective clothing and other safety measures; and
- Physical conditions of WSPs and reservoirs (e.g., plant growth, sludge build-up, odors, effluent colors).

Other health protection measures, such as produce washing, peeling, disinfection, and cooking, are more difficult to observe at the household level. Verification of crop contamination at the point of harvest or sale requires sampling and laboratory analysis. It is imperative, however, that the health protection measures be monitored and verified since they are central to the overall pathogen reduction required to protect public health. Table 7.6 lists the minimum monitoring requirements as recommended by WHO (2006).

7.2.1.5 Pathogen reduction in sludges produced in wastewater treatment

Pathogens removed by the various wastewater treatment processes are concentrated in sludges, which must be managed properly to avoid the re-release of these pathogens to the environment. Sludges produced by wastewater treatment processes, such as those from anaerobic and facultative ponds, upflow anaerobic sludge blanket (UASBs), secondary clarifiers for trickling filters, and anaerobic digesters, need to be dewatered and then (i) sanitarily disposed or (ii) reused in agriculture, which

Table 7.6 Minimum verification monitoring frequencies for health protection control measures.

Health Protection Measure	Minimum Verification Monitoring Frequency
Wastewater treatment	(a) Urban areas: one effluent sample every 2 weeks for *E. coli* and one sample per month for helminth eggs (b) Rural areas: one sample every 3–6 months for helminth eggs
Irrigation with root and leaf crops	Annual surveys to verify the irrigation method used and the type of crops grown
Spray irrigation	Annual surveys to verify the spray drift control methods and the extent of the buffer zone
Pathogen die-off	Annual local surveys to determine microbial quality of wastewater-irrigated crops at harvest and at various points of sale
Produce washing, disinfection, peeling, cooking with water	Annual local surveys to verify occurrence at household level of food preparation control measures and to assess the impact of food hygiene education programs

Source: Modified from WHO (2006).

presents high risks in resource-challenged areas where excreta-related infections are common. In WSP systems, dewatering can be effected in situ in facultative ponds that have been drained; otherwise, liquid waste sludges produced in all processes must be removed by gravity or pumping and sent to sludge drying beds or ponds for dewatering.

Unfortunately, the sanitary management of dewatered sludges has not been accomplished in practice, and there is a constant risk of re-contaminating the environment with the very same pathogens that the liquid stream processes were designed to remove. Figures 7.5 and 7.6 present common examples of

Figure 7.5 Sludge from a drained facultative pond being disposed on the pond's embankment. This pond in Tela, Honduras, had operated for 15 years, and no plans had ever been made for sludge removal, drying, and storage. When the pond finally needed desludging, the only space available for disposal was the embankment. As a result, the neighboring population, which had open access to the pond installation along with domestic animals, was put at high risk. Viable helminth eggs in the pond sludge after draining ranged from 0.4 to 25 eggs/g dry weight, and one sample taken on the embankment measured 5.1 eggs/g (Oakley *et al.*, 2012). The prevalence of soil-transmitted helminth infections in Honduras ranges from 20 to 50% (Figure 7.3).

Figure 7.6 Dewatered UASB sludges from two small treatment plants in Guatemala are distributed in bags to give to the local population as a fertilizer and a soil conditioner. This is a common practice in many small cities that should be prohibited: A significant fraction of all influent pathogens will concentrate in the sludge. As an example, concentrations of helminth eggs in UASB sludges have been reported up to 62.9 eggs/g dry weight (Oakley *et al.*, 2017). The prevalence of soil-transmitted helminth infections in Guatemala is >50% (Figure 7.3). These sludges should be well dried and buried onsite. (Top: Sololá; Bottom: Maria Tecún.)

this widespread problem for sludge dried in drained facultative ponds, and liquid sludges from UASB reactors were dewatered in sludge drying beds.

Table 7.7 presents the WHO recommendations for the treatment of dry sludges and excreta for pathogen reduction for reuse in agriculture (WHO, 2006). The most appropriate option for resource-challenged cities and peri-urban areas is drying and burial without reuse.

Table 7.7 Recommendations for treatment of dry sludges for pathogen reduction with or without reuse in agriculture.

Batch Treatment[1]	Criteria	Comment
Burial/long-term storage; ambient temperature: 2–35°C	Permanent; or ≥10 yr[2]	Best option for small cities to avoid recontamination of the environment with pathogens; facultative pond sludges can be buried for 10–15 yr, removed and used as a soil conditioner, with fresh dewatered sludge from pond placed in the same excavation.
Storage; ambient temperature: 2–20°C	1.5–2 yr	Eliminates bacterial pathogens if not rewetted; *E. coli* and *Salmonella* can regrow if rewetted; can reduce viral and protozoan pathogens below risk levels; soil-transmitted helminth ova persist at low numbers.
Storage; ambient temperature: >20–35°C	>1 yr	Substantial to total inactivation of viral, bacterial and protozoan pathogens; inactivation of hookworm and *Trichuris* eggs; inactivation of *Ascaris* at 1 yr.
Alkaline treatment[3]	pH >9 for >6 mo	If temperature >35°C and moisture content <25%; lower pH and wetter material will prolong time for complete elimination.
Composting[3]	Temperature >50°C for >1 week	Minimum requirement; longer time required if temperature requirement cannot be met.

Source: Modified from WHO (2006).
[1]New material cannot be added during the treatment cycle; during storage, material should be covered with a roof to avoid wetting and allow air circulation.
[2]With safety factor. *Ascaris* ova have remained viable up to 7 yrs in soil (Feachem et al., 1983).
[3]Not recommended. Unlikely to be successful in resource-challenged cities worldwide.

For sludge drying, storage, and burial, facultative ponds offer several advantages over other sludge-producing processes. The sludge only needs to be handled once every 10–15 years or longer and can be buried onsite directly after removal. When sludge removal is required a second time after 10–15 years, the buried sludge can be excavated and reused as a soil conditioner without concern for pathogens, and the fresh dewatered sludge can be disposed of in the same excavation.

All other processes, such as UASBs, sedimentation basins, and anaerobic digesters, require sludge drying beds, which consume the treatment plant area, and the sludge must be handled as a liquid, posing higher risks of contamination and infection of plant personnel. In addition, the liquid sludge must flow by gravity or be pumped to the drying beds on a regular basis, from days to weeks, and typically must be removed from the drying beds in bi-weekly to monthly intervals. Finally, the dried sludge must be taken to an onsite storage area or a site of final disposal. During all of this handling and processing, there will be viable pathogens in the sludge, and workers will be continuously at risk unless they are well trained and wear protective equipment, which is most often not the case. Figure 7.7 shows an example of a well-designed excavation for dewatered sludge disposal at a wastewater treatment plant in Costa Rica.

7.2.2 Case studies: pathogen reduction for wastewater and sludge use in agriculture
The following case studies from various countries present examples of past and recurrent problems, and successes, with wastewater and sludge reuse in agriculture. The use of night soil and fecal sludges is discussed for its relevance to the recurring problems of inadequate pathogen reduction methods, also a common problem in treatment plants, and for which wastewater treatment is supposed to resolve for both effluents and dewatered sludges.

Figure 7.7 An excavation for disposal of dewatered sludge at the Lagos de Lindora wastewater treatment plant, San Jose, Costa Rica. In areas where excreta-related disease prevalence is high, the reuse of sludges in agriculture should be prohibited and sludges should be dried and buried onsite. Long-term onsite storage for reuse of dewatered sludge for a minimum of 2 years for pathogen reduction (Table 7.7) should be included only in designs in areas where excreta-related disease prevalence is low or nonexistent; in most cases, it is not worth the potential risks.

7.2.2.1 Night soil use in China and East Asia: the original valorization of human excreta

The use of night soil, or human excreta, as a fertilizer in agriculture has a long history and is best documented in East Asia, where night soil use in agriculture is mentioned in texts dating back to 300 BCE (McNeill & Winiwarter, 2004). A Chinese agricultural instruction handbook describing night soil application as a fertilizer was published around 550 CE (Kawa *et al.*, 2019). Centuries later, in 1649, toilets that discharged to surface waters in Tokyo, Japan, were banned to maximize the collection of night soil for use in agriculture (McNeill & Winiwarter, 2004).

As recently as the early 1990s in China, approximately 110 million metric tons of night soil were produced every year, by 200 million people in 450 cities (Ling *et al.*, 1993). For the majority of these cities, untreated night soil was transported to peri-urban and rural areas and used as fertilizer in agriculture and also used as food in fish ponds (Ling *et al.*, 1993).

In high prevalence areas, untreated night soil contains significant concentrations of pathogenic bacteria, viruses, protozoan cysts, and helminth ova. The toll on the public health of night soil collectors, farmers, and the general public as a result of night soil use has been significant. Table 7.8 cites a few examples of helminth and viral diseases associated with night soil use in agriculture from China, Korea, and Vietnam. Figures 7.8 and 7.9 show examples of untreated night soil being collected, transported, and applied to crops in a peri-urban area of Chengdu, China, in 1991.

The use of night soil in China has greatly decreased to the present day as a result of urbanization, construction of sewers, development of wastewater treatment discharging to surface waters, and the widespread introduction of synthetic fertilizers (Kawa *et al.*, 2019). While the public health risks are greatly diminished, infections with soil-transmitted helminth (*Ascaris*, Hookworm, *Trichuris*) have not been eliminated as shown in Table 7.9 (see also Figure 7.3). While the prevalence has decreased

Table 7.8 Excreta-related diseases associated with night soil use in China and East Asia.

Excreta-Related Pathogens	Disease	Comments
Helminths		
Ascaris lumbricoides	Ascariasis	• 94% prevalence in vegetable farmers in China in the 1980s. • 24% prevalence in general population in Vietnam attributed to night soil use. • Found as a result paleoparasitological studies in various vegetable gardens in Hansong City, Korea, dating from the 1890s, with source attributed to night soil use.
Hookworm	Ancylostomiasis	• 65% prevalence in vegetable farmers in China in the 1980s. • 2% prevalence in general population in Vietnam attributed to night soil use.
Trichuris trichiura	Trichuriasis	• 93% prevalence in vegetable farmers in China in the 1980s. • 40% prevalence in general population in Vietnam attributed to night soil use. • Found as a result of paleoparasitological studies in various vegetable gardens in Hansong City, Korea, dating from the 1890s, with source attributed to night soil use.
Viruses		
Hepatitis A	Hepatitis A	• Hepatitis A outbreak in 1988 in Shanghai, with 2 million people infected from eating shellfish contaminated with night soil.

Source: Data from Ling *et al.* (1993), Pham-Duc *et al.* (2013), and Kim *et al.* (2014).

significantly, the population infected in 2019, more than 100 million persons, had an estimated 5.0×10^{-5} DALYs/person yr, which is 50 times greater than the WHO recommended value of 10^{-6} DALYs/person yr (WHO, 2006).

While the application of nutrients and organic matter in human excreta to agricultural fields, originally developed in East Asia, has been largely replaced (at least in China) with the use of synthetic fertilizers, developments in latrine technology, such as dehydration and urine diversion toilets, are promoted globally for the use of fecal sludges and urine in agriculture. These newer latrine technologies, together termed ecological sanitation, or ecosan, are supposed to eliminate the many problems associated with pit latrines and pour-flush toilets, such as pathogen and nutrient leaching to groundwater, and the constant need for additional space when pits need to be moved. Importantly, they are also promoted for being able to produce dehydrated sludges acceptable for use in agriculture for their nutrient value, replacing costly inorganic fertilizers in small-scale agriculture (Kumwenda *et al.*, 2017). Key questions to be addressed in the next case study are as follows: How well do ecosan technologies perform in practice for pathogen reduction so that dried fecal sludges can be safely used in agriculture? Should fecal sludge, or wastewater treatment plants sludges, be used in agriculture in areas where the prevalence of excreta-related infections, particularly helminth infections, is high, and where the estimated DALYs/person yr is much higher than the WHO guideline of 10^{-6}/DALYs/person yr?

7.2.2.2 Use of fecal sludge from ecosan toilets in Africa: Burkina Faso and Malawi
Burkina Faso and Malawi are two countries in Africa where the use of ecosan toilets for fecal sludge reuse in agriculture has been promoted in recent years (Kumwenda *et al.*, 2017).

Burkina Faso. Table 7.10 lists the prevalence and estimated DALYs/ person yr for soil-transmitted helminth infections in 2019. The prevalence in Burkina Faso reached a low of 18.6% in 2005 and

Figure 7.8 A night soil collector with cart in a peri-urban area of Chengdu, China, in 1991. The collector stopped at public latrines (bottom left), where fresh, liquid excreta was collected from an open tank on the outside of the latrines (bottom right). No protective equipment was used, and the collector was barefoot. While the use of nightsoil was historically an important resource for rural and peri-urban agriculture, it posed serious health risks for night soil collectors, field workers, and the consumers of field crops. Although in the 1970s–80s much work had been done to educate collectors and farmers and to develop night soil treatment methods, diffusion was still limited in the early 1990s (Ling, 1994), as evident in these photos in a major Chinese city. Historically, it has been suggested that consumers had been at less risk than night soil collectors and field workers as a result of the historical Chinese tradition of boiling water and cooking all food (Anderson, 1988).

has increased to 20% or slightly above ever since (GBD, 2021). The estimated DALYs/ person yr of 4.75×10^{-4} is very high: 475 times greater than the WHO guideline of 10^{-6}.

Figures 7.10 and 7.11 show a dehydration/urine separation toilet promoted in Burkina Faso, which was not operated properly by the very organization promoting it. The disposal of untreated fecal sludge on agricultural land, a common practice in Burkina Faso, is shown in Figure 7.12.

The unsafe management of excreta in Burkina Faso, coupled with the high prevalence of helminth infections and estimated DALYs/ person yr that are 475 times greater than the WHO guidelines,

Figure 7.9 Excreta from night soil carts is discharged into storage tanks built directly in the agricultural fields (top photo). In the bottom photo, field workers take excreta from the storage tanks and apply it directly to the soil, where contact with crops is unavoidable. The photos shown in Figures 7.8 and 7.9 were taken in 1991 in a peri-urban area near Sichuan University of Science and Technolgy in Chengdu, China. Night soil practice no longer exists in Chengdu and is now much rarer in China, perhaps only existing in more remote rural areas.

Table 7.9 Soil transmitted Helminths in China.

Year	Population Infected	Prevalence (%)	DALYs/person yr	WHO Guidelines DALYs/person yr
1990	297 764 740	26.2	7.30E-04	–
2019	114 271 784	8.4	5.00E-05	1.00E-06

Source: Data from the Global Burden of Disease Collaborative Network (GBD, 2021): <https://ghdx.healthdata.org/organizations/global-burden-disease-collaborative-network>.

Table 7.10 Soil transmitted Helminths in Burkina Faso in 2019.

Country	Population Infected	Prevalence (%)	DALYs/ person yr	# DALYs / person yr [1] 10^{-6} DALYs / person yr
Burkina Faso	4 441 699	20.0	4.75E-04	475

Source: Data from the Global Burden of Disease Collaborative Network (GBD, 2021): <https://ghdx.healthdata.org/organizations/global-burden-disease-collaborative-network>.
[1] WHO Guidelines (2006) recommend ≤10−6 DALYs/ person yr for reuse of fecal sludges.

raises questions whether human excreta should be used in agriculture. The results of investigations carries out in Malawi raise further concerns.

Malawi. Table 7.11 lists the prevalence and the estimated DALYs/person yr for soil-transmitted helminth infections for Malawi in 2019. The prevalence has gone down in recent years in Malawi, more likely as a result of deworming campaigns for high-risk groups than improved sanitation. The estimated 2.84×10^{-4} DALYs/person yr, however, is 248 times higher than the recommended WHO value of $\leq 10^{-6}$ DALYs/person yr for agricultural reuse.

Table 7.12 presents data for pathogen concentrations and estimated risks for agricultural reuse of the dehydrated fecal sludge, from 22 dehydration/urine separation toilets in Malawi (Kumwenda *et al.*,

Figure 7.10 An ecosan dehydration toilet with urine diversion operated at a nongovernmental organization education center in Burkina Faso. The design has two feces collection chambers in parallel that are operated as batch reactors: one chamber is in operation, while the second, filled chamber, is closed to use. In both chambers, the fecal sludge dehydrates through natural evaporation, ventilation, and the addition of absorbent materials. To be successful, water and urine should not enter the chambers. The chambers should be designed for a minimum sludge retention time of 1 year in the batch mode (WHO, 2006). Figure 7.11 shows the interior of the toilet.

Figure 7.11 Details of the dehydration/urine diversion toilet in Figure 7.10. Top: The interior of the toilet: the urination opening is flanked on each side by the defecation openings, with one chamber closed in the batch mode. Bottom left: The urine collection container; when full, it is replaced with an empty container. Bottom right: The active defecation chamber was almost full, and the accumulated material was wet, with maggots and foul odor. No absorption materials were being used, with the exception of some office stationery used as toilet paper. The operation of this toilet, by the very NGO promoting it, was inadequate to ensure the fecal sludge would be free of pathogens for reuse in agriculture after 1 year in a closed compartment.

2017). Samples of dehydrated fecal sludge were taken from dehydration compartments that had been sealed for 12 months, the minimum time recommended by WHO for temperatures ranging 20–35°C (Table 7.7). The samples were analyzed for select helminth and bacterial pathogen concentrations. The QMRA was then performed to estimate risks from key exposure pathways of sludge handling, including removal from dehydration compartments, field application, and walking barefoot in fields contaminated with sludge (Kumwenda *et al.*, 2017).

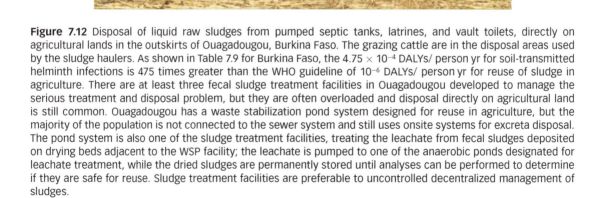

Figure 7.12 Disposal of liquid raw sludges from pumped septic tanks, latrines, and vault toilets, directly on agricultural lands in the outskirts of Ouagadougou, Burkina Faso. The grazing cattle are in the disposal areas used by the sludge haulers. As shown in Table 7.9 for Burkina Faso, the 4.75×10^{-4} DALYs/ person yr for soil-transmitted helminth infections is 475 times greater than the WHO guideline of 10^{-6} DALYs/ person yr for reuse of sludge in agriculture. There are at least three fecal sludge treatment facilities in Ouagadougou developed to manage the serious treatment and disposal problem, but they are often overloaded and disposal directly on agricultural land is still common. Ouagadougou has a waste stabilization pond system designed for reuse in agriculture, but the majority of the population is not connected to the sewer system and still uses onsite systems for excreta disposal. The pond system is also one of the sludge treatment facilities, treating the leachate from fecal sludges deposited on drying beds adjacent to the WSP facility; the leachate is pumped to one of the anaerobic ponds designated for leachate treatment, while the dried sludges are permanently stored until analyses can be performed to determine if they are safe for reuse. Sludge treatment facilities are preferable to uncontrolled decentralized management of sludges.

Table 7.11 Soil transmitted Helminths in Malawi in 2019.

Country	Population Infected	Prevalence (%)	DALYs/ person yr	# DALYs / person yr [1] $\overline{10^{-6} \text{DALYs / person yr}}$
Malawi	1 715 412	9.7	2.48E-04	248

Source: Data from the Global Burden of Disease Collaborative Network (GBD, 2021).
[1]WHO Guidelines (2006) recommend ≤10⁻⁶ DALYs/ person yr for reuse of fecal sludges.

Table 7.12 Concentrations of pathogens in sludges stored for 1 year in dehydration toilet compartments with estimated risks for sludge handlers and farm workers in Malawi.

Pathogen	Mean	Range	Estimated Risk (infections/person-yr)	Estimated risk [1] $\overline{10^{-4} \text{ Infections/person yr}}$
Helminths, viable eggs/g				
Ascaris lumbricoides	0.39	0–2.42	5.6E-01	5600
Hookworm	5.2	0–1727	4.4E-01	4400
Taenia	0.30	0–2.61	1.0E + 00	10 000
Pathogenic bacteria, CFU/g				
E. coli	859	0–5500	5.1E-01	5100
Salmonella	509	0–3200	8.9E-02	890

Source: Data from 22 systems (Kumwenda et al., 2017).
[1]The WHO acceptable risk for use of fecal sludge in agriculture is 10⁻⁴ infections/person yr in developing countries (WHO, 2006).

The data in Table 7.12 show that viable helminth eggs and pathogenic bacteria can survive a 1-year batch cycle, with very high concentrations observed at the high ends of the concentration ranges. No information was given on moisture content and temperature of the dehydrated sludges at sampling, which are important variables for pathogen survival. A plot of minimum, mean, and maximum air temperatures for Blantyre, Malawi, the location of the study, is shown in Figure 7.13. The minimum air temperature is ≤15°C for 6 months of the year, which could have a significant effect on sludge temperatures in dehydration compartments. Thus, the recommended batch cycle of 1 year in a dehydration compartment perhaps should be changed to 1.5–2 years as recommended by WHO and as presented in Table 7.7, and even then, *Ascaris* ova can still survive.

The estimated risks in Table 7.12 calculated from QMRA studies are all much higher than the WHO recommendation of 10⁻⁴ infections/ person yr, ranging from 2 to 3 orders of magnitude higher. In conclusion of their study, Kumwenda *et al.* (2017) recommended the following to reduce the health risks:

- Public health officers need to design effective interventions aimed at reducing the risks that users of ecosan toilets face.
- Promoters of ecosan toilets need to advocate for strict guidelines on sludge use.
- Users of ecosan toilets should properly store sludge; children should not be allowed to play where sludge is kept.
- Workers should use personal protective equipment when emptying dehydration.
- Compartments, and applying sludge to the fields.

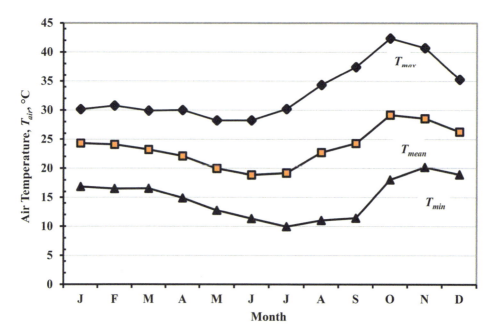

Figure 7.13 Monthly mean, minimum, and maximum air temperatures at 2 m for Blantyre, Malawi, in 2019. The minimum temperatures are insufficient to inactivate all bacterial, viral, protozoan, and helminth pathogens in 1 year in a dehydration toilet batch compartment (see Table 7.7). (*Source*: Data from NASA/LARC (https://power.larc.nasa.gov/data-access-viewer/).)

Commentary. Another way of reducing risks is not relying on ecosan toilets to produce a safe fecal sludge at the individual household level. The problems encountered with decentralized sludge management with onsite dry toilets are the same as sludge management at centralized wastewater treatment plants: process-produced sludges, from facultative ponds, UASB reactors, or ecosan toilets, contain pathogens that can be inactivated by treatment or removed from the environment by storage and/or burial. If the treatment is ineffective, as it usually is, even in wastewater treatment plants with operating personnel (Figures 7.5 and 7.6), storage and burial are the best alternatives. Sludge reuse in agriculture from decentralized systems is not a good alternative unless it is collected and taken to a central treatment facility, dried, stored, and monitored for pathogens.

In urban and peri-urban areas, such as in Ouagadougou, Burkina Faso, fecal sludge treatment facilities do exist, and trucks that pump latrines and septic tanks should discharge the sludges to the city's facilities. The discharge of sludges on agricultural lands (Figure 7.12) occurs in Ouagadougou when the sludge treatment facilities are overloaded and not accepting new loads (GFA, 2018). In rural areas, especially where the prevalence of excreta-related disease is high, it is probably best to use pit latrines if site conditions allow; if not, vault toilets or septic tanks, with vaults large enough to enable infrequent pumping, can be used. Pumped sludges would be taken to a fecal sludge treatment facility, a WSP system designed to accept fecal sludges, or a designated disposal site.

The following case study looks at the WSP system in Ouagadougou, Burkina Faso, and will address wastewater reuse along with the issue of peri-urban sludge management.

7.2.2.3 Wastewater reuse in Ouagadougou, Burkina Faso

The city of Ouagadougou WSP system, built in 2004, was designed for wastewater reuse in agriculture. The system consists of three anaerobic ponds in parallel, two facultative ponds in parallel, and three maturation ponds in series (Figures 7.14 and 7.15). (The anaerobic ponds were not designed for

Wastewater reuse in agriculture 219

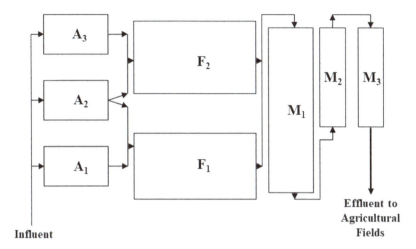

Figure 7.14 The city of Ouagadougou waste stabilization pond system consists of three anaerobic ponds in parallel, two facultative in parallel, and three maturation ponds in series.

Figure 7.15 The Ouagadougou stabilization pond system. Top photo: One of the anaerobic ponds, which have historically been underloaded as a result of a low number of sewer connections. Middle photo: One of the two facultative ponds in operation. Bottom photo: Maturation pond 3, which discharges into the irrigation canal. The anaerobic ponds should be covered for methane capture.

Figure 7.16 Left photo: The main irrigation canal receiving the final effluent. Middle: Smaller canals fed from the main canal conveying effluent to small plots. Right photo: Effluent is taken from the canals in buckets to irrigate crops. (*Source*: Right photo courtesy of Nitiema *et al.* 2013).

methane capture and are not covered; hopefully, this will be changed in the future to avoid methane emissions and have a sustainable energy source.) The system design focused on pathogen removal, with effluent reuse on 35 ha of land adjacent to the treatment plant; the reuse area is divided into small plots of approximately 50 m² each (Kpoda *et al.* 2022). One anaerobic pond also receives leachate from the nearby fecal sludge drying beds.

Figure 7.16 shows the main irrigation canal receiving the treated effluent, which feeds smaller canals passing through the 50 m² plots. Farmers use buckets to take water from the feeder canals to irrigate their plots. The farmers are mostly women with children, many of whom are barefoot, and do not use any protective equipment (Figure 7.17). There are no toilets in the entire 35 ha, and as a result, open defecation in the fields is common. There is also no source of potable water in the fields, and personal hygiene is practiced by washing hands and feet in the irrigation canal (Kpoda *et al.*, 2016).

Tables 7.13 and 7.14 present pathogen monitoring data that have been reported for the WSP system. The fecal coliform data in Table 7.13 show a $\log_{10}$ reduction of 5.33 for the entire system with a final effluent concentration of only 517 CFU/100 mL. This reduction is almost at the level of unrestricted irrigation based on the WHO guidelines. At this time period, however, the system was underloaded, and the actual hydraulic retention time was likely much longer than the design value of 31.5 d (Maiga, 2006). Unfortunately, currently, there are few data available to assess pathogen reduction based on the flowrate and hydraulic retention time.

Figure 7.17 Farmers are mainly women with children, most of whom are barefoot, and none use any protective equipment. There are no toilets onsite, and open defecation in the fields is common, with washing of hands and feet in the irrigation canals for lack of an onsite potable water supply.

Table 7.13 Fecal coliform monitoring of the Ouagadougou waste stabilization ponds.

Raw Wastewater (CFU/100 mL)	Pond Effluent (CFU/100 mL)				
	$(A_2, A_3)^1$	F_2	M_1	M_2	M_3
1.10E + 08	4.31E + 06	6.80E + 05	9.40E + 03	1.10E + 03	5.17E + 02
Log_{10} reduction/pond	1.41	0.80	1.86	0.93	0.33
Total Log_{10} reduction $= 1.41 + 0.80 + 1.86 + 0.93 + 0.33 = 5.33$					

Source: Data from Maiga (2006).
[1] Geometric mean of both pond effluents discharging to F_2; F_1 was not in operation.

The monitoring results of Kpoda et al. (2016) for the Ouagadougou WSP shown in Table 7.14 demonstrate that the system was performing well for helminth and protozoan pathogen reduction at the time of monitoring. Of the seven species of parasitic pathogens, not one viable egg or cyst was detected in the final effluent. The main irrigation canal, however, was contaminated with five of the monitored pathogens, with egg or cyst viabilities ranging from 34.4 to 89.9%, a surprising finding. The authors concluded that the pathogens were a result of open defecation throughout the many small plots in the wastewater reuse agricultural area (Figure 7.16). An epidemiological survey of 20 farmers, all of whom were female, found that 70% were infected with at least one of the pathogens (Kpoda et al., 2016).

Commentary. While pathogen reduction in the Ouagadougou WSP system was excellent and met the WHO guidelines for the few times it had been monitored, treatment alone was not sufficient to protect farmers working in their small plots. One would think that the designers of the system—an excellent example of wastewater reuse benefitting poor farmers in an arid climate—would have considered toilets and a potable water supply for farmers in the fields, but the sole focus only on treatment is, unfortunately, common. Integrated wastewater management includes all participants: treatment and reuse are part of the same water–nutrient cycle. At present, the farmers have abandoned their plots due to problems with effluent water quality, such as high salinity and pH (Y. Maiga, personal communication), possibly caused by industrial discharges into the pond system.

WSPs treating domestic wastewaters can also co-treat fecal sludges from urban and peri-urban areas if designed for it. As mentioned previously, the Ouagadougou WSP system treats leachate from the adjacent fecal sludge drying beds: The leachate is discharged directly into one of the anaerobic ponds designated for the leachate treatment. The fecal sludge drying beds, however, have experienced

Table 7.14 Pathogen detection and viability in the Ouagadougou waste stabilization pond effluent and main irrigation canal.

Pathogen	WSP Final Effluent		Main Irrigation Canal	
	% of Samples Detected	% Viable	% of Samples Detected	% Viability
Helminths				
Ascaris lumbricoides	0	n/a	4.3	48.8
Ascaris duodenalis	0	n/a	26.1	50.0
Taenia spp.	0	n/a	4.3	0
Hymenolepis nana	0	n/a	4.3	89.9
Protozoa				
Entamoeba coli	0	n/a	13.0	34.4
Entamoeba histolytica	0	n/a	4.3	0
Giardia lamblia	0	n/a	17.4	42.9

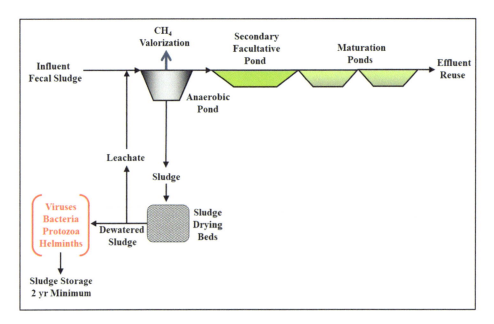

Figure 7.18 An anaerobic pond modified to treat fecal sludges discharged from trucks.

problems with slow infiltration rates of the sludge leachate, which requires longer sludge residence times on the beds; there is also a high cost of capital investment (GFA, 2018). Finally, the dewatered sludge has been stored onsite since the start of operation (2014) awaiting chemical analyses to assess its safety for reuse in agriculture (GFA, 2018).

An alternative option is to expand the Ouagadougou system with anaerobic ponds designed to treat fecal sludge by the direct discharge of liquid sludges into the ponds, a practice that is common in industry worldwide for high-strength wastes and also used for septage (Ingallinella et al., 2002). The ponds would be covered to capture methane, which then could be used for its heat value or to generate electricity. Pond effluents would be discharged to facultative ponds. Well-digested sludges would be pumped to drying beds, with leachates returned to the influent flow. The dried sludges should be stored for at least 2 years to meet the WHO guidelines shown in Table 7.7 for agricultural reuse. Figure 7.18 shows a conceptual design, and Figure 7.19 shows an anaerobic pond treating palm oil wastewater with methane capture and electricity generation at an industrial facility in Guatemala.

7.2.2.4 Fecal coliform mortality on wastewater-irrigated cattle fodder, Aurora II Sanitary Engineering Research Center, Guatemala City

A pilot-scale investigation by Ruano (2005) reported on the natural die-off of fecal coliforms on cattle fodder (*Pennisetum purpureum*) irrigated with primary treated wastewater at the Aurora II Sanitary Engineering Research Center, Regional School of Sanitary Engineering, University of San Carlos, Guatemala. Aurora II has various pilot scale wastewater treatment plants treating wastewater from an adjacent neighborhood. The objective of the study was (i) to determine the optimum time period after the cessation of irrigation to harvest the fodder crop with acceptable reductions of fecal coliforms on crop surfaces as a result of natural die-off and (ii) to optimize the crop yield (Ruano, 2005). The fodder is used to feed cattle raised by the university at Aurora II.

The fodder was irrigated with primary treated wastewater and monitored for 20 days at 5-day intervals after the cessation of irrigation. Gravity-fed sprinkler irrigation was used as a result of steep

Figure 7.19 A covered anaerobic pond treating high-strength palm oil wastewater designed to capture methane for electricity generation using 750 kW gas motors. A city the size of Ouagadougou could, in theory, have a similar facility for fecal sludge treatment at the existing waste stabilization pond site, with the added benefit of electricity generation from renewable energy. Well-digested in-pond sludges would be pumped to sludge drying beds, with the leachate returned to the headworks; dewatered sludges would be stored for a minimum of 2 years to satisfy the WHO guidelines for agricultural reuse. Monitoring would be necessary to ensure pathogen inactivation. (Agrocaribe palm oil facility, Puerto Barrios, Guatemala).

slopes at the irrigation site (Figure 7.20), and field worker safety precautions were used to protect the public from aerosols (see Table 7.2).

Figure 7.21 shows fecal coliform concentrations on plant surfaces as a function of time after irrigation. At 15 and 20 days after irrigation, the foliar concentrations were below the detection limit of 1000 CFU/100 g. Table 7.15 shows the results of fecal coliform reduction as a result of primary sedimentation and natural die-off on plant surfaces. The 0.38 $\log_{10}$ reduction for primary sedimentation, with an effluent of 4.6×10^5 MPN/100 mL, is far below the WHO guideline of 4.0 $\log_{10}$ reduction for restricted irrigation for the protection of field workers; primary-treated wastewater should not be used for irrigation, especially sprinkler irrigation, except in controlled conditions such as this research study.

The >3.79 $\log_{10}$ reduction of fecal coliforms on crop surfaces is significant and approaches a level where crops could be harvested by field workers. The mortality rate constant for fecal coliform

Figure 7.20 Gravity sprinkler irrigation of cattle fodder, *P. purpureum*, with primary treated wastewater on steep slopes at the Aurora II Research Center, Regional School of Sanitary Engineering, University of San Carlos, Guatemala City. Sprinkler irrigation is necessary on steep slopes, and protection measures for aerosols were implemented to protect field workers (see Table 7.2).

reduction can be estimated by assuming a first-order rate reaction as shown in Equation (7.1) (Feachem *et al.*, 1983).

$$C_t = C_0 e^{-kt} \tag{7.1}$$

C_t = concentration of fecal coliforms at time t (CFU/100 g)
C_0 = initial concentration (CFU/100 g)
k = mortality rate constant (d^{-1})
t = time (d)

Table 7.15 Fecal coliform reduction of by sedimentation and natural mortality on irrigated crops.

Sedimentation Basin		Log$_{10}$ reduction
Influent (MPN/100 mL)	Effluent (MPN/100 mL)	
1.1E + 06	4.6E + 05	0.38
Irrigated fodder		
$t = 0$ d	$t = 15$ d	
CFU/100 g	CFU/100 g	
6.20E + 06	<1.00E + 03	>3.79

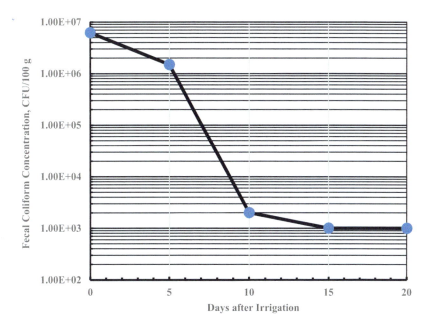

Figure 7.21 Fecal coliform inactivation on the surfaces of *P. purpurem*, a plant commonly used for cattle fodder in Guatemala. Fecal coliform concentrations were measured on plant surfaces at the cessation of irrigation and at 5-day intervals up to 20 days. At 15 and 20 days, the concentrations were below the detection limit of 1000 CFU/100 g. (*Source*: Data from Ruano (2005).)

Converting to $\log_{10}$ and rearranging Equation (7.1) to solve for k gives Equation (7.2):

$$k = \frac{2.3}{t} \log_{10}\left(\frac{C_0}{C_t}\right) \tag{7.2}$$

Using the data from Table 7.15, and allowing that the effluent concentration is less than 1000 CFU/100 g, gives the following value of k:

$$k = \frac{2.3}{t}\log_{10}\left(\frac{C_0}{C_t}\right) = \frac{2.3}{15\,\mathrm{d}}\log_{10}\left(\frac{6.20\mathrm{E}+06\,\mathrm{CFU}/100\,\mathrm{g}}{<1.00\mathrm{E}+03\,\mathrm{CFU}/100\,\mathrm{g}}\right) > 0.58\,\mathrm{d}^{-1}$$

The value of k is within the range of values for pathogen die-off listed in Table 7.4. The mean air temperature for the study was 19.5°C (Ruano, 2005); so this would be the estimated value of k_{20}.

Equations (7.1) and (7.2) do not consider the effect of solar radiation on fecal coliform mortality, which is a key mechanism for inactivation on plant surfaces. It is thus important to record the solar radiation at Aurora II to be able to compare it with other sites for possible design. The solar radiation data for Aurora II are tabulated in Table 7.16. March, April, and May were the months in which the data by Ruano (2005) were collected.

The range of values for Aurora II, 17 700–20 500 kJ/m² d, are higher than those for Belo Horizonte, Brazil, 14 000–20 600 kJ/m² d, discussed in Chapter 6 for *E. coli* removal using the von Sperling equation.

While the study of fecal coliform reduction is important for the protection of field workers harvesting crops, other pathogens, such as the beef tapeworm, Taenia saginata, are of critical concern for the

Table 7.16 Monthly solar radiation at Aurora II, Guatemala City.

Month	Solar Radiation (kJ/m² d)
January	17 700
February	19 200
March	20 500
April	20 200
May	19 200
June	17 000
July	18 400
August	18 900
September	16 800
October	16 500
November	16 600
December	16 800

Source: Data from CLIMWAT. Elevation = 1502 m.

continued transmission through the human excreta–beef tapeworm–consumption of meat infection cycle. Taenia species eggs have been reported to survive from 30 to 60 days on crops, and T. saginata eggs in the soil may remain infective for 5.5–9.5 years (Gonzalez & Thomas, 2018; WHO, 2006). For this reason, helminth eggs should be removed in wastewater treatment; and primary sedimentation alone is insufficient.

Figure 7.22 shows the cattle at Aurora II being fed the *P. purpurem* stalks without leaves as an added barrier of protection from the multiple barrier concept (Figure 7.2). It is assumed the majority of pathogens would be on leaf surfaces as a result of sprinkler irrigation.

Figure 7.21 presents the reduction in the crop yield as a function of time after irrigation. The highest yield, 360 metric tons/ha-yr, occurred at the time irrigation was stopped. At the time of

Figure 7.22 Cattle at Aurora II are fed the stalks of *P. purpurem* without leaves, which are more easily contaminated with sprinkler irrigation, to add an additional barrier against pathogen transmission.

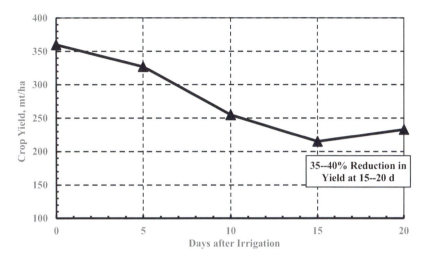

Figure 7.23 Reduction in crop yield (metric tons/ha) versus the number days after irrigation.

maximum fecal coliform reduction at 15 days, the reduction of crop yield was 35–40%, a reduction that would have to be accepted to enable safe harvesting and handling of the fodder. The alternative is to add pathogen reduction treatment processes beyond primary sedimentation. Aurora II also has a facultative-maturation WSP system that could be used and would produce an improved effluent for irrigation that would include helminth egg reduction (Figures 7.23 and 7.24).

Figure 7.24 The waste stabilization pond system at Aurora II consists of a facultative pond (left) followed by a maturation pond (right). The system would produce a high-quality effluent in terms of pathogen reduction for irrigation of the cattle fodder.

7.2.2.5 Wastewater reuse in the Municipality of Sololá, Lake Atitlán Basin, Guatemala

The municipality of Sololá, Guatemala, has two small wastewater treatment plants, built in 1995 and 1998, which were designed and constructed with the goals of reusing the wastewater effluent in agriculture and producing methane for use as a cooking fuel for the neighboring population. These two plants are still in operation and well maintained to the present day (November 2021). The goal of the project was best stated by Sánchez de León (2001):

> 'The project directly benefits more than 100 farmers who are given the opportunity to irrigate their crops with wastewater. This results in a savings in fertilizer costs, and as a result a decrease in the costs of production. On the other hand, all of the treated wastewater is not discharged to surface Cwaters and ditches, and consequently, does not enter Lake Atitlán, providing important protection for the environment.' (Sánchez de León, p 8, 2001) (Translated from Spanish)

The two plants at Sololá are identical designs with the configurations shown in Figure 7.25.

The design, construction, operation and maintenance, reuse of effluent in agriculture, and capture of methane for cooking of these two systems, operating for more than 20 years in a very poor region of the world, are among the best likely to be found anywhere. The treatment plants were never designed for pathogen removal, however, and neither have disinfection unit processes. As a result, serious health risks exist for farmers and their families, nearby communities, and consumers.

Table 7.17 lists the design data for population, flowrate, and maximum irrigated area for the two systems, along with the most commonly irrigated crops. The Lake Atitlán regulatory authority prohibits the irrigation of crops eaten raw with treated wastewater, but this regulation has never been enforced. Onions, carrots, and tomatoes are commonly eaten raw in Guatemala and may also be exported. (As examples of risks that have been documented from exported crops, Table 7.2 cites the outbreaks in the US of (i) cyclosporiasis linked to raspberries and snow peas imported from Guatemala and (ii) hepatitis A linked to green onions imported from Mexico.)

Figure 7.26 shows the San Antonio irrigation area and the health risks the farmers and their children face with unsafe irrigation practices, washing and handling of crops.

There are very few data on the microbiological quality of the two plants' effluents, and Table 7.18 presents the data reported by Sánchez de León (2001) for *E. coli*. The $\log_{10}$ reductions are very low, as would be expected, and far below the WHO guideline of 4.0 $\log_{10}$ reduction for labor-intensive restricted irrigation; the final effluent concentrations $>10^7$ MPN/100 mL are 4.0 $\log_{10}$ greater than the WHO guideline of 10^3 *E. coli*/100 mL for restricted irrigation (Table 7.5). There is no doubt that protozoan (oo)cysts and helminth eggs would pass through both treatment plants as has been reported in the literature for UASBs and trickling filters (Oakley & Mihelcic, 2019).

The prevalence of excreted-related parasite infections in Guatemala is high in both rural and urban areas as shown in Table 7.19. Contaminated crops eaten uncooked are surely a contributing factor. Protozoan (oo)cysts, which are infective upon excretion and can be transmitted through various pathways, have been found on wastewater-irrigated vegetable surfaces (WHO, 2006), and six outbreaks of cyclosporiasis in the US that have been linked to raspberries and snow peas imported from Guatemala suggest the wastewater irrigation pathway (Table 7.2). Helminth eggs must complete their life cycles, in the soil or on crops irrigated with poorly treated wastewater, to reach the infective stage; thus, helminth infections in urban areas are likely caused by contaminated produce that is eaten uncooked. Various studies have reported contamination of crops with both protozoan and helminth pathogens as a result of irrigation with inadequately treated wastewater (Amahmid *et al.*, 1999).

Disinfection with chlorine is often considered for improving the microbiological quality of effluents for reuse, or discharge to surface waters, in situations similar to Sololá. Chlorine is the disinfectant of choice because of the widespread availability and low cost. Helminth eggs and protozoan (oo)cysts, especially *Cryptosporidium*, however, are resistant to chlorine disinfection. Worse still, there are normally high concentrations of ammonium ions in treated wastewater effluents that react with chlorine to form chloramines, which are 100–400 times less effective than free available chlorine as a disinfectant in wastewater for *E. coli* reduction (Metcalf & Eddy/AECOM, 2014). Disinfection

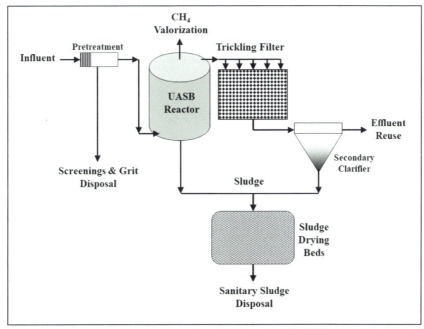

Figure 7.25 The San Antonio and San Bartolo (pictured) wastewater treatment plants consist of pretreatment followed by two UASBs in parallel, two trickling filters in parallel, and a secondary sedimentation basin; the final effluent is both used in agriculture and discharged to surface waters. The San Bartolo plant captures methane and distributes it to nearby dwellings for cooking. Sludge drying beds are used for the digested UASB sludges, with the dried sludges offered to farmers as a fertilizer/soil conditioner, a practice that should be prohibited (Figure 7.6). The final effluent used for irrigation is not disinfected and poses serious health risks.

Table 7.17 Reuse of wastewater in agriculture in the municipality of Sololá, Guatemala.

Wastewater Treatment Plant Effluent	Start-Up Year	Design Population	Q_{Design} (m³/d)	Maximum Irrigated Area (ha)	Irrigated Crops	Farmers Using Effluent
San Bartolo	1995	12000	2074	24	Onion, potato, carrots, beets, yuca, beans, tomato	>100
San Antonio	1998	7000	907	1.84	Onion, potato, beans, tomato	

Source: Data from Sánchez de León (2001).

with UV or ozone would work, but the complexities of operation and maintenance, higher costs, and high level of wastewater treatment required make them infeasible disinfectants for municipalities like Sololá.

Given the current, long-lasting situation, the following question must be asked: Is wastewater reuse in agriculture a good idea in a peri-urban area such as Sololá, where there is no infrastructure to implement the necessary public health measures for the safe use of wastewater in agriculture as developed by the World Health Organization (WHO, 2006)?

Figure 7.26 Top photo: Irrigated areas of the San Antonio wastewater treatment plant, which is on the left side. Irrigation is by gravity sprinklers, with onion fields on the right side. Left bottom photo: Farmers with their children wash freshly picked onions with plant effluent. Middle and right bottom photos: Onions are dried, bagged and trucked to local and countrywide markets and restaurants. Health risks are high for farmers and their families, local communities, and consumers. This situation has existed for both the San Antonio and San Bartolo reuse projects since the mid-1990s.

Table 7.18 *E. coli* reduction in the San Bartolo/San Antonio wastewater treatment plants.

Wastewater Treatment Plant	E. coli (MPN/100 mL) Influent	E. coli (MPN/100 mL) Effluent	Log$_{10}$ reduction
San Bartolo	1.60E + 07	1.20E + 07	0.12
San Antonio	6.10E + 08	2.20E + 07	1.44

Source: Data from Sánchez de León (2001).

Table 7.19 Prevalence of excreta-related parasite infections in Guatemala.

Parasite	Prevalence in Select Communities (%) Rural	Prevalence in Select Communities (%) Urban
Protozoa		
Entamoeba histolytica	22	21
Giardia duodenalis	30	15
Cryptosporidium parvum		32
Cyclospora cayetanensis	2.3–6.7[1]	
Helminths		
Ascaris lumbricoides	67	49
Trichuris trichiura	20	14
Hookworm	39.5[1]	

Source: Data from Bern *et al.* (1999), Jensen *et al.* (2009), Laubach *et al.* (2004), and Oakley (2005).
[1]Reported data were for both rural and urban communities.

Commentary. The continued use over 20 years of inadequately treated wastewater for agricultural reuse, with farmers and their families, neighboring communities, and countrywide consumers all at risk, with a high prevalence of excreta-related infections countrywide, and with various outbreaks of cyclosporiasis in the US linked to produce imported from Guatemala, is, unfortunately, not an uncommon occurrence in many parts of the world. While the problem is a complex mix of technical and social issues, involving many actors and organizations, it is the responsibility of design engineers and public health professionals in cases such as this, who caused the problem by allowing reuse of unsafe wastewater, to make sound engineering and public health recommendations to protect public health. Improved practices implementing the multiple barrier approach to meet the WHO guidelines for restricted irrigation at the San Antonio and San Bartolo sites should have included the following measures:

(1) Public health requirements dictate that wastewater reuse be clearly designated for restricted irrigation, with irrigation fields fenced with clear warning signs.
(2) Field workers need training in health risks, use of protective equipment in the fields, personal hygiene upon leaving the fields, and washing of harvested crops with disinfectant solutions for bacterial/viral pathogens, and soap for helminth eggs (WHO, 2006). Protective equipment should be included as part of the operation and maintenance budget of the treatment plant.
(3) Waiting periods to allow natural die-off of pathogens before farmers enter the fields are necessary. The case study at Aurora II in Guatemala City is an excellent example of waiting period determination for crop harvesting. Table 7.20 presents the monthly solar radiation data at Sololá, where the monthly radiation values are all higher than those at Aurora II (Table 7.16) with the exception of May, which is very close (18 900 versus 19 200 kJ/m² d). Assuming

Table 7.20 Monthly solar radiation at Sololá.

Month	Solar Radiation (kJ/m² d)
January	18 972
February	20 484
March	21 204
April	20 556
May	18 900
June	18 252
July	20 052
August	19 620
September	17 568
October	17 172
November	18 108
December	18 216

Source: Data from NASA POWER. Elevation = 2100 m.

similar conditions of temperature apply, the waiting period for a 4.0 log$_{10}$ reduction is roughly estimated as follows:

$$t_{4.0\,\log_{10},\text{Reduction}} = \frac{2.3}{k}\log_{10}\left(\frac{C_0}{C_t}\right) = \left(\frac{2.3}{0.58\,\text{d}^{-1}}\right)\log_{10}\left(\frac{1.0\,\text{E}+07\,\text{CFU}/100\,\text{mL}}{1.0\,\text{E}+03\,\text{CFU}/100\,\text{mL}}\right) = 15.8\,\text{d}$$

There is a university laboratory in the city of Sololá, and field studies similar to those done at Aurora II could, if resources were made available, be performed to determine accurate *E. coli* mortality rate constants and waiting periods for different crops and seasons.

(4) Sprinkler irrigation, with increased risks from aerosols, should be changed to gravity flow border irrigation using the sloped, terraced fields at both San Antonio and San Bartolo, shown in Figure 7.27. The rocks on the terraces' borders could easily be sealed with clay soil.

(5) Helminth egg exposure to field workers and their children and contamination of crop surfaces are very serious problems needing abatement. The two wastewater treatment plants do not

Figure 7.27 The sloped terraced fields irrigated with sprinklers at the San Antonio (left) and San Bartolo sites could easily be converted to gravity flow border irrigation: the terraces are gently sloped and have rock borders able to be sealed with a clay soil. Postirrigation waiting periods would allow natural bacterial/viral pathogen die-off and contribute to the protection of field workers and their families.

remove helminth eggs, and the verification monitoring or the treatment technology options in Table 7.5 cannot be used. Although the irrigation of crops eaten raw has been prohibited at both sites for 20 years (Sánchez de León, 2001), the regulation has never been enforced. Onions are a major crop and source of income for farmers in Sololá, which is unlikely to change. Onions are a root crop, however, that can have helminth eggs on plant surfaces from both soil and irrigation water. Worse still, the practice at San Antonio after harvesting onions is to wash them with the wastewater treatment plant effluent (Figure 7.26). (An example of the risks from pathogens: Green onions imported from Mexico have been linked to the largest outbreak of hepatitis A in the US history from imported produce and were likely irrigated or washed with wastewater, as mentioned in Table 7.5) The following measures could at least improve the current situation:

- Wastewater effluent should never be used to wash produce. A potable water source needs to be provided in the vicinity of the harvest and washing areas.
- A 1–2 $\log_{10}$ reduction of helminth eggs on crop surfaces can be obtained by washing in a weak detergent solution and rinsing thoroughly with potable drinking water (WHO, 2006). Farmers would have to be trained, and this could be difficult since in many cultures the use of detergents on crops eaten raw would not be accepted (WHO, 2006).
- All packaged produce should be labeled that it has been irrigated with wastewater and should be washed with detergents and disinfected if eaten raw.

(6) Irrigation at Sololá is only possible for 6 months of the year, and wastewater effluents are discharged to surface waters during this period. If reservoirs were well designed and constructed, increased pathogen reduction and 100% helminth egg removal could be obtained. The San Antonio treatment plant has a small reservoir that could be used for a pilot project by filling, storing the water during the 6-month rainy season, and monitoring for *E. coli* and helminth egg reduction at monthly intervals. It is unlikely that reservoirs could be constructed, however, because of the volume needed to store 6 months of flow, coupled with the steep terrain throughout the irrigation area.

What is the likelihood that any of the aforementioned measures could be implemented anytime soon? The past 20 years of operation are not conducive to optimism.

7.2.2.6 Pathogen reduction for wastewater reuse in agriculture at the Campo Espejo waste stabilization pond system, Mendoza, Argentina

The Campo Espejo WSP system was built in 1976 and upgraded in 1996 specifically to meet the 1989 WHO guidelines for wastewater use in agriculture (Category A in Table 7.1). The pond system consists of 12 modules of one facultative pond followed by two maturation ponds in series (Figure 7.28). The final effluent is discharged to a distribution canal and then conveyed to a 3000 ha controlled irrigation area called an ACRE (Áreas de Cultivos Restringidos Especiales or Special Restricted Crop Area). Excess irrigation and drainage waters are discharged downstream to a collection canal, mixed with river water, and used for irrigation of an additional 7000 ha (Bartone, 2012).

The institutions responsible for wastewater treatment and reuse for the city of Mendoza and the surrounding agricultural area are as follows:

- The provincial water and sanitation agency, Ente Provincial del Agua y Saneamiento (EPAS), regulates the water and sanitation services in the Province of Mendoza. The EPAS Resolution 35/96 established standards for irrigation in ACREs, including effluent requirements for helminths and fecal indicator bacteria from the 1989 WHO guidelines.
- The metropolitan water and sewerage company, Obras Sanitarias de Mendoza (OSM), is in charge of operation and maintenance of the Campo Espejo system.
- The Departamento General de Irrigación (DGI) is responsible for the management of water resources and the Campo Espejo ACRE, including monitoring the microbiological quality of irrigation water, irrigated produce, and the health of agricultural workers.

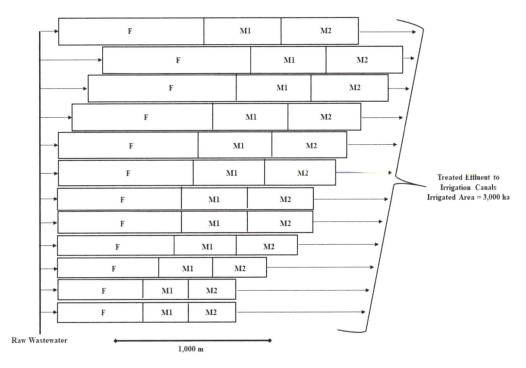

Figure 7.28 The Campo Espejo waste stabilization pond system comprised 12 modules of one facultative pond followed by two maturation ponds in series. The total water surface area is 259 ha, with a total hydraulic retention time of 31.5 d at the design flowrate of 129 600 m3/d (Barbeito Anzorena, 2001). The system was built in 1976 and upgraded in 1996 specifically to meet the 1989 WHO guidelines for wastewater use in agriculture (Table 7.1). Approximately 3000 ha are presently irrigated with the effluent, which also meets the newer 2006 WHO guidelines for restricted and unrestricted irrigation.

Table 7.21 lists the permitted activities and crops that can be grown on the Campo Espejo ACRE as developed by the DGI. Crops grown on the ACRE include alfalfa, artichokes, garlic, grapes, peaches, pears, squash, tomatoes, and poplar trees (Bartone, 2012). Figures 7.29 and 7.30 present a few examples of the Campo Espejo pond system and ACRE in operation.

Table 7.21 Resolution No. 400/03: wastewater reuse with secondary effluent on ACREs[1,2]

(1) Sprinkler irrigation is prohibited.
(2) Permitted crops and procedures:
• Pasture and green fodder crops fed to animals do not need special handling.
• Edible portions of plants eaten uncooked must not come into contact with treated wastewater.
• Crops consumed cooked that may have contact with wastewater should be harvested 30 days after irrigation.
• Crops for human consumption that have a rind or husk that prevent treated wastewater from contacting the edible portion can be in contact with wastewater.

[1] Departamento General de Irrigación, Mendoza, Argentina (https://www.irrigacion.gov.ar/web/).
[2] Secondary effluent in this case is waste stabilization pond effluent in the Mendoza area.

Wastewater reuse in agriculture 235

Figure 7.29 Top photo: One of the 12 secondary maturation ponds in parallel at the Campo Espejo waste stabilization pond system. Based on the design in the mid-1990s to meet the 1989 WHO guidelines for agricultural reuse, each primary and secondary maturation pond has a hydraulic retention time of 6.15 days at the design flowrate of 129 600 m^3/d or 10 800 m^3/d per pond. Bottom photo: One of the several final effluent discharge canals that feed the network of smaller irrigation canals for the entire 3000 ha Campo Espejo ACRE, all by gravity flow. The effluent also meets the 2006 WHO guidelines for restricted irrigation: For crops eaten cooked that have contact with wastewater, a 30-day waiting period after cessation of irrigation is required before harvest to allow pathogen die-off from natural mortality; for crops eaten uncooked, the contact with treated wastewater is strictly prohibited (e.g., fruit trees).

Table 7.22 presents several of the original design parameters relevant to pathogen reduction for the Campo Espejo system. Barbeito Anzorena (2001) presented the performance estimates in Table 7.23 for a flowrate of 155 520 m^3/d, which is 120% of the design flowrate. How the design values shown in Table 7.23 for fecal bacteria mortality rate constants, and dispersion numbers, were developed, unfortunately, was not explained. An estimate of helminth egg and fecal coliform/*E. coli* reduction using the data from Table 7.23 is presented as follows.

Figure 7.30 Top photo: A farmer diverts irrigation water from a Campo Espejo ACRE effluent feeder canal to his furrow-irrigated fields, which contain onions, squash, and peppers. Bottom photos: Only the onion seeds, which do not have contact with the treated wastewater, are harvested. The regulations governing the ACRE prohibit sprinkler irrigation, and the most commonly used method is furrow irrigation. Edible portions of plants eaten uncooked cannot come into contact with treated wastewater, which eliminates rooted onions. If treated wastewater comes into contact with crops eaten cooked, they must not be harvested for 30 days after the last irrigation to allow natural die-off.

Table 7.22 Select design parameters for the Campo Espejo WSP system.

Mean flowrate, m³/d	129 600
Water temperature, °C	10
Influent BOD₅, mg/L	160
Influent fecal coliforms, MPN/100 mL	1.07E + 07
Pond evaporation/infiltration, cm/d	0.52

Source: From Barbeito Anzorena (2001).

Helminth egg reduction. The data in Table 7.23 can be used to estimate 100% helminth egg reduction for each pond based on the overflow rate and the total system using Equation (5.28). The reduction in the facultative pond would be:

$$\text{OFR} = \frac{Q_m}{A_f} \leq 0.12\,\text{m}^3/\text{m}^2\,\text{d} \tag{5.28}$$

Q_m = mean flowrate (m³/d = 155 520 m³/d)
A_f = area of facultative pond (m² = 1 294 200 m²)

$$\text{OFR} = \frac{Q_m}{A_f} = \frac{155\,520\,\text{m}^3/\text{d}}{1\,294\,200\,\text{m}^2} = 0.120\,\text{m}^3/\text{m}^2\,\text{d} \leq 0.12\,\text{m}^3/\text{m}^2\,\text{d}$$

The overflow rates for all ponds and the total of ponds are tabulated in Table 7.24.

Table 7.23 Performance estimates at flowrate = 155 520 m³/d (120% of design value).

Parameter	Facultative	Maturation 1	Maturation 2
Total area (12 ponds in parallel), m²	1 294 200	646 900	646 900
Depth m	2.17	2.17	2.07
Total volume (12 ponds in parallel), m³	2 043 533	1 021 766	1 021 766
Hydraulic retention time, d	13.14	6.57	6.57
Bacteria mortality rate constant, k_b, d⁻¹	0.3683	0.4297	0.6080
Dispersion number, d	0.0550	0.0986	0.1095
Effluent fecal coliform concentration (range of 12 ponds in parallel), MPN/100 mL	7.63E + 04– 9.47E + 04	3.48E + 03–4.98E + 03	1.59E + 02–2.62E + 02

Source: From Barbeito Anzorena (2001).

Table 7.24 Overflow rates for the Campo Espejo WSP system for Q_{mean} = 155 520 m³/d.

Pond	Area (m²)	OFR (m³/m² d)
F	1 294 200	0.120
M1	646 900	0.240
M2	646 900	0.240
Total	2 588 000	0.060

While the facultative pond exactly meets the maximum value for OFR, the value for the total pond area is well below the maximum value. Equation (5.30) can be used to estimate the 95% LCL $\log_{10}$ reduction for the facultative ponds:

$$\log_{10\ 95\%\ \text{LCL}} = -\log_{10}[1 - (1 - 0.41\exp(-0.49 t_{V/Q} + 0.0085(t_{V/Q})^2))] \tag{5.30}$$

where $t_{V/Q} = 13.14\,\text{d}$.

$$\log_{10\ 95\%\ \text{LCL}} = -\log_{10}[1 - (1 - 0.41\exp(-0.49(13.1) + 0.0085(13.13)^2))] = 2.55$$

The values of $\log_{10\ 95\%\ \text{LCL}}$ for each pond and the total system are shown in Table 7.25. The results suggest that 100% helminth egg removal should not be a problem even at 120% of the design flowrate.

The Wehner–Wilhem equation for dispersed flow is used to calculate the $\log_{10}$ reduction of fecal coliforms using the mortality rate constants and dispersion numbers from Table 7.23. For the facultative pond:

$k_{b,T} = 0.3683\,\text{d}^{-1}$

$t_{V/Q} = 13.14\,\text{d}$

$d = 0.055$

(1) Determine a in the Wehner–Wilhem equation (Equation (5.32)):

$$a = \sqrt{1 + 4 k_{b,T} \cdot t_{V/Q} \cdot d} = \sqrt{1 + 4(0.3683\,\text{d}^{-1})(13.14\,\text{d})(0.055)} = 1.4369$$

(2) Determine the effluent fecal coliforms/*E. coli* concentration using Equation (5.33):

$$N_{\text{effluent}} = N_{\text{influent}} \frac{4a \cdot e^{1/2d}}{((1+a)^2 e^{a/2d} - (1-a)^2 e^{-a/2d})} =$$

$$= (1.07\,\text{E}+07) \left(\frac{4(1.4369) \cdot e^{1/2(0.055)}}{((1+1.4369)^2 e^{1.4369/2(0.055)} - (1-1.4369)^2 e^{-1.4369/2(0.055)})} \right)$$

$$= 1.95\,\text{E}+05\,\text{MPN}/100\,\text{mL}$$

(3) Determine the $\log_{10}$ reduction of fecal coliforms/*E. coli*:

$$\log_{10,\text{Reduction}} = \log_{10}(1.07\,\text{E}+07) - \log_{10}(1.95\,\text{E}+05) = 1.74$$

The calculated values for effluent concentration from each pond and the $\log_{10}$ reductions are tabulated in Table 7.26. The values are higher than those in Table 7.23 but still meet the WHO guideline of a 4.0 $\log_{10}$ reduction for *E. coli*.

Table 7.25 The 95% lower confidence limit for Helminth egg reduction.

Ponds	$t_{V/Q}$ (d)	$\text{Log}_{10,95\%,\text{LCL}}$
F	13.14	2.55
M1	4.9	1.34
M2	5.3	1.41
Total	23.3	5.3

Table 7.26 Estimated $\log_{10}$ reduction of fecal indicator bacteria using design values of $k_{b,T}$ and d in the Wehner–Wilhem equation.

Pond	$t_{V/Q}$ (d)	$k_{b,10\ °C}$	d	a	$N_{influent}$	$N_{effluent}$	$\log_{10}$ Reduction
F	13.14	0.3683	0.055	1.4369	1.07E + 07	1.95E + 05	1.74
M1	6.57	0.4297	0.099	1.4538	1.95E + 05	1.89E + 04	1.01
M2	6.57	0.6080	0.110	1.6582	1.89E + 04	8.77E + 02	1.33
Total	26.28						4.09

Daily monitoring of the Campo Espejo pond system for 1 year in 2000 showed that it was performing better than the design calculation prediction. Table 7.27 presents a summary of the data for helminth egg and *E. coli* reduction. Helminth eggs were never detected in the final effluent, and the *E. coli* geometric mean effluent concentration was 74.8 MPN/100 mL, giving a 5.17 $\log_{10}$ reduction. There are few WSP systems that have consistently had a performance such as this one. Unfortunately, the mean flowrate was not reported, but it was surely no greater than the design flowrate of 129 600 m³/d.

Table 7.28 presents results for *E. coli* reduction for the design flowrate of 129 600 m³/d. The hydraulic retention times were calculated from the volumes of the ponds. For the facultative pond, for example,

$$t_{V/Q,F} = \frac{V_F}{Q_{mean}} = \frac{2\,043\,533\,\text{m}^3}{129\,600\,\text{m}^3/\text{d}} = 15.77\,\text{d}$$

The overall $\log_{10}$ reduction is close to the measured values in Table 7.29. A recent presentation by an engineer from the DGI in Mendoza stated that the current flowrate for Campo Espejo was approximately 128 000 m³/d, which is close to the design value of 129 600 m³/d (Rauek, 2020).

Commentary. Pathogen reduction at the Campo Espejo pond system has consistently met the verification monitoring requirements of ≤1 helminth egg/L and ≤10³ *E. coli*/100 mL (Table 7.5), which meet the WHO guideline for restricted irrigation; the unrestricted irrigation guidelines could

Table 7.27 Results of daily monitoring of Helminth Eggs and *E. coli*, Campo Espejo WSP system, January 1–December 31, 2000.

Parameter	Influent	Effluent	$\log_{10}$ Reduction
Helminth eggs, eggs/L	26	<1	>1.41
E. coli, MPN/100 ml	1.01E + 07	7.48E + 01	5.17

Source: Data from Barbeito Anzorena (2001). Although not stated in the report, it is assumed the helminth egg data are represented in arithmetic means and the *E. coli* data in geometric means.

Table 7.28 Estimated $\log_{10}$ reduction of fecal indicator bacteria using design values of $k_{b,T}$ and d in the Wehner–Wilhem equation for the design flowrate = 129 600 m³/d.

Pond	$t_{V/Q}$ (d)	$k_{b,10\ °C}$	d	a	$N_{influent}$	$N_{effluent}$	$\log_{10}$ Reduction
F	15.77	0.3683	0.055	1.5092	1.07E + 07	1.00E + 05	2.03
M1	7.88	0.4297	0.099	1.5284	1.00E + 05	6.57E + 03	1.18
M2	7.88	0.6080	0.110	1.7606	6.57E + 03	1.88E + 02	1.54
Total	31.54						4.75

Table 7.29 Mean air temperature and solar radiation for Mendoza and Aurora II.

Month	Mendoza Mean Air Temperature (°C)	Mendoza Solar Radiation (kJ/m² d)	Aurora II Guatemala City Mean Air Temperature (°C)	Aurora II Guatemala City Solar Radiation (kJ/m² d)
January	25.2	23 300		17 700
February	23.8	23 100		19 200
March	21.2	17 400	19.5	20 500
April	16.4	14 400		20 200
May	12.1	10 800		19 200
June	8.8	9 000		17 000
July	7.9	10 000		18 400
August	10.5	13 200		18 900
September	13.5	16 700		16 800
October	17.7	21 100		16 500
November	21.4	25 300		16 600
December	24.3	23 200		16 800

Source: Data from CLIMWAT and NASA Power.

be met if additional pathogen reduction measures from Table 7.4, such as pathogen die-off and produce washing, were implemented. The DGI regulations, however, do not permit unrestricted irrigation; resolution No. 400/03 in Table 7.21 is actually stricter that the WHO guidelines by requiring a 30-day waiting period before harvest after irrigation of crops eaten cooked.

An estimate for pathogen reduction by natural die-off at the Campo Espejo ACRE during the 30-day waiting period can be estimated by comparing the solar insolation and air temperature with the data from Aurora II in Guatemala, shown in Table 7.29. For the months November–February, during the peak of the irrigation season in Mendoza, the mean air temperatures and solar radiation data are all higher than those reported at Aurora II. Equation (7.2) can be used to estimate the $\log_{10}$ reduction at Mendoza assuming the value of k is the same for the climate and crop conditions.

Equation (7.2) is used to estimate $\log_{10}$ reduction:

$$k = \frac{2.3}{t} \log_{10}\left(\frac{C_0}{C_t}\right) \tag{7.2}$$

Rearranging Equation (7.2), and inserting $k = 0.58$ d^{-1} and $t = 30$ d, gives the following $\log_{10}$ reduction:

$$\log_{10} \text{Reduction} = \log_{10}\left(\frac{C_0}{C_t}\right) = \frac{k \cdot t}{2.3} = \frac{(0.58\,\text{d}^{-1})(30\,\text{d})}{2.3} = 7.56$$

The 7.56 $\log_{10}$ reduction is double the 3.79 $\log_{10}$ reduction at Aurora II in 15 d, as would be expected. The actual reduction would depend also on the plant foliage and its exposure to sunlight. Nevertheless, the 30-d waiting period gives a substantial safety factor for field worker protection and is laudable given the numerous problems existing elsewhere, such as presented in the case studies of Ouagadougou and Sololá.

The Campo Espejo WSP and agricultural reuse (ACRE) system, which has been in operation for over 25 years, is one of the best examples of integrated wastewater management prioritizing reuse in agriculture, anywhere. Its success is due to various environmental and historical circumstances (Foresi, 2016):

- Agriculture in the arid climate of Mendoza Province depends almost exclusively on irrigation, and the demand for irrigation water represents more than 80% of all water demand.
- Farmers began using wastewater for irrigation over 50 years ago as they learned its value not only for the water but also for the nutrients and organic matter it contains.
- The metropolitan water and sewerage company, OSM, has had a 20-year contract for operation and maintenance of the Campo Espejo system, with the requirement of continually meeting the effluent standards of ≤ 1 helminth egg/L and $\leq 1 \ 10^3$ *E. coli*/100 mL.
- The DGI, which was founded 1916, has long focused on the importance of using treated wastewater as an important water resource for irrigation and has played the key role in developing public health guidelines for the protection of farmers and the public, managing the Campo Espejo ACRE and monitoring the microbiological quality of irrigation water, irrigated produce, and the health of agricultural workers.

7.3 PHYSICAL–CHEMICAL WATER QUALITY FOR IRRIGATION WITH WASTEWATER

7.3.1 Physical–chemical guidelines for interpretation of water quality for irrigation

Wastewater reuse in agriculture requires, in addition to pathogen reduction concerns, an assessment of the physicochemical water quality to ensure it is adequate for crop irrigation with limited potential problems. Normally treated domestic wastewater, as long as the raw water domestic water source is from surface water, and that there are no industrial wastes entering the wastewater treatment plant, should not pose any serious problems. Table 7.30 presents the physical–chemical water quality guidelines for potential irrigation problems that will be discussed later: salinity, infiltration, specific ion toxicity, and miscellaneous effects from specific parameters.

7.3.1.1 Salinity

Salinity is a quantitative measure of the total dissolved solids (TDS) in water or soil Metcalf and Eddy/AECOM (2006). Salinity in soil or water can reduce water availability to crops to an extent where yields are affected (Ayers & Westcot, 1985). The salinity of water is typically measured as electrical conductivity, EC_W, in units of decisiemens per meter (dS/m), which can be measured rapidly in the field. The relationship between EC_W and TDS is expressed by Equation (7.3):

$$\frac{\text{For } EC_w < 5.0 \, \text{dS/m} : \text{TDS} = 640 \, EC_w}{\text{For } EC_w > 5.0 \, \text{dS/m} : \text{TDS} = 800 \, EC_w} \tag{7.3}$$

As salinity increases in irrigation water, the osmotic gradient between soil water and root cells decreases; as a result, plants must use more energy to concentrate solutions in root cells by taking up water from the soil. This results in plant growth reduction similar in appearance to that of drought conditions (Metcalf & Eddy/AECOM, 2006). Leaching is the key control for salinity problems when wastewater is used for irrigation. Leaching of applied wastewater is defined by the following equations (Ayers & Westcot, 1985):

$$\text{Leaching fraction (LF)} = \frac{\text{Electrical conductivity of irrigation water}}{\text{Electrical conductivity of drainage water}} \tag{7.4}$$

$$LF = \frac{EC_W}{EC_{DW}} \tag{7.5}$$

LF = leaching fraction expressed as a decimal, ranges from 0.10 to 0.40
EC_W = electrical conductivity of irrigation water (dS/m)
EC_{DW} = electrical conductivity of drainage water (dS/m)

Table 7.30 Physical–chemical guidelines for interpretation of water quality for irrigation.

Potential Irrigation Problem	Units	Degree of Restriction on Use		
		None	Slight to Moderate	Severe
Salinity (affects crop water availability)				
EC_W (electrical conductivity)	dS/m	<0.7	0.7–3.0	>3.0
TDS	mg/L	<450	450–2000	>2000
Infiltration				
(affects rate of water infiltration into the soil; evaluate using EC_W and SAR, sodium adsorption ratio, together)				
SAR = 0–3 and EC_W =		>0.7	0.2–0.7	<0.2
SAR = 3–5 and EC_W =		>1.2	0.3–1.2	<0.3
SAR = 6–12 and EC_W =		>1.9	0.5–1.9	<0.5
SAR = 12–20 and EC_W =		>2.9	1.3–2.9	<1.3
SAR = 20–40 and EC_W =		>5.0	2.9–5.0	<2.9
Specific ion toxicity				
(affects sensitive crops)				
Na^+ (sodium)				
Surface irrigation	SAR	<3	3–9	>9
Sprinkler irrigation	mg/L	<70	>70	
Cl^- (chloride)				
Surface irrigation	mg/L	<140	140–350	>350
Sprinkler irrigation	mg/L	<3	>3	
B (boron)	mg/L	<0.7	0.7–3.0	>3.0
Miscellaneous effects (affects susceptible crops)				
TN	mg/L	<5.0	5–30	>30
HCO_3^- (bicarbonate)[1]	mg/L	<90	90–520	>520
H_2S (hydrogen sulfide)	mg/L	<0.5	0.5–2.9	>2.9
pH		Normal range: 6.5–8.4		

Source: From Ayers and Westcot (1985) and WHO (2006).
[1] Expressed as HCO_3 and not as $CaCO_3$ as is customary.

After many successive irrigations, the salinity in the soil reaches an equilibrium concentration based on the salinity of the applied water: a high leaching fraction (LF = 0.4) results in less salt accumulation than a low leaching fraction (LF = 0.1). If the irrigation water salinity and the LF are known, the salinity of the drainage water, and the root zone salinity, can be estimated with Equations (7.5) and (7.6) (Ayers & Westcot, 1985):

$$EC_{sw} = 3\,EC_W \tag{7.6}$$

$$EC_e = 1.5\,EC_W \tag{7.7}$$

EC_{sw} = electrical conductivity of soil water (dS/m)
EC_e = electrical conductivity of soil in the root zone (dS/m)

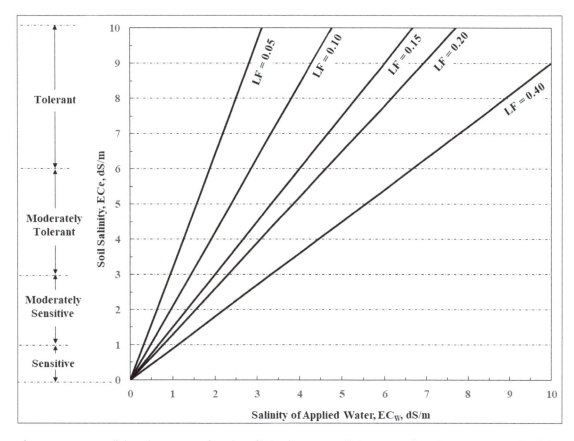

Figure 7.31 Crop salinity tolerance as a function of irrigation water salinity, EC_w, and resultant root zone soil salinity, EC_e, for various LFs. As the irrigation water salinity, EC_w, increases, the leaching factor must be increased to maintain a constant value of soil salinity to stay within the crop tolerances. If upper crop tolerance ranges are exceeded, crop yields will decrease commensurately. (*Source*: Modified from Metcalf & Eddy/AECOM (2006).)

Equations (7.6) and (7.7) are only valid for values of LF from 0.15 to 0.20. (For other values of LF, see Table 3, page 18, in the study by Ayers and Westcot (1985).)

Figure 7.31 shows the relationship for the electrical conductivity of the applied irrigation water, EC_w, and the conductivity of the root zone soil, EC_e, for various values of leaching factor (LF). The crop classifications at the left show that moderately sensitive crops, for example, can be irrigated with LF = 0.10 up to a value of EC_w = 1.5 dS/m and with LF = 0.15 up to a value of EC_w = 2.0 dS/m. Above the root zone soil threshold of EC_e = 3.0 dS/m for moderately sensitive crops, the crop yield will begin to decrease. If the yield decrease is not acceptable, the LF would have to be increased.

7.3.1.2 Infiltration: sodium adsorption ratio

When sodium is the dominant cation in irrigation and soil water, clay soil particles can disperse and swell, with the resulting soil particles plugging large pore spaces, reducing the rates that water and air can enter the soil. Water ponding at the soil surface is often observed when infiltration becomes restricted (Metcalf & Eddy/AECOM, 2006).

The sodium adsorption ratio (SAR) is used to express the conditions where sodium-induced reduction in infiltration can occur. The SAR equation is the ratio of sodium to calcium and magnesium cations as shown in Equation (7.8):

$$\text{SAR} = \frac{[\text{Na}^+]}{\sqrt{(0.5) \cdot ([\text{Ca}^{2+}] + [\text{Mg}^{2+}])}} \tag{7.8}$$

SAR = sodium adsorption ratio (unitless)
[Na⁺] = sodium concentration (meq/L)
[Ca²⁺] = calcium concentration (meq/L)
[Mg²⁺] = magnesium concentrations (meq/L)

The calculated value of SAR is used with the measured values of EC_W in Table 7.30 to determine the degree of restriction on wastewater reuse. Figure 7.32 can also be used to estimate the relative reductions in the rate of infiltration as a function of SAR and EC_W.

7.3.1.3 Specific ion toxicity

Toxicity occurs when certain ions are taken up by plants with the soil water and accumulate in the leaves during evapotranspiration, resulting in tissue damage to the plant. If damage is severe, crop yields can be reduced. Most annual crops are not sensitive to toxicity from sodium, chloride, and boron at the concentrations presented in Table 7.30; the majority of tree crops and woody perennial plants are however (Ayers & Westcot, 1985).

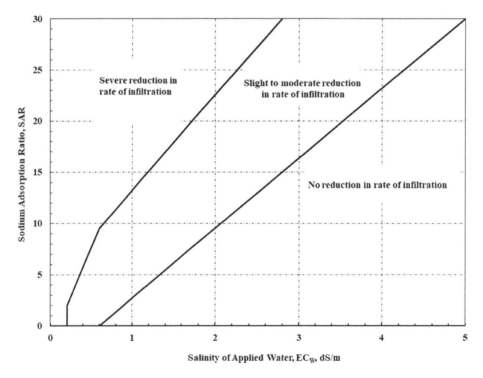

Figure 7.32 Relative rate of water infiltration as influenced by irrigation water salinity (EC_W) and the SAR. (*Source:* Redrawn from Ayers and Westcot (1985).)

Sodium. High sodium concentrations can cause leaf burn and dead tissue along the outside edges of leaves; these symptoms appear after several days or weeks after the sodium accumulation in plant tissue reaches toxic concentrations. Sodium toxicity is a potential problem with high values of SAR. Sodium tolerance varies with crops, and many do not show sodium toxicity.

Chloride. Chloride toxicity from irrigation water is the most common specific ion toxicity. Chloride ions are not adsorbed by soils and move readily with soil water in plant transpiration to accumulate in leaves. If the chloride concentration exceeds the plant threshold tolerance, leaf burn and drying of leaf tissue occurs and can eventually lead to defoliation (Ayers & Westcot, 1985).

Boron. Boron is an essential element for plants, but is toxic to various plants at concentrations above those listed in Table 7.30. Concentrations of boron in surface waters are not normally high enough to cause toxicity problems at the concentrations shown in Table 7.30, but boron concentrations in well waters can be above the threshold values for toxicity. Wastewater used in agriculture where the potable water source is groundwater should always be examined for boron concentrations.

7.3.1.4 Miscellaneous effects

Total nitrogen (TN). Application of nitrogen in excess, in applied fertilizer or in irrigated wastewater, of crop needs can cause overstimulation of growth, delayed maturity, and lower quality harvests. Sensitive crops can be affected at TN concentrations as low as 5 mg/L, while most crops exhibit toxic effects at levels above 30 mg/L (Ayers & Westcot, 1985). While wastewater effluents from larger wastewater treatment plants will usually have TN concentrations less than 30 mg/L, this may not be the case in small cities or peri-urban areas, where TN concentrations can range up to 60 mg/L or more. TN problems can also occur if the annual loading of nitrogen in wastewater exceeds the allowable crop uptake in kg TN/ha yr. Table 7.31 shows the ranges of crop uptake for forage, field, and forest crops for the climates of the US. High TN concentrations in wastewater used for crop irrigation can be controlled by dilution with surface or groundwater, and annual TN loading rates can be controlled by irrigating for crop nitrogen needs rather than crop water needs.

Bicarbonate. High concentrations of bicarbonate and calcium ions can cause the precipitation of calcium carbonate on leaves or fruit when sprinkler irrigation is used as shown with the following equation:

$$Ca^{2+} + 2\,HCO_3^- = CaCO_{3\downarrow} + 2\,H^+$$

While there is no toxicity, calcium carbonate deposits on vegetable leaves and fruit cause problems requiring treatment before marketing of produce. Fortunately, $CaCO_3$ precipitation is not a problem with surface irrigation.

Hydrogen sulfide (H_2S). Hydrogen sulfide is toxic to plants and can cause corrosion problems in irrigation systems and pumps. Adequately treated wastewater for agricultural irrigation should not have significant concentrations of H_2S in the final effluent.

pH. The pH of wastewater effluent used for irrigation is not usually a problem since treated wastewater effluents typically contain sufficient alkalinity in the form of HCO_3^- and CO_3^{2-}. Only low salinity waters with $EC_W < 0.2$ dS/m have had pH problems (Ayers & Westcot, 1985), but this would be a rare occurrence for wastewater effluents.

Table 7.31 Ranges of nitrogen uptake for major crop types in the US.[1]

Crop Type	Total Nitrogen Uptake (kg TN/ha yr)
Forage crops	130–675
Field crops	125–250
Forest crops	110–400

Source: USEPA (1981).

7.3.2 Case studies of physicochemical characteristics of wastewater used in agriculture

7.3.2.1 Physicochemical parameter concentrations in wastewater effluent in the US and Germany

Table 7.32 lists physicochemical data for treated wastewater effluents used for agricultural irrigation in the US and Germany as reported by Ayers and Westcot (1985). Data for a hypothetical high-strength domestic raw wastewater are also shown for comparison.

The data in Table 7.32 can be assessed for possible irrigation problems using Table 7.30, and the results are shown in Table 7.33.

Commentary. The results from Table 7.33 show that treated wastewater from various cities in the US and Germany, and even hypothetical high-strength raw domestic wastewater in the US, have physicochemical characteristics enabling wastewater irrigation with none or slight to moderate reduction in use based on the criteria of Ayers and Westcot (1985). Unless groundwater with high concentrations of certain physicochemical constituents is used as the source water in a city or peri-urban area, treated domestic wastewater should normally not pose any significant restrictions for irrigation if surface irrigation is used.

7.3.2.2 Salinity, infiltration, and sodium ion toxicity potential problems, Cochabamba, Bolivia

To reuse the WSP effluent from the proposed redesign of the Cochabamba system discussed in Chapter 6, it is necessary to assess the available physiochemical data presented in Table 7.34 for potential irrigation problems from Table 7.30.

Salinity. Both the values of $EC_W = 1.75$ and $TDS = 808$ mg/L fall into the range of slight to moderate degree of restriction on use from Table 7.30.

Infiltration. The value of SAR is calculated with Equation (7.8), where all concentrations are expressed in meq/L:

$$SAR = \frac{[Na^+]}{\sqrt{((0.5) \cdot ([Ca^{2+}] + [Mg^{2+}]))}} \tag{7.8}$$

Conversion of mg/L to meq/L:

$[Na^+] = (139\,mg/L)/(23\,mg/meq) = 6.04\,meq/L$

$[Ca^{2+}] = (47\,mg/L)/(20\,mg/meq) = 2.35\,meq/L$

$[Mg^{2+}] = (16.7\,mg/L)/(12.2\,mg/meq) = 1.37\,meq/L$

Table 7.32 Physicochemical analyses of wastewater effluents used for irrigation in Germany and the US.

Location	EC$_W$ (dS/m)	TDS (mg/L)	pH	Na (mg/L)	Cl (mg/L)	HCO$_3$ (mg/L)	SAR
Braunschweig, Germany	1.11	710	7.1	78.2	127.6	280.6	1.8
Bakersfield, CA	0.88	563	7.0	108.1	106.4	219.6	4.1
Santa Rosa, CA	0.70	448		89.7	117.0	164.7	2.9
Tuolumne, CA	0.35	224		27.6	42.5	79.3	1.2
Fresno, CA	0.69	442	7.2	25.3	70.9	219.6	3.1
High-Strength Raw Domestic Wastewater, US[1]	1.75	1121	–	–	118	–	–

Source: Data from Ayers and Westcot (1985).
[1]Typical concentrations in the US for high-strength domestic wastewater (Metcalf & Eddy/AECOM, 2014).

Table 7.33 Potential irrigation problems for treated wastewater effluents used for irrigation in Germany and the US.

Potential Irrigation Problem	Degree of Reduction in Use — None	Degree of Reduction in Use — Slight to Moderate
Salinity		
EC_W	Tuolumne, Fresno	Braunschweig, Bakersfield, Santa Rosa, High-strength WW
TDS	Santa Rosa, Tuolumne, Fresno	Braunschweig, Bakersfield, High-strength WW
Infiltration		
SAR and EC_W	Braunschweig	Bakersfield, Santa Rosa, Tuolumne, Fresno
Specific ion toxicity		
Na+ (surface irrigation)	Braunschweig, Santa Rosa, Tuolumne	Bakersfield, Fresno
Cl− (surface irrigation)	Braunschweig, Bakersfield, Santa Rosa, Tuolumne, Fresno, High-strength WW	
Miscellaneous effects		
HCO_3^-	Tuolumne	Braunschweig, Bakersfield, Santa Rosa, Fresno, High-strength WW
pH		Braunschweig, Bakersfield, Fresno

Table 7.34 Physicochemical analyses of wastewater effluent, Cochabamba WSP system, Bolivia.

Location	EC_W (dS/m)	TDS (mg/L)	pH	Na+ (mg/L)	Ca^2 (mg/L)	Mg^{2+} (mg/L)
WSP Final Effluent	1.75	808	7.90	139	47	16.7

Source: Data from Coronado *et al.* (2001), and the Centro de Aguas y Saneamiento Ambiental (CASA), Universidad Mayor de San Simón, courtesy of C. Oporto.

$$\therefore \text{SAR} = \frac{[\text{Na}^+]}{\sqrt{((0.5) \cdot ([\text{Ca}^{2+}] + [\text{Mg}^{2+}]))}}$$

$$= \frac{[6.04\,\text{meq/L}]}{\sqrt{((0.5) \cdot ([2.35\,\text{meq/L}] + [1.37\,\text{meq/L}]))}} = 4.43$$

From Table 7.30 and Figure 7.32, there would be no restriction on use.
Specific ion toxicity for Na⁺.

- The SAR value of 4.43 gives a slight to moderate restriction on use for surface irrigation.
- The Na⁺ concentration of 139 mg/L also gives a slight to moderate restriction on use for sprinkler irrigation.

Table 7.35 summarizes the results, which show that there should be none or slight to moderate reduction in use irrigating with the Cochabamba WSP effluent.

7.3.2.3 Potential salinity and pH irrigation problems, Ouagadougou WSP system effluent

Kpoda *et al.* (2022) reported the electrical conductivity and pH data on the Ouagadougou WSP effluent that are presented in Table 7.36.

Salinity. The median value of EC_W for the dry season is within the range of slight to moderate restriction on irrigation use from Table 7.30. The maximum value measured of 12 dS/m, however, is unacceptable for irrigation. The wet season median value of 10.5 dS/m is unacceptable for irrigation.

pH. The median, minimum, and maximum pH values in Table 7.36 are all above the normal range in Table 7.30.

Table 7.35 Potential irrigation physicochemical water quality problems, Cochabamba WSP system.

Potential Irrigation Problem	Degree of Reduction in Use	
	None	Slight to Moderate
Salinity		
EC_W		X
TDS		X
Infiltration		
SAR and EC_W	X	
Specific ion toxicity		
Na+ (surface irrigation)		X
Na+ (sprinkler irrigation)		X

Table 7.36 Electrical conductivity and pH of final effluent, Ouagadougou WSP system.

Season	EC_W (dS/m)			pH		
	Median	Minimum	Maximum	Median	Minimum	Maximum
Dry season	2.50	1.73	12.00			
Wet season	10.50	2.00	13.53			
Yearlong				10.2	8.7	11.0

Source: Data from Kpoda *et al.* (2015). Samples were collected twice a month for 12 months.

Commentary. The Ouagadougou effluent is unacceptable for irrigation, which is the reason it had been abandoned by farmers as mentioned previously in the pathogen reduction case study. The high pH and EC_W values are the results of industrial wastewater discharged into the pond system, which is at least two-third of the total flowrate (Kpoda *et al.*, 2022). Unfortunately, another project to help poor farmers failed as a result of poor planning and management. If the industrial wastewater could be diverted, or pretreated, the quality of the wastewater would improve for agricultural reuse.

doi: 10.2166/9781789061536_0251

Chapter 8
Design of wastewater irrigation systems with valorization of nutrients

8.1 INTRODUCTION
8.1.1 Types of irrigation systems
Irrigation systems can be generally classified as gravity surface flow, pressurized surface application, and pressurized subsurface application. Table 8.1 shows the basic features of each of the most commonly used types.

The gravity surface flow irrigation methods in Table 8.1 all have efficiencies comparable to sprinkler and subsurface drip irrigation, but at lower cost and simplicity of operation. Subsurface drip irrigation, which has the highest cost and requires a high-quality effluent, is not a possible solution for resource-poor urban and peri-urban areas. Sprinkler irrigation presents higher pathogen risks as a result of aerosols, but can be an appropriate irrigation method in steep terrain if public health measures are taken, as discussed in the case study of Aurora II in Chapter 7.

Figure 8.1 shows examples of border and furrow irrigation as commonly used worldwide.

8.1.2 Design equations and parameters for irrigation requirements
Irrigation requirements are a function of the positive difference of monthly evapotranspiration and precipitation, irrigation efficiency, and leaching factor as shown in Equation (8.1):

$$\text{Irrigation requirement} = f(\text{Positive net evapotranspiration, leaching factor, efficiency}) \quad (8.1)$$

Equation (8.1) is put into mathematical form in Equation (8.2) (Ayers & Westcot, 1985):

$$L_{w,i} = \frac{(\text{ET}_{C,i} - P_{\text{eff},i})}{(1 - \text{LF})E} \quad (8.2)$$

where $L_{w,i}$ is the hydraulic loading rate based on irrigation requirements for month i (mm/mo), $\text{ET}_{C,i}$ is the crop evapotranspiration requirement for month i (mm/mo), $P_{\text{eff},i}$ is the effective precipitation for month i (mm/mo), LF is the leaching fraction expressed as a decimal, and E is the efficiency of irrigation system expressed as a decimal.

The hydraulic loading rate, $L_{w,i}$, is the principal parameter for determining the irrigation area.

© 2022 The Author. This is an Open Access book chapter distributed under the terms of the Creative Commons Attribution Licence (CC BY-NC-ND 4.0), which permits copying and redistribution for noncommercial purposes with no derivatives, provided the original work is properly cited (https://creativecommons.org/licenses/by-nc-nd/4.0/). This does not affect the rights licensed or assigned from any third party in this book. The chapter is from the book *Integrated Wastewater Management for Health and Valorization: A Design Manual for Resource Challenged Cities*, Stewart M. Oakley (Author)

Table 8.1 Characteristics of commonly used irrigation systems.

Irrigation Method	Use Factors	Efficiency
Gravity surface flow		
Furrow irrigation	Low cost Survey leveling may be required Medium level of health protection	70–85
Border irrigation	Low cost Leveling required Medium level of health protection	65–90
Flood irrigation	Low cost Some leveling may be required Medium level of health protection	70–85
Sprinkler irrigation	Medium to high cost Leveling not required; can be used on steep terrain Low level of health protection, especially with aerosols	70–80
Subsurface drip irrigation	High cost Highest level of health protection Requires high-quality effluent produced with expensive, technologically complex treatment methods	70–90

Source: Developed from Metcalf and Eddy/AECOM (2006).

Figure 8.1 Examples of the most commonly used gravity surface irrigation systems in small cities. Top photo: A harvested plot that uses border irrigation with earthen berms with an arrow showing the flow direction. Bottom photo: Furrow irrigation of various plots with irrigation water from the small canal in the center. Both photos are from peri-urban areas of Cochabamba, Bolivia.

8.1.2.1 Crop evapotranspiration and reference evapotranspiration

Crop evapotranspiration is defined by Equation (8.3):

$$ET_{C,i} = K_C ET_{0,i} \tag{8.3}$$

where $ET_{C,i}$ is the crop evapotranspiration for month i (mm/mo), K_C is the crop coefficient, and $ET_{0,i}$ is the reference evapotranspiration for month i (mm/mo).

The monthly reference evapotranspiration, $ET_{0,i}$, is defined as the evapotranspiration of a well-irrigated reference surface, resembling the evaporation of an extension surface of green grass of uniform height, actively growing and adequately watered (Allen et al., 1998). It is determined by climatic factors only, calculated from weather data without consideration of crop and soil factors. Data for $ET_{0,i}$, are available from CLIMWAT/CROPWAT.

The crop coefficient, K_C, varies with crop characteristics rather than climate. K_C changes with the four principal crop growth stages of crops shown in Figure 8.2. At the initial stages of growth, K_C values are low; during the growth stage, K_C increases, reaching and maintaining its maximum value during the longest period, the mid-season stage. Finally, K_C decreases during the late season/harvest stage. More frequent wetting during all growth stages yields higher values of K_C.

For the preliminary design of irrigation systems, the reference evapotranspiration, $ET_{0,i}$, can be used in lieu of the crop evapotranspiration, $ET_{C,i}$, which is equivalent to having $K_C = 1.0$ for all crop growth stages, resulting in Equation (8.4):

$$L_{w,i} = \frac{(ET_{0,i} - P_{\text{eff},i})}{(1 - LF)E} \tag{8.4}$$

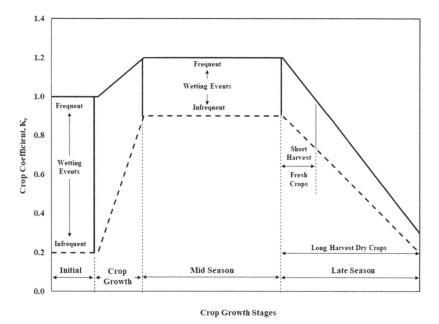

Figure 8.2 Typical ranges of the crop coefficient, K_c, for the four crop growth stages. The values of K_c after the initial growth stage can be close to 1.0 the majority of the time if frequent wetting events occur, and crops are harvested fresh (wet). (*Source*: Developed from Allen et al. (1998).)

Table 8.2 Reference evapotranspiration and rainfall data for CLIMWAT/CROPWAT Station Peña Plata, Guatemala.

Month	ET₀ (mm/d)	Rain (mm)	Eff rain (mm)
January	3.96	17	16.5
February	4.55	34	32.2
March	4.73	60	54.2
April	4.42	153	115.5
May	3.91	401	165.1
June	3.77	597	184.7
July	3.71	501	175.1
August	3.70	512	176.2
September	3.50	688	193.8
October	3.18	520	177
November	3.51	137	107
December	3.82	46	42.6
Total		**3666**	**1440**

8.1.2.2 Effective precipitation

Effective precipitation, P_{eff}, is the water available to plants that is not lost through runoff and deep percolation and is the form of precipitation that must be used in Equation (8.4). The precipitation data in CROPWAT are reported for both total precipitation and effective precipitation and are termed rain and effective rain in CROPWAT.

Table 8.2 gives an example of how reference evapotranspiration and precipitation data are reported in CROPWAT from a station in Guatemala. Note that the annual effective rainfall is less than half the total rainfall, which is typical of wet climates. Also, the ET_0 data are represented as mm/d, which must be converted to mm/mo to use in Equation (8.4).

8.1.2.3 Leaching factor and irrigation efficiency

A value of 0.15–0.20 is used for the leaching factor for preliminary design as discussed in Section 7.3.1 on physicochemical water quality for irrigation.

Irrigation efficiencies are listed in Table 8.1, and the ranges of all, with the exception subsurface drip irrigation, are close.

8.1.2.4 Irrigation area

The determination of irrigation area varies for the three possible irrigation methods:

(1) **Without a reservoir, changing the irrigation area each month in relation to changing crop water requirements.**

This method is applicable in dry climates where $ET_{0,i} > P_{eff,i}$ for all or most months of the year; it is especially applicable in resource-limited cities where the cost of reservoir construction is beyond the economic means. The Campo Espejo ACRE irrigation system discussed in Chapter 7 is a good example of this method. The irrigated area is calculated with Equation (8.5):

$$A_{w,i} = \frac{V_{final,i}}{L_{w,i}/1000} \tag{8.5}$$

where $A_{w,i}$ is the irrigation area for each month i, with positive net ET_0 (m²/mo), $V_{w,\text{final},i}$ is the effluent wastewater volume for month i, with positive net ET_0 (m³), $L_{w,i}$ is the hydraulic loading for each month i with positive ET_0 (mm/mo), and 1000 is the conversion of mm to m.

(2) **With a reservoir, irrigating a fixed area during the growing season, and storing wastewater during the wet season.**

Equation (8.6) is used to calculate the irrigation area:

$$A_w = \frac{\sum_{i=1}^{12} V_{w,i}}{\sum_{i=1}^{12} L_{w,i}/1000} \tag{8.6}$$

where A_w is the total irrigation area (m²), $V_{w,i}$ is the effluent wastewater volume for month i (m³), and $L_{w,i}$ is the hydraulic loading based on irrigation requirements for month i (mm). Note: Negative values of $L_{w,i}$ are not included in summations.

(3) **Without a reservoir, irrigating year-round during both the dry and wet seasons.**

During the dry season, method 1 is used. During the wet season, land application is used (USEPA, 1981), where wastewater is applied to the soil at predetermined percolation rates, which are much lower than those used during the irrigation season. This method is preferable to surface water discharge, and a large fraction of nutrients can be adsorbed in the soil to be used by plants during the growing season. The leaching of nitrate nitrogen to groundwater is a potential problem, however, that must be assessed. Each of the three methods will be discussed in detail in the respective design sections.

8.1.2.5 Nutrient loading rates

Estimated loading rates of total nitrogen (TN) and total phosphorus (TP) in kg/ha yr should be calculated for all designs after the irrigated area is determined for the following design issues:

- Are nutrient loadings with the range required for adequate crop growth?
- If nutrient loadings are below the minimum required for crop growth, can the irrigated area be lowered so nutrient loadings meet crop needs?
- If nutrient loadings of TN are above the maximum allowed to avoid crop toxicity, what remediation measures can be taken? (e.g., increase irrigated area, use different crops).

Table 8.3 presents the ranges of TN and TP uptake rates for forage crops and field crops. The nitrogen loading rate is calculated using Equation (8.7):

$$L_{\text{TN}} = \frac{0.001 \cdot (V_{w,\text{yr}})(C_{\text{TN}})}{A_w} \tag{8.7}$$

Table 8.3 Nutrient uptake rates for forage and field crops (USEPA, 1981).

Crop Type	Range of Uptake Rates (kg/ha yr)	
	Nitrogen	Phosphorus
Forage crops	130–675	20–85
Field crops	125–200	10–30
Forest crops	110–400	–

where L_{TN} is the TN loading (kg TN/ha yr), $V_{w,yr}$ is the annual wastewater application of pond effluent (m³/yr), C_{TN} is the concentration of total N in applied wastewater (mg/L), and A_w is the irrigated area (ha).

The value of $V_{w,yr}$ for the effluent can be significantly lower than the influent in large waste stabilization pond (WSP) systems as a result of evaporation and infiltration losses and should be used instead.

The phosphorus loading is calculated in a similar fashion with Equation (8.8):

$$L_{TP} = \frac{0.001 \cdot (V_{w,yr})(C_{TP})}{A_w} \tag{8.8}$$

where L_{TP} is the total phosphorous loading (kg TP/ha yr) and C_{TP} is the concentration of total P in applied wastewater (mg/L).

8.1.2.6 Carbon emissions saved by using wastewater in lieu of synthetic fertilizers

If wastewater irrigation can satisfy crop nutrient requirements as determined by Equations (8.7) and (8.8), the carbon offsets in metric tons of carbon dioxide equivalents per year, $t\ CO_{2,equiv}$/yr, should be reported to emphasize the importance of wastewater reuse in agriculture in regards to climate change caused by synthetic fertilizer use; it is particularly important to raise awareness of design engineers, wastewater infrastructure organizations, and local governments of the issue and the value of wastewater nutrients.

Table 8.4 lists emission factors for fertilizer production for the most commonly used nitrogen and phosphorus fertilizers. The mean values for greenhouse gas production of nitrogen and phosphorus fertilizers are then used in Equations (8.9) and (8.10).

Nitrogen fertilizer production

$$t\ CO_{2,equiv,Saved,TN}/yr = 0.001 \cdot (V_{w,yr})(C_{TN})(2.522\ kg\ CO_{2,equiv}/kg\ N) \tag{8.9}$$

where $t\ CO_{2,equiv,saved,TN}$/yr = metric tons of CO_2 emissions/year saved as a result of use of biogenic nitrogen in wastewater rather than synthetic fertilizers; $V_{w,yr}$ is the annual wastewater application of pond effluent (m³/yr), and C_{TN} is the concentration of TN in effluent applied to crops (mg/L).

Phosphorus fertilizer production

$$t\ CO_{2,equiv,Saved,TP}/yr = 0.001 \cdot (V_{w,yr})(C_{TP})(0.472\ kg\ CO_{2,equiv}/kg\ P) \tag{8.10}$$

Table 8.4 Greenhouse gas emission factors for N and P fertilizer production.

Fertilizer	N-P-K	kg CO$_{2,equiv}$/kg N	kg CO$_{2,equiv}$/kg P$_2$O$_5$	kg CO$_{2,equiv}$/kg P
Urea	46-0-0	4.019		
		1.326		
		0.913		
		1.707		
TSP[1]	0-48-0		1.083	0.473
			0.354	0.155
DAP[2]	18-46-0	4.612	1.883	0.822
		2.556	1.000	0.437
	Mean values	**2.522**		**0.472**

Source: Developed from Wood and Cowie (2004).
[1]Triple superphosphate.
[2]Diammonium phosphate.

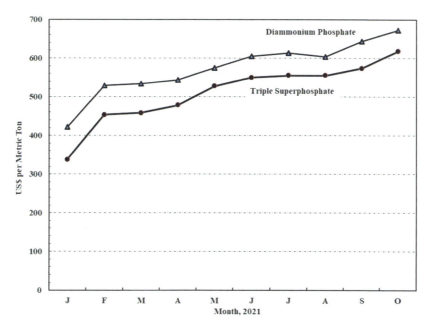

Figure 8.3 Increase in cost of two fertilizers commonly used worldwide. The cost of diammonium phosphate increased by 170%, and TSP by 190%, from January through October 2021. (*Source*: www.indexmundi.com/commodities.)

$t\ CO_{2,equiv,saved,TP}/yr$ = metric tons of CO_2 emissions/year saved as a result of use of biogenic phosphorus in wastewater rather than synthetic fertilizers; $V_{w,yr}$ is the annual wastewater application of pond effluent, m³/yr); and C_{TP} is the concentration of TP in effluent applied to crops (mg/L).

8.1.2.7 *Valorization of nitrogen and phosphorus*
The monetary value of TN and TP as fertilizers in irrigation water should be calculated to demonstrate the economic advantages of wastewater reuse in addition to those of environmental concerns. Fertilizer prices almost doubled during the first 10 months of 2021 (Figure 8.3), after a prior 5-year period with little change in prices (www.indexmundi.com/commodities). While farmers are aware of the economic value of nutrients, specialists in funding agencies, regulators, and government officials at national and local levels, and design engineers continue to focus on treatment systems for discharge to surface waters, and an economic argument for reuse with valorization of TN and TP should be included in reuse designs.

8.2 WASTEWATER IRRIGATION WITHOUT RESERVOIRS IN DRY CLIMATES
Reservoirs add sizeable costs to integrated wastewater management projects and are usually beyond the resources of resource-limited urban and peri-urban areas worldwide. In dry climates, wastewater irrigation can be implemented without the need for water storage in reservoirs. This is effected by changing the irrigated area on a monthly basis as calculated by the available wastewater volumes and hydraulic loading rates in Equations (8.4) and (8.5). The Campo Espejo ACRE discussed in Section 7.2.2 is an excellent example of this method that has been in operation for almost 30 years.

8.2.1 Design procedure for irrigation without reservoirs

In dry climates, where evaporation can exceed precipitation for every month of the year, significant volumes of water can be lost by evaporation in WSP, yielding lower effluent volumes available for irrigation than those estimated with the mean influent wastewater flowrate. In these circumstances, the following steps 1–4 are tabulated by month in the design procedure and are used at the beginning of the design to more accurately estimate water volumes available for irrigation.

(1) **Calculate initial monthly volumes of wastewater using the mean design flowrate**

$$V_{ww,i} = Q_{mean} \left(\# \text{days/month}_i \right) \tag{8.11}$$

where $V_{ww,i}$ is the initial volume of wastewater based on influent flowrate for month i (m³) and Q_{mean} is the design mean flowrate (m³/d).

(2) **Determine the monthly water evaporation rate from pan evaporation data**
Obtain long-term monthly pan evaporation data from local meteorological stations and convert to water surface evaporation with Equation (8.11) (Yihdego & Webb, 2018).

$$e_{water,i} = 0.70 e_{pan,i} \tag{8.12}$$

where $e_{water,i}$ is the evaporation from water surface for month i (mm), $e_{pan,i}$ is the pan evaporation for month i (mm), and 0.70 is the conversion factor from pan to free water surface evaporation.

(3) **Calculate the volume of water lost each month due to evaporation**

$$V_{lost,i} = \frac{A_{ws}}{e_{water,i}/1000} \tag{8.13}$$

where $V_{lost,i}$ is the volume of water lost in month i (m³), A_{ws} is the total water surface area of WSP system (m²), and 1000 is the conversion of mm to m.

(4) **Calculate the final volume of water remaining for irrigation each month**

$$V_{final,i} = V_{initial,i} - V_{lost,i} \tag{8.14}$$

$V_{final,i}$ is the final irrigation water volume for month i (m³).

(5) **Tabulate monthly data for P_{eff} and ET_0**
Obtain long-term average monthly site data for effective precipitation, P_{eff}, and reference evapotranspiration, ET_0, from CLIMWAT/CROPWAT or local meteorological stations. Monthly ET_0 values from CROPWAT are in units if mm/d and must be converted to mm/mo.

(6) **Select values for leaching factor, LF, and irrigation efficiency, E**
Generally, LF = 0.15 and E = 0.75 should be used as discussed in Chapter 7 unless local conditions require other values.

(7) **Calculate the monthly hydraulic loading rate using Equation (8.4)**

$$L_{w,i} = \frac{(ET_{0,i} - P_{eff,i})}{(1 - LF)E} \tag{8.4}$$

(8) **Calculate the irrigation areas for every month using Equation (8.5)**

$$A_{w,i} = \frac{V_{final,i}}{L_{w,i}/1000} \tag{8.5}$$

(9) **Determine the annual N and P loading rates using Equations (8.7) and (8.8)**

$$L_{TN} = \frac{0.001 \cdot (V_{w,yr})(C_{TN})}{A_w} \tag{8.7}$$

$$L_{TP} = \frac{0.001 \cdot (V_{w,yr})(C_{TP})}{A_w} \qquad (8.8)$$

(10) **Use Equations (8.15) and (8.16) to estimate the metric tons of $CO_{2,equiv}$ saved by using the nutrients in wastewater rather than synthetic fertilizers.**

Nitrogen: MT $CO_{2,equiv,Saved,TN}/yr = 0.001 \cdot (V_{w,yr})(C_{TN})(2.522 \text{ kg } CO_{2,equiv}/\text{kg N})$ (8.15)

Phosphorus: MT $CO_{2,equiv,Saved,TP}/yr = 0.001 \cdot (V_{w,yr})(C_{TP})(0.472 \text{ kg } CO_{2,equiv}/\text{kg P})$ (8.16)

(11) **Valorization of nitrogen and phosphorus**
The annual value of the nitrogen and phosphorus applied in irrigation should be calculated based on current global market costs of fertilizers.

8.2.2 Design example: pond redesign for wastewater irrigation, Cochabamba, Bolivia

The Cochabamba WSP effluent from the redesign in Sections 5.4 and 6.4 will be reused in agriculture without a reservoir using the methodology from Section 8.2.1. The following data on redesign from Table 6.21 are applied:

Cochabamba WSP Redesign	
Mean influent flowrate, m³/d	13,200
Water surface area, m²	
• Facultative ponds	21,600
• Maturation ponds	13,600
• Total	35,200

Steps 1–4 in the design procedure will be used to complete columns 3 through 8 for the month of January in the following table.

1	2	3	4	5	6	7	8
Month	No. of Days	Initial Volume Wastewater $V_{initial}$ (m³)	e_{pan} (mm)	e_{water} (mm)	Water Surface Area (m²)	Volume Wastewater Lost V_{lost} (m³)	Volume Wastewater Remaining V_{final} (m³)
January	31						

(1) **Calculate the initial monthly volume of wastewater using the mean design flowrate (Equation (8.11)) for column 3:**

$$V_{ww,1} = Q_{mean}(\#\text{days/month}_1) = (13,200 \text{ m}^3/\text{d})(31 \text{ d}) = 409,200 \text{ m}^3$$

(2) **Determine the monthly water evaporation rate from pan evaporation data (columns 4 and 5)**
Long-term monthly pan evaporation data from the Cochabamba airport meteorological station for four consecutive years are presented in Table 8.5. The 4-year mean pan evaporation value for January (Table 8.5) is expressed as follows:

$e_{pan,1} = 108.0 \text{ mm}$

By using Equation (8.12), the water surface evaporation is expressed as follows:

$e_{water,1} = 0.70 \, e_{pan,1} = (0.70)(108.0 \text{ mm}) = 76 \text{ mm}$

Table 8.5 Monthly pan evaporation data and calculated water surface evaporation, Cochabamba, Bolivia, 2008–2012.

Month	Pan Evaporation, e_{pan} (mm)					
	2008	2009	2010	2011	2012	Mean (2008–2012)
January	87	121.7	130.6	106.6	93.9	108.0
February	117.5	103.8	109.6	80.9	111.9	104.7
March	97.2	99.2	125.8	114	89.4	105.1
April	102.6	116.9	136.2	103.2	95.9	111.0
May	114	115.3	114.6	103	113	112.0
June	79.3	97.4	105.2	95.8	97.5	95.0
July	106.5	114.6	113	101.4	104.1	107.9
August	119.9	143	148.1	137	112.1	132.0
September	156.1	149.5	164	144.6	156.6	154.2
October	167.6	180.2	173.1	143.2	166.2	166.1
November	136.8	146.7	196.1	155.7	141.4	155.3
December	103	138.2	145.9	120.1	93.7	120.2
Annual	1387	1526	1662	1405	1376	1471

Source: Data accessed from Servicio Nacional de Meteorología e Hidrología (SENAMHI), Bolivia (http://senamhi.gob.bo/index.php/sismet).

(3) **Calculate the volume of water lost each month due to evaporation with Equation (8.13).**

$$V_{lost,1} = \frac{A_{ws}}{e_{water,1}/1000} = \frac{35,200 \text{ m}^2}{(76 \text{ mm})/(1000 \text{ mm/m})} = 26,611 \text{ m}^3$$

(4) **Calculate the final volume of water remaining for irrigation each with Equation (8.14).**

$$V_{final,1} = V_{ww,1} - V_{lost,1} = 409,200 \text{ m}^3 - 26,611 \text{ m}^3 = 382,589 \text{ m}^3$$

Table 8.6 presents the results for the entire year using the same procedure. The final volumes of wastewater after evaporation in column 8 will be used in Equation (8.5) to estimate the monthly irrigation areas.

The percent of annual wastewater volume lost to evaporation is:

$$\%V_{wastewater} \text{ lost to evaporation} = \frac{V_{lost,yr}}{V_{ww,yr}}(100\%) = \frac{362,578 \text{ m}^3}{4,818,000 \text{ m}^3}(100\%) = 7.52\%$$

Generally, in dry climates, losses to evaporation should be in this range as long as hydraulic retention times are not excessively long, with corresponding higher water surface areas.

Steps 5–8 in the design procedure are used to calculate the values in columns 9–14 in the following table for the month of January.

1	2	8	9	10		11	12	13		14
Month	No. of Days	Volume Wastewater Remaining V_{final} (m³)	P_{eff} (mm)	ET_0 (mm)		$ET_0 - P_{eff}$ (mm)	L_w (mm)	Irrigated Area (A_w)		Volume Irrigation Water V_w (m³)
				mm/d	m			m²	ha	
January	31	382,589								

Design of wastewater irrigation systems with valorization of nutrients 261

Table 8.6 Volume of wastewater remaining after evaporation losses, Cochabamba WSP.

1	2	3	4	5	6	7	8
Month	No. of Days	Initial Volume Wastewater V_{ww} (m³)	e_{pan} (mm)	e_{water} (mm)	Water Surface Area (m²)	Volume Wastewater Lost V_{lost} (m³)	Volume Wastewater Remaining V_{final} (m³)
January	31	409,200	108.0	76	352,000	26,611	382,589
February	28	369,600	104.7	73	352,000	25,798	343,802
March	31	409,200	105.1	74	352,000	25,897	383,303
April	30	396,000	111.0	78	352,000	27,350	368,650
May	31	409,200	112.0	78	352,000	27,597	381,603
June	30	396,000	95.0	67	352,000	23,408	372,592
July	31	409,200	107.9	76	352,000	26,587	382,613
August	31	409,200	132.0	92	352,000	32,525	376,675
September	30	396,000	154.2	108	352,000	37,995	358,005
October	31	409,200	166.1	116	352,000	40,927	368,273
November	30	396,000	155.3	109	352,000	38,266	357,734
December	31	409,200	120.2	84	352,000	29,617	379,583
Total		4,818,000				362,578	4,455,422

(5) **Tabulate monthly data for P_{eff} and ET_0.**
From CLIMWAT/CROPWAT for the month of January:

$P_{eff,1} = 94.5$ mm

$ET_{0,1} = (4.58 \text{ mm/d})(31 \text{ d}) = 142$ mm

(6) **Select values for leaching factor, LF, and irrigation efficiency, E.**
Assume LF = 0.15 and E = 0.75.

(7) **Calculate the monthly hydraulic loading rate using Equation (8.4).**
For January:

$$L_{w,1} = \frac{(ET_{0,1} - P_{eff,1})}{(1-LF)E} = \frac{(142 \text{ mm} - 94.5 \text{ mm})}{(1-0.15)(0.75)} = 74.5 \text{ mm}$$

(8) **Calculate the irrigation areas for every month using Equation (8.5).**
For January:

$$A_{w,1} = \frac{V_{final,1}}{L_{w,1}/1000} = \frac{382,589 \text{ m}^3}{(74.5 \text{ mm})/(1000 \text{ mm/m})} = 5,136,907 \text{ m}^2 = 514 \text{ ha}$$

Table 8.7 presents the results for the entire year, with monthly irrigation areas ranging from 165 to 536 ha. In addition to the savings in avoiding the cost of storage reservoir construction, this design method offers the flexibility in easily changing irrigation areas if wastewater volumes change significantly as a result of population growth or other factors.
Table 8.8 presents a summary of all data calculations from an Excel sheet.

(9) **Determine the annual N and P loading rates for the median annual irrigation area using Equations (8.7) and (8.8).**
Median irrigation area = 334 ha from Table 8.8.

$$\text{Nitrogen loadings: } L_{TN} = \frac{0.001 \cdot (V_{w,yr})(C_{TN})}{A_w} = \frac{0.001 \cdot (4,455,422 \text{ m}^3/\text{yr})(25 \text{ mg/L})}{334 \text{ ha}}$$

$$= 333 \text{ kg TN/ha yr}$$

Table 8.7 Monthly hydraulic loadings and irrigated areas for Cochabamba, Bolivia.

1	8	9	10	11	12	13		14
Month	Volume Wastewater Remaining V_{final} (m³)	P_{eff} (mm)	ET_0 (mm)	$ET_0 - P_{eff}$ (mm)	L_w (mm)	Irrigated Area A_w m²	ha	Volume Irrigation Water V_w (m³)
January	382,589	94.5	142	47	74.5	5,136,907	514	382,589
February	343,802	80.6	122	41	64.2	5,356,152	536	343,802
March	383,303	65.2	123	58	90.8	4,222,497	422	383,303
April	368,650	24.9	101	76	119.5	3,084,175	308	368,650
May	381,603	5.9	81	75	117.2	3,256,654	326	381,603
June	372,592	2.0	65	63	98.5	3,782,283	378	372,592
July	382,613	2.0	73	71	111.6	3,427,713	343	382,613
August	376,675	5.0	101	96	150.7	2,499,796	250	376,675
September	358,005	8.9	131	122	190.7	1,876,877	188	358,005
October	368,273	15.6	158	142	223.0	1,651,129	165	368,273
November	357,734	37.4	154	117	182.7	1,957,558	196	357,734
December	379,583	78.5	142	63	99.1	3,830,679	383	379,583
	4,455,422	420.5	1391	286	1523			4,455,422

$$\text{Phosphorus loadings: } L_{TP} = \frac{0.001 \cdot (V_{w,yr})(C_{TP})}{A_w} = \frac{0.001 \cdot (4,455,422 \text{ m}^3/\text{yr})(11.3 \text{ mg/L})}{334 \text{ ha}}$$
$$= 151 \text{ kg TP/ha yr}$$

The TN loading is within the range for forage and forest crops in Table 8.3, while the TP loading is far above the maximum values for both forage (85 kg/ha yr) and field crops (30 kg/ha yr). While excess phosphorus can be stored in the soil without harm to crops, excess nitrogen causes toxic effects. More data from continuous monitoring are needed to have more reliable estimates of nutrient loadings.

(10) **Use Equations (8.9) and (8.10) to estimate the metric tons of $CO_{2,equiv}$ saved by using the nutrients in wastewater rather than synthetic fertilizers.**

Nitrogen: $t \, CO_{2,equiv,Saved,TN}/\text{yr} = 0.001 \cdot (V_{w,yr})(C_{TN})(2.522 \text{ kg } CO_{2,equiv}/\text{kg N})$
$$= 0.001 \cdot (4,455,422 \text{ m}^3/\text{yr})(25 \text{ mg/L})(2.522 \text{ kg } CO_{2,equiv}/\text{kg N})$$
$$= 280,914 \text{ kg } CO_{2,equiv}/\text{yr}$$
$$= 281 \text{ t kg } CO_{2,equiv}/\text{yr}$$

Phosphorus: $t \, CO_{2,equiv,Saved,TP}/\text{yr} = 0.001 \cdot (V_{w,yr})(C_{TP})(0.472 \text{ kg } CO_{2,equiv}/\text{kg P})$
$$= 0.001 \cdot (4,455,422 \text{ m}^3/\text{yr})(11.3 \text{ mg/L})(0.472 \text{ kg } CO_{2,equiv}/\text{kg P})$$
$$= 23,763 \text{ kg } CO_{2,equiv}/\text{yr}$$
$$= 23.7 \text{ t } CO_{2,equiv}/\text{yr}$$

Total metric tons of $CO_{2,equiv}$ saved $= 281 + 24 = 305 \text{ t } CO_{2,equiv}/\text{yr}$.

This example overestimates the TP contribution since farmers using synthetic fertilizers would not apply an excess amount so far above crop maximum values for growth. Once again,

Table 8.8 Wastewater irrigation design for WSP system, Cochabamba, Bolivia.

1	2	3	4	5	6	7	8	9	10	11	12	13		14
Month	No. of Days	Initial Volume Wastewater V_{ww} (m³)	e_{pan} (mm)	e_{water} (mm)	Water Surface Area (m²)	Volume Wastewater Lost V_{lost} (m³)	Volume Wastewater Remaining V_{final} (m³)	P_{eff} (mm)	ET_0 (mm)	$ET_0 - P_{eff}$ (mm)	L_w (mm)	Irrigated Area A_w m²	ha	Volume Irrigation Water V_w (m³)
January	31	409,200	108.0	76	352,000	26,611	382,589	94.5	142	47	74.5	5,136,907	514	382,589
February	28	369,600	104.7	73	352,000	25,798	343,802	80.6	122	41	64.2	5,356,152	536	343,802
March	31	409,200	105.1	74	352,000	25,897	383,303	65.2	123	58	90.8	4,222,497	422	383,303
April	30	396,000	111.0	78	352,000	27,350	368,650	24.9	101	76	119.5	3,084,175	308	368,650
May	31	409,200	112.0	78	352,000	27,597	381,603	5.9	81	75	117.2	3,256,654	326	381,603
June	30	396,000	95.0	67	352,000	23,408	372,592	2.0	65	63	98.5	3,782,283	378	372,592
July	31	409,200	107.9	76	352,000	26,587	382,613	2.0	73	71	111.6	3,427,713	343	382,613
August	31	409,200	132.0	92	352,000	32,525	376,675	5.0	101	96	150.7	2,499,796	250	376,675
September	30	396,000	154.2	108	352,000	37,995	358,005	8.9	131	122	190.7	1,876,877	188	358,005
October	31	409,200	166.1	116	352,000	40,927	368,273	15.6	158	142	223.0	1,651,129	165	368,273
November	30	396,000	155.3	109	352,000	38,266	357,734	37.4	154	117	182.7	1,957,558	196	357,734
December	31	409,200	120.2	84	352,000	29,617	379,583	78.5	142	63	99.1	3,830,679	383	379,583
Total		4,818,000				362,578	4,455,422	420.5	1391	286	1523	Median	334	4,455,422

Table 8.9 Select fertilizer prices for October 2021.

1	2	3
Parameter	Diammonium Phosphate (18-46-0)	Triple Superphosphate 44–46% P_2O_5
US$/metric ton	$672.90	$618.00
% Total N	18	
kg Total N/metric ton	180	
Value of TN, US$/kg N	$3.74	
% P_2O_5	46	45
kg Total P_2O_5/metric ton	460	450
kg Total P/metric ton	201	196.5
Value of TP, US$/kg P	$3.35	$3.14

Source: Data from indexmundi.com/commodities.

this example is based on very few data, and rigorous monitoring is required to have reliable estimates for both nutrient loadings and $CO_{2,equiv}$ saved.

(11) **Valorization of nitrogen and phosphorus**

Table 8.9 lists fertilizer prices for three common fertilizers used globally: diammonium phosphate (DAP), urea, and triple superphosphate (TSP). The potential value of N and P as fertilizers in wastewater, in US$/kg of nutrient, is calculated as follows using prices calculated from DAP fertilizer prices.

Column 2: Diammonium phosphate (18-46-0)

DAP has 18% TN and 46% P_2O_5 per metric ton of fertilizer.

(1) **Value of TN (US$/kg)**

kg N/MT = (0.18)(1000 kg/t) = 180 kg/t

Value of N, US$/kg = $\frac{\$672.90/t}{180 \text{ kg N/t}}$ = US$3.74/kg N

(2) **Value of TP (US$/kg)**

kg P_2O_5/t = (0.46)(1000 kg/t) = 460 kg P_2O_5/t
kg P/t = (0.437 kg P/kg P_2O_5)(460 kg P_2O_5/t) = 201 kg P/t

Value of P, US$/kg = $\frac{\$672.90/t}{201 \text{ kg TP/t}}$ = US$3.35/kg P

(3) **Total potential fertilizer value of N and P in wastewater**

Mass of TN in wastewater effluent = 0.001(25 mg/L)(4,455,422 m^3/yr) = 111,386 kg TN/yr
Value of TN as fertilizer = (US$3.74/kg TN)(111,386 kg TN/yr) = US$416,582/yr

Mass of TP in wastewater effluent = 0.001(11.3 mg/L)(4,455,422 m^3/yr) = 50,346 kg TP/yr
Value of TP as fertilizer = (US$3.35/kg TP)(50,346 kg TP/yr) = US$168,660/yr

Total potential value as fertilizers = US$416,582/yr + US$168,660/yr = US$585,242/yr

Design of wastewater irrigation systems with valorization of nutrients

Figure 8.4 A farmer pumps sewage-contaminated river water from the Río Rocha to irrigate various crops on a site approximately 1 km downstream of the Cochabamba WSP system. The Río Rocha is seriously contaminated throughout its course through the city, with fecal coliform concentrations reported as too numerous to count (TNTC) (Coronado et al., 2001). A safe, reliable irrigation water supply is critically needed throughout the peri-urban areas of Cochabamba to the present day.

This example illustrates the potential value of wastewater nutrients when wastewater is used in agriculture rather than treated—oftentimes to remove nutrients—and discharged to surface waters losing all potential benefits.

An example of the dire irrigation situation in Cochabamba is shown in Figure 8.4. A reuse project such as this one, redesigning an abandoned system to use a portion of the city's wastewater flow for reuse, is preferable to a current proposal to construct a new wastewater treatment plant for surface water discharge.

8.2.3 Case study design example: wastewater irrigation at the Campo Espejo ACRE

This case study design example will use the preceding design procedure for wastewater irrigation without a reservoir for the Campo Espejo ACRE in Mendoza and compare the results with available data from Campo Espejo. The following data for the Campo Espejo WSP system from Section 7.2.2 apply:

Campo Espejo WSP System	
Mean influent flowrate, m³/d	129,600
Water surface area, m²	
Facultative + maturation ponds, m²	2,590,000

Steps 1–4 in the design procedure are used to complete columns 3 through 8 for the month of January in the following table.

1	2	3	4	5	6	7	8
Month	No. of Days	Initial Volume Wastewater $V_{initial}$ (m³)	e_{water}/l (mm/d)	e_{water}/l (mm)	Water Surface Area (m²)	Volume Wastewater Lost V_{lost} (m³)	Volume Wastewater Remaining V_{final} (m³)
January	31		5.2		2,590,000		

(1) **Calculate the initial monthly volume of wastewater using the mean design flowrate (Equation (8.11)) for column 3:**

$$V_{ww,1} = Q_{mean}(\#\,days/month_1) = (129{,}600\ m^3/d)(31\ d) = 4{,}017{,}600\ m^3$$

(2) **Determine the monthly water evaporation/infiltration rate (column 5).**
The design evaporation/infiltration (e_{water}/I) rate for the Campo Espejo WSP system was reported as 5.2 mm/d for the entire year, which will be used in the design calculations (Barbeito Anzorena, 2001). For January:

$$(e_{water}/I)_1 = (5.2\ mm/d)(31\ d) = 161\ mm$$

(3) **Calculate the volume of water lost each month due to evaporation with (Equation (8.13)) (column 7).**

$$V_{lost,1} = \frac{A_{ws}}{(e_{water}/I)_1/1000} = \frac{2{,}590{,}000\ m^2}{(161\ mm)/(1000\ mm/m)} = 417{,}508\ m^3$$

(4) **Calculate the final volume of water remaining for irrigation each with (Equation (8.14)) (column 8).**

$$V_{final,i1} = V_{ww,1} - V_{lost,1} = 4{,}017{,}600\ m^3 - 417{,}598\ m^3 = 3{,}600{,}092\ m^3$$

Table 8.10 presents the results for the entire year using the same procedure. The final volumes of wastewater after evaporation in column 8 will be used in Equation (8.5) to estimate the monthly irrigation areas.

The percent of annual wastewater volume lost to evaporation is:

$$\%\ V_{wastewater}\ \text{lost to evaporation} = \frac{V_{lost,yr}}{V_{ww,yr}}(100\%) = \frac{4{,}915{,}820\ m^3}{42{,}388{,}180\ m^3}(100\%) = 11.6\%$$

Table 8.10 Volume of wastewater remaining after evaporation losses, Campo Espejo WSP system.

1	2	3	4	5	6	7	8
Month	No. of Days	Initial Volume Wastewater V_{ww} (m³)	e_{evap}/I (mm/d)	e_{evap}/I (mm)	Water Surface Area (m²)	Volume Wastewater Lost V_{lost} (m³)	Volume Wastewater Remaining V_{final} (m³)
January	31	4,017,600	5.2	161	2,590,000	417,508	3,600,092
February	28	3,628,800	5.2	146	2,590,000	377,104	3,251,696
March	31	4,017,600	5.2	161	2,590,000	417,508	3,600,092
April	30	3,888,000	5.2	156	2,590,000	404,040	3,483,960
May	31	4,017,600	5.2	161	2,590,000	417,508	3,600,092
June	30	3,888,000	5.2	156	2,590,000	404,040	3,483,960
July	31	4,017,600	5.2	161	2,590,000	417,508	3,600,092
August	31	4,017,600	5.2	161	2,590,000	417,508	3,600,092
September	30	3,888,000	5.2	156	2,590,000	404,040	3,483,960
October	31	4,017,600	5.2	161	2,590,000	417,508	3,600,092
November	30	3,888,000	5.2	156	2,590,000	404,040	3,483,960
December	31	4,017,600	5.2	161	2,590,000	417,508	3,600,092
Total		47,304,000				4,915,820	42,388,180

Table 8.11 CLIMWAT/CROPWAT Data for Campo Espejo ACRE, Mendoza, Argentina.

1	2	8	9	10		11	12	13		14
Month	No. of Days	Volume Wastewater Remaining V_{final} (m³)	P_{eff} (mm)	ET_0 (mm)		$ET_0 - P_{eff}$ (mm)	L_w (mm)	Irrigated Area A_w		Volume Irrigation Water V_w (m³)
				mm/d	m			m²	ha	
January	31	3,600,092	35.3	5.03						3,600,092

This value is within the range expected for a large system in a dry climate such as Mendoza. Steps 5–8 in the design procedure are used to calculate the monthly ET_0, the hydraulic loading rate, L_w, and the irrigated area, A_w, in columns 10–13 in Table 8.11 for the month of January.

(5) **Tabulate monthly data for P_{eff} and ET_0 (columns 9 and 12).**
From CLIMWAT/CROPWAT for the month of January:

$P_{eff,1} = 35.3$ mm

$ET_{0,1} = (5.03$ mm/d$)(31$ d$) = 156$ mm

(6) **Select values for leaching factor, LF, and irrigation efficiency, E.**
Assume LF $= 0.15$ and $E = 0.75$.

(7) **Calculate the monthly hydraulic loading rate using Equation (8.4) (column 12).**
For January:

$$L_{w,1} = \frac{(ET_{0,1} - P_{eff,1})}{(1 - LF)E} = \frac{(156 \text{ mm} - 35.3 \text{ mm})}{(1 - 0.15)(0.75)} = 189.2 \text{ mm}$$

(8) **Calculate the irrigation area using Equation (8.5) (column 13).**
For January:

$$A_{w,1} = \frac{V_{final,1}}{L_{w,1}/1000} = \frac{3,600,092 \text{ m}^3}{(189.2 \text{ mm})/(1000 \text{ mm/m})} = 19,025,694 \text{ m}^2 = 1903 \text{ ha}$$

Table 8.12 presents the results for the entire year, with monthly irrigation areas ranging from 1760 to 8178 ha. The irrigated areas and irrigation water volumes in bold are the formal irrigation season in Mendoza (September–March).

For seven contiguous months of the year, September–March, the calculated irrigated areas range from 1760 to 3086 ha, with a mean value of 2298 ha, which is in the range of the Campo Espejo ACRE of 3000 ha. From April to August, the winter in Mendoza, the range of irrigated areas increase from 3629 to 8176 ha. Campo Espejo has only allotted 200 ha for winter irrigation, but there are 24,400 ha available for winter ACREs to be used in the future as the Departamento General de Irrigación (DGI) gradually promulgates a no discharge regulation for all domestic wastewater (Franci, 1999; Rauek, 2020).

(9) **Determine the annual N and P loading rates for the mean irrigation area and total water volume from September through March using Equations (8.7) and (8.8)**.
From Table 8.12, the mean irrigation area and water volume from September–March (in bold) are:

Mean irrigation area (ha)	2298
Water volume (m³/7 mo)	24,619,984

Table 8.12 Monthly irrigated areas without a reservoir for Campo Espejo ACRE.

1	2	8	9	10	11	12	13		14
Month	No. of Days	Volume Wastewater Remaining V_{final} (m³)	P_{eff} (mm)	ET_0 (mm)	$ET_0 - P_{eff}$ (mm)	L_w (mm)	Irrigated Area A_w m²	ha	Volume Irrigation Water V_w (m³)
January	31	3,600,092	35.3	156	121	189.2	19,025,604	1903	3,600,092
February	28	3,251,696	40.0	132	92	143.7	22,630,526	2263	3,251,696
March	31	3,600,092	29.8	104	74	116.6	30,864,156	3086	3,600,092
April	30	3,483,960	12.6	74	61	96.0	36,291,250	3629	3,483,960
May	31	3,600,092	7.9	49	41	64.0	56,292,829	5629	3,600,092
June	30	3,483,960	6.0	34	28	43.3	80,471,902	8047	3,483,960
July	31	3,600,092	8.2	36	28	44.0	81,761,975	8176	3,600,092
August	31	3,600,092	4.8	59	54	84.9	42,422,526	4242	3,600,092
September	30	3,483,960	11.9	85	73	114.0	30,550,543	3055	3,483,960
October	31	3,600,092	17.0	121	104	162.5	22,155,214	2216	3,600,092
November	30	3,483,960	19.9	146	126	198.0	17,599,243	1760	3,483,960
December	31	3,600,092	27.5	155	128	200.0	18,000,460	1800	3,600,092
	Total	42,388,180	220.9	1149	930	1456	Mean	2298	Total 24,619,984

Source: Data from CLIMWAT/CROPWAT.

$$\text{Nitrogen loading: } L_{TN} = \frac{0.001 \cdot (V_{w,yr})(C_{TN})}{A_w} = \frac{0.001 \cdot (24{,}619{,}984 \text{ m}^3/\text{yr})(17 \text{ mg/L})}{2298 \text{ ha}}$$
$$= 182 \text{ kg TN/ha yr}$$

$$\text{Phosphorus loading: } L_{TP} = \frac{0.001 \cdot (V_{w,yr})(C_{TP})}{A_w} = \frac{0.001 \cdot (24{,}619{,}984 \text{ m}^3/\text{yr})(13.9 \text{ mg/L})}{2298 \text{ ha}}$$
$$= 149 \text{ kg TP/ha yr}$$

The TN loading is within the range for all crops in Table 8.3, while the TP loading, as discussed for Cochabamba, is far above the maximum values for both forage (85 kg/ha yr) and field crops (30 kg/ha yr). Excess phosphorus, however, can be stored in the soil without harm to crops. Again, more data from continuous monitoring are needed to have more reliable estimates of nutrient concentrations to calculate loadings.

Commentary. The Campo Espejo ACRE is an excellent example of wastewater irrigation without a reservoir in dry climates that has operated successfully for over 25 years. Winter irrigation of select crops is feasible and is intended to be implemented in the near future as soon as available land can be designated as part of the Campo Espejo ACRE. It is planned that all wastewater used for irrigation in the Mendoza area eventually will meet the zero-discharge regulation of the DGI.

Table 8.13 presents the entire design as it appears on an Excel sheet.

Design of wastewater irrigation systems with valorization of nutrients

Table 8.13 Monthly irrigated areas for Campo Espejo ACRE.

1	2	3	4	5	6	7	8	9	10	11	12	13		14
Month	No. of Days	Initial Volume Wastewater V_{ww} (m³)	e_{evap}/l (mm/d)	e_{evap}/l (mm)	Water Surface Area (m²)	Volume Wastewater Lost V_{lost} (m³)	Volume Wastewater Remaining V_{final} (m³)	P_{eff} (mm)	ET_0 (mm)	$ET_0 - P_{eff}$ (mm)	L_w (mm)	Irrigated Area A_w m²	ha	Volume Irrigation Water V_w (m³)
January	31	4,017,600	5.2	161	2,590,000	417,508	3,600,092	35.3	156	121	189.2	19,025,604	1903	3,600,092
February	28	3,628,800	5.2	146	2,590,000	377,104	3,251,696	40.0	132	92	143.7	22,630,526	2263	3,251,696
March	31	4,017,600	5.2	161	2,590,000	417,508	3,600,092	29.8	104	74	116.6	30,864,156	3086	3,600,092
April	30	3,888,000	5.2	156	2,590,000	404,040	3,483,960	12.6	74	61	96.0	36,291,250	3629	3,483,960
May	31	4,017,600	5.2	161	2,590,000	417,508	3,600,092	7.9	49	41	64.0	56,292,829	5629	3,600,092
June	30	3,888,000	5.2	156	2,590,000	404,040	3,483,960	6.0	34	28	43.3	80,471,902	8047	3,483,960
July	31	4,017,600	5.2	161	2,590,000	417,508	3,600,092	8.2	36	28	44.0	81,761,975	8176	3,600,092
August	31	4,017,600	5.2	161	2,590,000	417,508	3,600,092	4.8	59	54	84.9	42,422,526	4242	3,600,092
September	30	3,888,000	5.2	156	2,590,000	404,040	3,483,960	11.9	85	73	114.0	30,550,543	3055	3,483,960
October	31	4,017,600	5.2	161	2,590,000	417,508	3,600,092	17.0	121	104	162.5	22,155,214	2216	3,600,092
November	30	3,888,000	5.2	156	2,590,000	404,040	3,483,960	19.9	146	126	198.0	17,599,243	1760	3,483,960
December	31	4,017,600	5.2	161	2,590,000	417,508	3,600,092	27.5	155	128	200.0	18,000,460	1800	3,600,092
Total		47,304,000				4,915,820	42,388,180	220.9	1149	475	1456			42,388,180

Data for P_{eff} and ET_0 from CLIMWAT/CROPWAT.
Source: Data from Barbeito Anzorena (2001).
Mean flowrate = 129,600 m³/d; evapotranspiration/infiltration (e_{evap}/l) = 5.2 mm/d; water surface area = 2,590,000 m².

8.3 WASTEWATER IRRIGATION WITH STABILIZATION RESERVOIRS IN WET CLIMATES

8.3.1 Wastewater stabilization reservoirs

In wet climates, or in climates with cold winters, wastewater must be stored during the months when irrigation is not possible. The wastewater storage can also be designed as part of the treatment system, where the residence time distribution in the reservoir contributes to pathogen reduction, and biochemical oxygen demand and total suspended solids removal, hence the term stabilization reservoir. There are numerous configurations for stabilization reservoir design (Juanico, 1996), and the following three to be discussed in the following sections are commonly used in different parts of the world:

(1) Single reservoir
(2) Batch stabilization reservoir with secondary reservoir
(3) Batch stabilization reservoirs in parallel

8.3.1.1 Single reservoir

A single reservoir shown in Figure 8.5 receives a continuous inflow of treatment plant effluent throughout the year. During the nonirrigation season, the reservoir is filled. After the irrigation season begins, as older, stabilized water is removed from the reservoir, fresh influent wastewater from the treatment plant enters, and gradually becomes an increasingly higher percentage of the reservoir volume, what has been called the percentage of fresh effluents (Juanico & Shelef, 1994). If the treatment plant effluent is not of high quality, the reservoir effluent will gradually deteriorate, potentially limiting irrigation potential to the restricted category. If the treatment plant effluent is of high quality in terms of pathogens, such as a well-designed WSP effluent, there should be no deterioration in the reservoir effluent. Under these conditions, a single reservoir is the best solution as a result of ease of operation, and the required reservoir volume, which is the lowest of all alternatives.

8.3.1.2 Batch stabilization reservoir with secondary reservoir

In this system shown in Figure 8.6, a batch stabilization reservoir is filled during the nonirrigation season and allowed to rest for 2–3 months before the irrigation season begins. The resting period

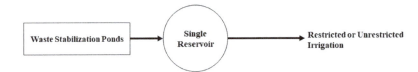

Figure 8.5 A single reservoir receives continuous inflow of treatment plant effluent throughout the year.

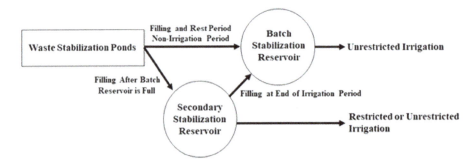

Figure 8.6 A batch stabilization reservoir with a secondary reservoir can combine restricted with unrestricted irrigation possibilities, depending on the quality of the wastewater treatment plant effluent. (*Source*: After Libhaber and Orozco-Jaramillo (2012).)

Design of wastewater irrigation systems with valorization of nutrients 271

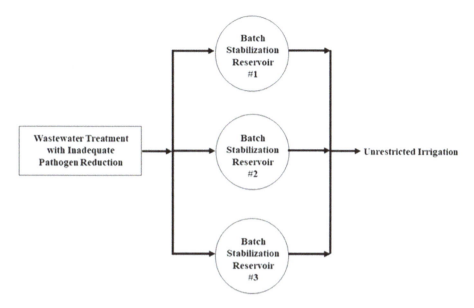

Figure 8.7 An example of three batch stabilization reservoirs operated in parallel. Each reservoir is operated on a staggered, fill-rest-irrigate cycle. The rest cycle should be a minimum of 2–3 months to obtain a high-quality effluent for unrestricted irrigation. The total volume of reservoirs is equal to 1 year's flow volume. (*Source*: After Mara (2003).)

allows for increased pathogen reduction, which is especially important if the wastewater treatment plant effluent is of low quality. The secondary stabilization reservoir is operated with the continuous inflow during the irrigation season, operating in parallel with the batch stabilization reservoir; the effluent can be used for restricted or unrestricted irrigation depending of the quality of the treatment plant effluent. At the end of the irrigation season, the effluent of the secondary reservoir is used to fill the larger batch stabilization reservoir.

8.3.1.3 Batch stabilization reservoirs in parallel

Batch stabilization reservoirs were developed to improve irrigation water quality from inadequately treated wastewater effluents, or even raw wastewater (Juanico, 1996). Figure 8.7 shows an example of three batch stabilization reservoirs operated in parallel. In this design, each reservoir is operated on a staggered, fill-rest-irrigate cycle as presented in Table 8.14. For each month, the total flow volume is partitioned among the three reservoirs by trial and error until all volumes balance for every month (last column in Table 8.14). The rest cycle should be a minimum of 2–3 months to obtain a high-quality effluent for unrestricted irrigation, with a >3.5 $\log_{10}$ reduction of fecal coliforms (Juanico, 1996). The total volume of reservoirs is equal to 1 year's flow volume. This configuration is the most expensive to build and most difficult to operate, but it produces the highest quality effluent for agricultural reuse.

8.3.2 Design procedure for single reservoirs

Steps 1–3 are identical to steps 5–7 used in Section 8.2.1.

(1) **Tabulate monthly data for P_{eff} and ET_0.**
Obtain long-term average monthly site data for effective precipitation, P_{eff}, and reference evapotranspiration, ET_0, from CLIMWAT/CROPWAT or local meteorological stations. Monthly ET_0 values from CROPWAT are represented in units of mm/d and must be converted to mm/mo.

Table 8.14 Example of monthly flow volume partitioning.

Three Batch Stabilization Reservoirs in Parallel for a 6-Month Irrigation Period						
Month	No. of Days	$L_{w,i}$ (mm)	Batch Stabilization Reservoirs Decimal Fraction (DF) of Monthly Volume			$\sum_{n=1}^{3} DF_{n,i}$
			DF_1	DF_2	DF_3	
May	31	0.0	0.333	0.50	0.167	1.00
June	30	0.0	0.333	0.50	0.167	1.00
July	31	0.0	0.333	0.50	0.167	1.00
August	31	0.0	Rest	0.75	0.25	1.00
September	30	0.0	Rest	0.75	0.25	1.00
October	31	0.0	Rest	Rest	1.00	1.00
November	30	60.7	Irrigate	Rest	1.00	1.00
December	31	101.1	Irrigate	Rest	1.00	1.00
January	31	174.1	1.00	Irrigate	Rest	1.00
February	28	151.8	1.00	Irrigate	Rest	1.00
March	31	184.5	0.50	0.50	Irrigate	1.00
April	30	160.9	0.50	0.50	Irrigate	1.00
Months of inflow			4	4	4	12

(2) **Select values for leaching factor, LF, and irrigation efficiency, E.**
Generally, LF = 0.15 and E = 0.75 should be used as discussed in Chapter 7 unless local conditions require other values.

(3) **Calculate and tabulate the monthly hydraulic loading rate using Equation (8.4).**

$$L_{w,i} = \frac{(ET_{0,i} - P_{eff,i})}{(1-LF)E} \tag{8.4}$$

(4) **Calculate the monthly volume of wastewater available with Equation (8.11).**

$$V_{ww,i} = Q_{mean}(\#\,days/month_i) \tag{8.11}$$

where $V_{ww,i}$ is the volume of wastewater for month i (m³) and Q_{mean} is the mean design flowrate (m³/d).

(5) **Determine the irrigation area by summing the monthly volumes of wastewater available, $V_{ww,i}$, and hydraulic loading rates, $L_{w,i}$, and solving Equation (8.17) to determine the irrigation area (USEPA, 1981). Negative values of $L_{w,i}$ are not included in the summation.**

$$A_w = \frac{\sum_{i=1}^{12} V_{ww,i} + \Delta V_S}{\sum_{i=1}^{12} L_{w,i}/1000} \tag{8.17}$$

where A_w is the irrigation area (m²), $V_{ww,i}$ is the wastewater volume for month i (m³), ΔV_S is the net gain or loss in stored wastewater volume in reservoir due to precipitation, evaporation, and infiltration (m³/yr), and $L_{w,i}$ = hydraulic loading based on irrigation requirements for month i (mm).

Note: Negative values of $L_{w,i}$ are not included in summations.

For preliminary design, ΔV_S is typically neglected, but should be included in the final design since the net gain or loss in large reservoirs can be significant (USEPA, 1981).

(6) **Calculate the monthly water volume requirements using Equation (8.18).**

$$V_{\text{required},i} = (L_{w,i}/1000)(A_w) \qquad (8.18)$$

where $V_{\text{required},i}$ is the volume of water required for irrigation for month i (m³)

(7) **Calculate the monthly change in storage with Equation (8.19).**

$$\Delta S_i = V_{\text{required},i} - V_{\text{ww},i} \qquad (8.19)$$

where ΔS_i is the change in storage for month i (m³).

(8) **Calculate the monthly cumulative storage with Equation (8.20).**

$$\sum_i \Delta S = \Delta S_{i-1} + \Delta S_i \qquad (8.20)$$

where ΔS_{i-1} is the change in storage from the previous month (m³).

(9) **Determine the maximum volume of water to be stored in the reservoir.**
The maximum volume is found by inspection of the monthly cumulative storage values.

8.3.3 Case study design example: irrigation with a single reservoir, Sololá, Guatemala

The wastewater treatment and reuse project at San Antonio, Sololá, discussed in Section 7.2.2, has an inadequately sized reservoir originally constructed for a pilot reuse project of 1.84 ha. Irrigation is only possible during the 6-month dry season, and a single reservoir will be designed to enable 100% wastewater storage during the wet season. Because the treatment plant was not designed for pathogen reduction, wastewater reuse will be for restricted irrigation, with a waiting period for natural pathogen die-off in the fields as discussed in Section 7.2.2.

The pertinent data for the San Antonio irrigation system is shown in the following table (Sánchez de León, 2001).

San Antonio Wastewater Reuse System	
Mean influent flowrate, m³/d	907
Pilot project irrigated area, ha	1.84
Existing reservoir volume, m³	48

A photo of the existing reservoir is shown in Figure 8.8.

Table 8.15 lists the precipitation and evapotranspiration data for Sololá, along with the calculated monthly results for the hydraulic loading, $L_{w,i}$, and the available wastewater volume, $V_{ww,i}$. In the last column, the irrigation area is calculated using Equation (8.17). The order of the months is arranged with the first 6 months filling the reservoir, and the last 6 months withdrawing water for irrigation.

Using the design procedure for a single reservoir, the calculations for the first month, May, are as follows:

(1) **Tabulate monthly data for P and ET_0.**

The long-term average monthly site data for precipitation, P, and reference evapotranspiration, ET_0, are tabulated from CLIMWAT/CROPWAT in columns 3 and 4. Effective precipitation data were not available. Column 5: The monthly ET_0 value for May = (3.87 mm/d)(31 d) = 120 mm

Column 6: $ET_{0,1} - P_1 = 120$ mm $- 179.4$ mm $= -59$ mm

(2) **Select values for leaching factor, LF, and irrigation efficiency, E.**
LF = 0.15 and $E = 0.75$.

Figure 8.8 The existing reservoir at the San Antonio, Sololá wastewater reuse project has an estimated maximum volume of 50 m³ and was designed for a pilot irrigation project of 1.84 ha.

Table 8.15 Irrigation area for Sololá, Guatemala.[1]

1		2	3	4	5	6	7	8	9
Month		No. of Days	P (mm)	ET_0 (mm/d)	ET_0 (mm)	$ET_0 - P$ (mm)	L_w (mm)	Volume Wastewater Available V_{ww} (m³)	A_w (m²)
May	1	31	179.4	3.87	120	−59		28,117	
June	2	30	304.9	3.53	106	−199		27,210	
July	3	31	174.4	3.33	103	−71		28,117	
August	4	31	202.4	3.27	101	−101		28,117	
September	5	30	279.7	3.11	93	−186		27,210	
October	6	31	164.6	3.22	100	−65		28,117	
November	7	30	45.6	2.81	84	39	60.7	27,210	
December	8	31	10.9	2.43	75	64	101.1	28,117	
January	9	31	3.4	3.69	114	111	174.1	28,117	
February	10	28	4.6	3.62	101	97	151.8	25,396	
March	11	31	19.7	4.43	137	118	184.5	28,117	
April	12	30	41.4	4.8	144	103	160.9	27,210	
		Total	1431		1280	531	833	331,055	397,371

[1]Mean flowrate = 907 m³/d; data for P and ET_0 from CLIMWAT/CROPWAT; LF = 0.15; E = 0.75.

(3) **Calculate and tabulate the monthly hydraulic loading rate using Equation (8.4).**
For May,

$$L_{w,1} = \frac{(ET_{0,1} - P_1)}{(1-LF)E} = \frac{(120 \text{ mm} - 179.4 \text{ mm})}{(1-0.15)(0.75)} = 0$$

Only positive values can be used in column 7.

(4) **Calculate the monthly volume of wastewater available with Equation (8.11).**

$$V_{ww,1} = Q_{mean}(\#\text{days/month}_i) = (907 \text{ m}^3/\text{d})(31 \text{ d}) = 28,117 \text{ m}^3$$

Design of wastewater irrigation systems with valorization of nutrients 275

(5) **Determine the irrigation area by summing the monthly volumes of wastewater available, $V_{ww,i}$, and hydraulic loading rates, $L_{w,i}$, and solving Equation (8.17) to determine the irrigation area (USEPA, 1981). Negative values of $L_{w,i}$ are not included in the summation.**

$$A_w = \frac{\sum_{i=1}^{12} V_{ww,i} + \Delta V_S}{\sum_{i=1}^{12} L_{w,i}/1000} = \frac{331{,}055 \text{ m}^3}{(833 \text{ mm})/(1000 \text{ mm/m})} = 397{,}371 \text{ m}^2 = 39.7 \text{ ha}$$

This area is substantially larger than the 1.84 ha originally used as a pilot project.

Table 8.16 lists the calculated monthly results the volume of irrigation water, $V_{w,i}$, the change in storage, ΔS_i, and the cumulative storage, $\Sigma \Delta S$.

(6) **Calculate the monthly water volume requirements using Equation (8.13).**
For May–October:

$$V_{w,1-6} = (L_{w,1-6}/1000)(A_w) = 0$$

For November:

$$V_{w,7} = (L_{w,7}/1000)(A_w) = (60.7 \text{ mm})/(1000 \text{ mm/m})(397{,}371 \text{ m}^2) = 24{,}123 \text{ m}^3$$

(7) **Calculate the monthly change in storage with Equation (8.14).**
For May:

$$\Delta S_1 = V_{w,1} - V_{ww,1} = 28{,}117 \text{ m}^3 - 0 = 28{,}117 \text{ m}^3$$

For November:

$$\Delta S_7 = V_{w,7} - V_{ww,7} = 27{,}210 \text{ m}^3 - 24{,}123 \text{ m}^3 = 3087 \text{ m}^3$$

Table 8.16 Storage volumes and reservoir volume for Sololá, Guatemala.

1		8	9	10	11	12
Month		Volume WW Available V_{ww} (m³)	A_w (m²)	Volume Irrigation Water V_w (m³)	Change in Storage ΔS (m³)	Cumulative Storage $\Sigma \Delta S$ (m³)
May	1	28,117		0	28,117	0
June	2	27,210		0	27,210	27,210
July	3	28,117		0	28,117	55,327
August	4	28,117		0	28,117	83,444
September	5	27,210		0	27,210	110,654
October	6	28,117		0	28,117	138,771
November	7	27,210		24,123	3087	**141,858**
December	8	28,117		40,161	−12,044	129,814
January	9	28,117		69,183	−41,066	88,748
February	10	25,396		60,313	−34,917	53,831
March	11	28,117		73,322	−45,205	8626
April	12	27,210		63,953	−36,743	−28,117
Totals: 331,055			397,371	331,055		

(8) **Calculate the monthly cumulative storage with Equation (8.20).**
The last contiguous month when wastewater is taken out of the reservoir is the end of the irrigation season, which is April in Table 8.16. The reservoir is thus considered empty in May at the start of the storage season.
The cumulative storage for June is

$$\sum_{i=1}^{2} \Delta S = \Delta S_{2-1} + \Delta S_2 = 0 + 27{,}210 \text{ m}^3 = 27{,}210 \text{ m}^3$$

The subsequent calculations continue until the final month of April, which gives a cumulated storage of −28,117 m³, which closes the cumulative storage to 0 in May:

$$\sum_{i=12}^{1} \Delta S = \Delta S_{12-1} + \Delta S_1 = -27{,}210 \text{ m}^3 + 27{,}210 \text{ m}^3 = 0$$

(9) **Determine the maximum volume of water to be stored in the reservoir.**
The maximum volume to be stored in the reservoir is 141,858 m³ as found by inspection of column 12. The required storage volume is 42.8% of the annual volume of wastewater produced for irrigation:

$$\frac{\text{Required storage volume}}{\text{Annual volume of irrigation water}} = \left(\frac{141{,}858 \text{ m}^3}{331{,}055 \text{ m}^3}\right)(100\%) = 42.8\%$$

Table 8.17 presents the entire design as it appears on an Excel sheet.

Commentary. This integrated wastewater reuse design for a population of 7000 persons, with a potential of irrigating 39.7 ha, could significantly aid impoverished farmers and foster sustainable agriculture with wastewater reuse in a region where synthetic fertilizers are seriously overused. A reservoir would also eliminate at least 6 months of wastewater discharge into Lake Atitlán, which is undergoing serious contamination from wastewater pathogens and nutrients (Oakley & Saravia, 2021). There are several limiting factors, however, that must be addressed if proposed projects such as this one will ever have a chance of execution.

- With a required minimum reservoir volume of 141,858 m³, a reservoir with a mean depth of 10 m would require and area of 14,186 m², or 1.4 ha, in an area within the Lake Atitlán basin where land sells for approximately US$100/m².
- The cost of construction of the reservoir, apart from the cost of land, could range from US$5 to $10/m³, for a total cost ranging from US$70,000 to US$1,400,000.
- Finally, and most importantly, the quality of the effluent from the San Antonio wastewater treatment plant is unacceptable in terms of pathogen reduction as discussed in Section 7.2.2, and a continuous flow reservoir cannot provide an acceptable irrigation effluent meeting the WHO guidelines for restricted irrigation every month of irrigation. In this case, pathogen reduction by natural die-off, washing, and cooking (Table 7.4) would have to be used as discussed in Section 7.2.2. A continuous flow reservoir, with all its limitations, is still preferable to discharging 6 months of wastewater into the seriously threatened Lake Atitlán.

8.3.4 Design procedure for three batch stabilization reservoirs in parallel
In this design shown in Table 8.14, each reservoir is operated on a fill-rest-irrigate cycle. The monthly volumes of wastewater available are partitioned among the three reservoirs by trial and error until the monthly totals are equal to the wastewater available (last column in Table 8.14), with each reservoir

Design of wastewater irrigation systems with valorization of nutrients 277

Table 8.17 Irrigation area and reservoir volume for San Antonio, Sololá, Guatemala.[1]

1	2	3	4	5	6	7	8	9	10	11	12
Month	No. of Days	P (mm)	ET_0 (mm/d)	ET_0 (mm)	$ET_0 - P$ (mm)	L_w (mm)	Volume Wastewater Available V_{ww} (m^3)	A_w (m^2)	Volume Irrigation Water V_w (m^3)	Change in Storage ΔS (m^3)	Accumulation in Storage $\Sigma\Delta S$ (m^3)
May	1	31	179.4	3.87	120	−59		28,117		0	28,117	0
June	2	30	304.9	3.53	106	−199		27,210		0	27,210	27,210
July	3	31	174.4	3.33	103	−71		28,117		0	28,117	55,327
August	4	31	202.4	3.27	101	−101		28,117		0	28,117	83,444
September	5	30	279.7	3.11	93	−186		27,210		0	27,210	110,654
October	6	31	164.6	3.22	100	−65		28,117		0	28,117	138,771
November	7	30	45.6	2.81	84	39	60.7	27,210		24,123	3087	141,858
December	8	31	10.9	2.43	75	64	101.1	28,117		40,161	−12,044	129,814
January	9	31	3.4	3.69	114	111	174.1	28,117		69,183	−41,066	88,748
February	10	28	4.6	3.62	101	97	151.8	25,396		60,313	−34,917	53,831
March	11	31	19.7	4.43	137	118	184.5	28,117		73,322	−45,205	8626
April	12	30	41.4	4.8	144	103	160.9	27,210		63,953	−36,743	−28,117
Total		1431		1280	531	833	331,055	397,371	331,055		

[1]Mean flowrate = 907 m^3/d (Sánchez de León, 2001); data for P and ET_0 from CLIMWAT/CROPWAT; LF = 0.15; E = 0.75.

having the same allotted time for rest and irrigation cycles, if possible. The rest cycles should be at least 2–3 months for pathogen reduction measured as *E. coli*.

For a given month, the volumes partitioned to each reservoir are determined with Equation (8.21).

$$V_{1,i} = DF_{1,i}(V_{ww,i})$$
$$V_{2,i} = DF_{2,i}(V_{ww,i})$$ (8.21)
$$V_{3,i} = DF_{3,i}(V_{ww,i})$$

where $V_{1,i}, V_{2,i}, V_{3,i}$ are the volumes of water partitioned to each reservoir during month i (m³), $DF_{1,i}, DF_{2,i}, DF_{3,i}$ are decimal fractions of wastewater partitioned to each reservoir in month i, and $V_{ww,i}$ is the volume of wastewater available during month i (m³).

The decimal fractions are determined by trial and error as presented in Table 8.14.

The sum of the decimal fractions for each month must equal 1.0 as shown in Equation (8.22).

$$\sum_{n=1}^{3} DF_{n,i} = 1.0$$ (8.22)

The total volume of the wastewater stored in month i, also must equal the volume of wastewater available from Equation (8.23):

$$\sum_{n=1}^{3} V_{w,n,i} = V_{ww,i}$$ (8.23)

where $V_{w,n,i}$ is the wastewater stored in reservoir n, during month i (m³).

8.3.5 Case study design example: batch stabilization reservoirs, Sololá, Guatemala

To improve irrigation water quality, it is proposed to design a system of three batch stabilization reservoirs in parallel for the San Antonio irrigation system in Sololá, Guatemala. The design of a single, continuous flow reservoir is presented in Section 8.3.3.

The precipitation, evapotranspiration, hydraulic loading rate, available wastewater volume, irrigation area, and volume of irrigation water data from Table 8.17 apply. Table 8.18 presents the calculated design data using the decimal fractions developed in Table 8.14.

Example calculations for the first month, May, are shown below using Equations (8.21) and (8.23):
Using Equation (8.21):

Reservoir 1 (column 11): $V_{1,1} = DF_{1,1}(V_{ww,1}) = (0.333)(28,117 \text{ m}^3) = 9372 \text{ m}^3$

Reservoir 2 (column 12): $V_{2,1} = DF_{2,1}(V_{ww,1}) = (0.50)(28,117 \text{ m}^3) = 14,059 \text{ m}^2$

Reservoir 3 (column 13): $V_{3,1} = DF_{3,1}(V_{ww,1}) = (0.167)(28,117 \text{ m}^3) = 4686 \text{ m}^3$

From Equation (8.23), the sum of the three volumes put in each reservoir (column 14) equals the volume of available wastewater for the month (column 8):

$$\sum_{n=1}^{3} V_{w,n,1} = 9372 \text{ m}^3 + 14,059 \text{ m}^2 + 4686 \text{ m}^3 = 28,117 \text{ m}^3 = V_{ww,1}$$

The annual maximum volumes of water stored in each reservoir in Table 8.18, 108,991, 110,881, and 111,183 m³, are within a narrow range and add up to the annual total volume of wastewater produced of 331,055 m³.

In this design, each reservoir uses all of its stored water in 2 months and begins refilling immediately afterward. After filling, reservoirs 2 and 3 have a 3-month rest period before stored water is used for

Table 8.18 Three batch stabilization reservoir design for San Antonio, Sololá, Guatemala.

1		8	10	11		12		13		14
Month	n	Volume WW Available V_{ww} (m³)	Volume Irrigation Water V_w (m³)	Reservoir Number						Volume WW Available V_w (m³)
				1		2		3		
				DF_1	V_1 (m³)	DF_2	V_2 (m³)	DF_3	V_3 (m³)	
May	1	28,117	0	0.333	9372	0.50	14,059	0.167	4686	28,117
June	2	27,210	0	0.333	9070	0.50	13,605	0.167	4535	27,211
July	3	28,117	0	0.333	9372	0.50	14,059	0.167	4686	28,118
August	4	28,117	0	Rest		0.75	21,088	0.25	7029	28,118
September	5	27,210	0	Rest		0.75	20,408	0.25	6803	27,211
October	6	28,117	0	Rest		Rest		1.00	28,117	28,118
November	7	27,210	24,123	Irrigate		Rest		1.00	27,210	27,211
December	8	28,117	40,161	Irrigate		Rest		1.00	28,117	28,118
January	9	28,117	69,183	1.00	28,117	Irrigate		Rest		28,117
February	10	25,396	60,313	1.00	25,396	Irrigate		Rest		25,396
March	11	28,117	73,322	0.50	14,059	0.50	14,059	Irrigate		28,118
April	12	27,210	63,953	0.50	13,605	0.50	13,605	Irrigate		27,211
Totals		331,055	331,055	4.00	108,991	4.00	110,881	4.00	111,183	331,055

irrigation; reservoir 3 has a 2-month rest period, and the effluent quality may be slightly lower in terms of pathogen reduction. But still should be much better than a continuous flow single reservoir.

Commentary. The operational characteristics of a batch reservoir can also influence effluent quality. If water is withdrawn from the bottom of the reservoir, settled helminth eggs can be swept up with the effluent in rising currents; bacterial die-off in the hypolimnion is also much lower than the sunlit epilimnion (Juanico, 1996). If possible, water should be drawn off close to the surface, and a minimum depth of water should always be left in the reservoir during irrigation season to avoid resuspension of settled helminth eggs.

The construction of this batch stabilization reservoir system would be double the cost of the single reservoir design as it requires storing 1 year's volume of wastewater. The operation would also be more difficult, requiring accurate flow measurements to partition water inflows into the three reservoirs on a monthly basis. In resource-challenged areas in wet climates, a better approach would be to design a WSP system that has adequate pathogen reduction, and use a single, continuous flow reservoir, which is the easiest to operate with the lowest cost of construction.

8.4 WASTEWATER IRRIGATION WITHOUT RESERVOIRS IN WET CLIMATES USING LAND APPLICATION

A viable alternative to reservoir construction in wet climates is the use of land application of treated wastewater during the wet season, when precipitation is greater than evapotranspiration. Wastewater is applied during month i based on the following water balance equation in units of mm/mo (USEPA, 1981):

$$\text{Applied wastewater}_i = (\text{Evapotranspiration} - \text{Precipitation})_i + \text{Percolation}_i \tag{8.24}$$

In land application, wastewater is applied to the soil at predetermined percolation rates, unlike the irrigation season where wastewater is applied according to crop needs. This method is preferable to surface water discharge, as a significant fraction of nutrients and organic matter can be adsorbed by the soil to be used by plants during the growing seasons. The leaching of nitrate nitrogen to groundwater is a potential problem; however, that must be assessed in areas where groundwater is used as a drinking water source (USEPA, 1981).

8.4.1 Design equations for land application during the wet season

This procedure uses the method presented in Section 8.2.1 during the irrigation season to determine monthly irrigation areas, with the following equations used during the wet season (USEPA, 1981).

8.4.1.1 Hydraulic loading rate based on soil permeability

Equation (8.25) is used to calculate the hydraulic loading rate for the months when total precipitation exceeds evapotranspiration: $(ET_{0,i} - P_i) < 0$. Total precipitation is used rather than effective precipitation to add an additional safety factor.

$$L_{w(p),i} = (ET_{0,i} - P_i) + P_{ww,i} \quad (8.25)$$

where $L_{w(p),i}$ is the hydraulic loading rate based on soil permeability for month i (mm), $ET_{0,i}$ is the reference evapotranspiration for month i (mm), P_i is the total precipitation for month i (mm), and $P_{ww,i}$ is the design wastewater percolation rate for month i (mm); varies from 4.0 to 10% of clear water soil permeability or saturated hydraulic conductivity.

The design monthly wastewater percolation rate, $P_{ww,i}$, is determined from field studies measuring the clear water permeability for the restricted soil layer using Equation (8.26):

$$P_{ww,i} = P_{cw}(d_i)(CF_{ww}) \quad (8.26)$$

where $P_{ww,i}$ is the wastewater percolation rate for month i (mm), P_{cw} is the clear water permeability for soil (mm/d), d_i is the number of days in month i, and CF_{ww} is the correction factor for wastewater (0.04–0.10).

The ranges of clear water permeability for the most restrictive layer in a soil profile are shown in Table 8.19. For most designs, the land application system should be designed as a slow rate system, with clear water permeabilities measured in the field ranging from 36 to 360 mm/d.

The correction factor for wastewater, CF_{ww}, in Equation (8.26) ranges from 4 to 10% of the measured clear water permeabilities. Lower values of CF_{ww} should be used for variable or poorly defined soil conditions.

The annual design wastewater percolation rate without considering nitrogen loading limits can range from 500 to 6000 mm/yr for slow rate systems (Table 8.19). Design percolation rates should be as low as possible for the site conditions and land area available to foster the adsorption of nutrients and organic matter into the soil.

Table 8.19 Clear water permeability of most restrictive layer in a soil profile.

Land Application System	Restrictive Layer Classification	Clear Water Permeability (mm/d)	Design Wastewater Percolation Rate[1] (mm/yr)
Unsuitable	Very slow	<36	–
Slow rate	Slow	36–122	500–4000
	Moderately slow	122–360	4000–6000
Rapid infiltration	Moderate	360–1224	6000–45,000
	Moderately rapid	1224–3648	45,000–150,000
	Rapid	3648–12,000	150,000–200,000

Source: From USEPA (1981).
[1]Nitrogen limits are not considered.

8.4.1.2 Land area requirements without reservoir varying irrigated area each month

Monthly land area requirements are calculated using Equation (8.27):

$$A_{w,i} = \frac{V_{w,i}}{L_{w(p),i}/1000} \tag{8.27}$$

where $A_{w,i}$ is the irrigation area for each month i, with $ET_{0,i} - P_i < 0$ (m²), $V_{w,i}$ is the available wastewater volume for each month with $ET_{0,i} - P_i < 0$ (m³), and 1000 is the conversion factor (mm to m).

8.4.1.3 Hydraulic loading based on nitrogen limits

In areas where groundwater may be used for drinking water, Equation (8.28) is used to calculate the maximum annual hydraulic loading based on nitrogen limits.

$$L_{w(N)} = \frac{(C_p)(P_{yr} - ET_{0,yr}) + U(100)}{(1-f)(C_N) - C_p} \tag{8.28}$$

where $L_{w(N)}$ is the hydraulic loading based on nitrogen limits (mm/yr); C_p is the TN concentration in percolating water, assume 10 mg/L; P_{yr} is the annual precipitation (mm/yr); $ET_{0,yr}$ is the annual reference evapotranspiration (mm/yr); U is the TN uptake by crops (kg TN/ha yr); 100 is the combined conversion factor; f is the decimal fraction of nitrogen removed by denitrification and volatilization; and C_N is the TN concentration in applied wastewater (mg/L).

Equation (8.28) assumes that the maximum concentration of nitrate nitrogen in the receiving groundwater does not exceed 10 mg/L.

8.4.2 Design procedure for irrigation without reservoirs in wet climates

In wet climates, irrigation is separated into the dry and wet season. During the dry season, irrigation design uses the procedure outlined in Section 8.2.1 assuming wastewater loss due to evaporation is negligible. Wet season design uses Equations (8.25)–(8.28).

Dry season design:

(1) **Calculate initial monthly volumes of wastewater using the mean design flowrate.**

$$V_{ww,i} = Q_{mean}(\#\,days/month_i) \tag{8.11}$$

where $V_{ww,i}$ is the initial volume of wastewater based on influent flowrate for month i (m³) and Q_{mean} is the design mean flowrate (m³/d).

(2) **Tabulate monthly data for P_{eff} and ET_0.**
Obtain long-term average monthly site data for effective precipitation, P_{eff}, and reference evapotranspiration, ET_0, from CLIMWAT/CROPWAT or local meteorological stations. Monthly ET_0 values from CROPWAT are in units if mm/d and must be converted to mm/mo.

(3) **Select values for leaching factor, LF, and irrigation efficiency, E.**
Generally, LF = 0.15 and E = 0.75 should be used as discussed in Chapter 7 unless local conditions require other values.

(4) **Calculate the monthly hydraulic loading rate using Equation (8.4).**

$$L_{w,i} = \frac{(ET_{0,i} - P_{eff,i})}{(1 - LF)E} \tag{8.4}$$

(5) **Calculate the irrigation areas for every month using Equation (8.5).**

$$A_{w,i} = \frac{V_{final,i}}{L_{w,i}/1000} \tag{8.5}$$

Wet season design:

(6) **Calculate the monthly hydraulic loading rate based on soil permeability using Equations (8.25) and (8.26).**

$$L_{w(p),i} = (ET_{0,i} - P_i) + P_{ww,i} \qquad (8.25)$$

$$P_{ww,i} = P_{cw}(d_i)(CF_{ww}) \qquad (8.26)$$

(7) **Calculate the monthly land application areas using Equation (8.27).**

$$A_{w,i} = \frac{V_{w,i}}{L_{w(p),i}/1000} \qquad (8.27)$$

(8) **Check the annual hydraulic loading based on TN limits with Equation (8.28).**

$$L_{w(N)} = \frac{(C_p)(P_{yr} - ET_{0,yr}) + U(100)}{(1-f)(C_N) - C_p} \qquad (8.28)$$

8.4.3 Case study design example: crop irrigation and land application, Sololá, Guatemala

As a result of the high costs and improbability of constructing a single continuous flow reservoir, or three batch stabilization reservoirs in parallel, designed in Sections 8.3.3 and 8.3.4, it is proposed to design a wastewater irrigation and land application system for the San Antonio reuse project in Sololá, Guatemala. The precipitation, evapotranspiration, and wastewater volume data from Table 8.17 apply. Table 8.20 presents the results for both dry season and wet season design.

Dry season design.

Calculations are shown for the first irrigation month of November.

(1) **Calculate the initial monthly volume of wastewater using the mean design flowrate (Equation (8.11)) for column 7.**

$$V_{ww,1} = Q_{mean}(\#\,days/month_1) = (907\ m^3/d)(30\ d) = 27{,}210\ m^3$$

(2) **Tabulate monthly data for P_{eff} and ET_0.**
From CLIMWAT/CROPWAT for the month of January:
P_{eff} data are not reported and total precipitation will be used.

$$P_1 = 45.6\ mm$$
$$ET_{0,1} = (2.81\ mm/d)(30\ d) = 84\ mm$$
$$ET_{0,1} - P_1 = 84\ mm - 45.6\ mm = 39\ mm$$

(3) **Select values for leaching factor, LF, and irrigation efficiency, E.**
Assume $LF = 0.15$ and $E = 0.75$.

(4) **Calculate the monthly hydraulic loading rate using Equation (8.4).**
For November:

$$L_{w,1} = \frac{(ET_{0,1} - P_1)}{(1 - LF)E} = \frac{(84\ mm - 45.6\ mm)}{(1 - 0.15)(0.75)} = 60.7\ mm$$

(5) **Calculate the irrigation areas for every month using Equation (8.5).**
For November:

$$A_{w,1} = \frac{V_{final,1}}{L_{w,1}/1000} = \frac{27{,}210\ m^3}{(60.7\ mm)/(1000\ mm/m)} = 448{,}227\ m^2 = 45\ ha$$

Table 8.20 Irrigation and land application design for Sololá, Guatemala.[1]

		1	2	3	4	5	6	7	8	9		Irrigation and Land Applied Water V_w (m³)
	Month		P (mm)	ET₀ (mm/d)	ET₀ (mm)	P_{ww} (mm)	ET₀ − P (mm)	L_w (mm)	Wastewater Volume V_{ww} (m³)	A_w m²	ha	
Dry season crop irrigation[2]	November	1	45.6	2.81	84	39		60.7	27,210	448,227	45	27,210
	December	2	10.9	2.43	75	64		101.1	28,117	278,203	28	28,117
	January	3	3.4	3.69	114	111		174.1	28,117	161,497	16	28,117
	February	4	4.6	3.62	101	97		151.8	25,396	167,321	17	25,396
	March	5	19.7	4.43	137	118		184.5	28,117	152,381	15	28,117
	April	6	41.4	4.80	144	103		160.9	27,210	169,068	17	27,210
	Total	127			657		833					
					531	$L_{w(p)}$ (mm)						
Wet season land application[3]	May	7	179.4	3.87	120	−59	310	250.6	28,117	112,212	11	28,117
	June	8	304.9	3.53	106	−199	300	101.0	27,210	269,406	27	27,210
	July	9	174.4	3.33	103	−71	310	238.8	28,117	117,728	12	28,117
	August	10	202.4	3.27	101	−101	310	209.0	28,117	134,550	13	28,117
	September	11	279.7	3.11	93	−186	300	113.6	27,210	239,525	24	27,210
	October	12	164.6	3.22	100	−65	310	245.2	28,117	114,660	11	28,117
	Total	1305			624		1158					
	Total				331,055		331,055					

[1]Q_{mean} = 907 m³/d; data for P and ET₀ from CLIMWAT/CROPWAT.
[2]Crop irrigation; LF = 0.15; E = 0.75.
[3]Land application: P_{cw} = 100 mm/d; CF_{ww} = 0.10.

Table 8.20 presents the results for the 6-month irrigation season, November through April, with monthly irrigation areas ranging from 15 to 45 ha.

Wet season design:

Calculations are shown for the first land application month of May.

(6) **Calculate the monthly hydraulic loading rate based on soil permeability using Equations (8.25) and (8.26).**

No clear water permeability data are available at the San Antonia irrigation site. For preliminary design, a value of 100 mm/d, which is in the slow restrictive layer category in Table 8.19, will be used. A correction factor for wastewater of 0.10 is assumed.

$P_{cw} = 100$ mm/d

$CF_{ww} = 0.10$

Using Equation (8.26) for the month of May,

$P_{ww,7} = P_{cw}(d_7)(CF_{ww}) = (100 \text{ mm/d})(31 \text{ d})(0.10) = 310 \text{ mm}$

The hydraulic loading rate for the month of May is calculated with Equation (8.25):

$L_{w(p),7} = (ET_{0,7} - P_7) + P_{ww,7} = (120 \text{ mm} - 179.4 \text{ mm}) + 310 \text{ mm} = 250.6 \text{ mm}$

(7) **Calculate the monthly land application areas using Equation (8.27).**

The monthly volume of wastewater for the month of May is

$V_{ww,7} = Q_{mean}(\# \text{days/month}_1) = (907 \text{ m}^3/\text{d})(31 \text{ d}) = 28{,}117 \text{ m}^3$

The irrigated area for May is

$A_{w,7} = \dfrac{V_{w,7}}{L_{w(p),7}/1000} = \dfrac{28{,}117 \text{ m}^3}{250.6 \text{ mm}/1000 \text{ mm/m}} = 112{,}212 \text{ m}^2 = 11.2 \text{ ha}$

Table 8.20 presents the results for the 6-month land application period, May through October, with monthly irrigation areas ranging from 11 to 27 ha.

(8) **Check the annual hydraulic loading based on TN limits with Equation (8.28).**

Equation (8.28) is used for the wet season only. The following parameters are assumed:

$U = 50$ kg TN/ha yr during the wet season

$f = 0.20$

$C_N = 25$ mg/L

$L_{w(N)} = \dfrac{(C_p)(P_{yr} - ET_{0,yr}) + U(100)}{(1-f)(C_N) - C_p} = \dfrac{(10 \text{ mg/L})(1305 \text{ mm} - 624 \text{ mm}) + (50)(100)}{(1-0.2)(25 \text{ mg/L}) - 10 \text{ mg/L}}$

$= 1181$ mm/yr $> L_{w(p)} = 1158$ mm/yr (Table 8.20, Column 6)

The annual hydraulic loading based on nitrogen limits is slightly higher than the calculated hydraulic loading rate for soil percolation. Field tests on soil permeability, nitrogen uptake by crops or forests during the wet season, denitrification rates in the soil, and TN concentrations in wastewater would have to be performed to ensure that TN concentrations would not exceed the drinking water limit of 10 mg/L TN, none of which are likely to happen in a resource-limited city such as Sololá. Groundwater is not used as a drinking water source in Sololá, however, so nitrogen limits would not be an issue in this particular case.

Commentary. Assuming the values of $P_{cw} = 100$ mm/d and $CF_{ww} = 0.10$ represent actual field conditions, the monthly irrigated areas during the wet season (11–27 ha) are near the range of those during the irrigation season (15–45 ha). It would be best to have a designated wet season irrigation area to avoid excessive nitrogen accumulation in the soil that could affect crop growth during the irrigation season. Rudimentary field studies would have to be performed to better estimate values of P_{cw} and CF_{ww}. This approach is a viable alternative to discharging to surface waters in the wet season that enter Lake Atitlán, or proposing reservoirs that will likely never be built.

doi: 10.2166/9781789061536_0287

Chapter 9
Physical design and aspects of construction

9.1 INTRODUCTION

The process designs discussed in previous chapters eventually have to be implemented into a commensurate physical design. The physical design is as fundamental to the successful operation of a pond system as the process design and can fundamentally affect the treatment efficiency. Poorly designed and constructed inlet and outlet structures, for example, can cause extensive hydraulic short-circuiting that causes very poor performance. Physical design includes detailed site selection; adapting original pond dimensions to actual ones compatible with site topography; design of exterior and interior embankments, including the interior revetment and freeboard depth; design of pond inlet, outlet, and pond interconnection structures; the construction of fencing to prevent unauthorized access to the installation; and the construction of facilities for operators and a security guard. Table 9.1 presents recommended guidelines for the physical design and construction of waste stabilization ponds. Examples of well-designed and constructed pond systems are shown in Figure 9.1.

9.2 SITE SELECTION

The site to be selected for construction of pond systems should be located with respect to topography, existing and projected population centers, and wind direction. The selected site should have a flat topography to minimize earth movement and must be above the flood plain. The site should also be selected to take advantage of gravity flow to avoid the use of pumping, which requires maintenance and consumption of electrical energy: two pond systems in Honduras were abandoned due to problems with operation and maintenance of their pumping stations (Oakley, 2005). Also, consideration should be given to stormwater diversion and, if necessary, the construction of a stormwater diversion system to protect the ponds from embankment erosion and sediment loadings (see Figures 9.2 and 9.3).

It is recommended that a waste stabilization pond system be located downwind, at a distance greater than 200 m, and preferably greater than 500 m, from the population it serves (existing and projected). This objective is to alleviate the public's potential concerns of odors and unsightly conditions, which are unwarranted if the system is well designed and operated, but completely warranted if the opposite occurs as a result of bad design and poor operation and maintenance. The site should be located at a distance greater than 2 km from an airport, since ponds attract birds that can pose risks to air traffic (Mara *et al.*, 1992).

© 2022 The Author. This is an Open Access book chapter distributed under the terms of the Creative Commons Attribution Licence (CC BY-NC-ND 4.0), which permits copying and redistribution for noncommercial purposes with no derivatives, provided the original work is properly cited (https://creativecommons.org/licenses/by-nc-nd/4.0/). This does not affect the rights licensed or assigned from any third party in this book. The chapter is from the book *Integrated Wastewater Management for Health and Valorization: A Design Manual for Resource Challenged Cities*, Stewart M. Oakley (Author).

Table 9.1 Recommended physical design and construction guidelines.

Parameter	Recommended Guideline
Land selection	
Topography	Flat terrain where stormwater runoff and flooding can be avoided, and gravity flow can be used; pumping should be avoided if possible
Population distance	≥ 200 m and preferably ≥ 500 m
Distance from an airport	≥ 2 km
Orientation to the wind	Inlet–outlet axis of ponds should be oriented perpendicular to the prevailing winds to minimize hydraulic short-circuiting
Geotechnical investigations	
Slope and embankment design	Generally, 3/1 (horizontal/vertical) for interior slopes and 1.5/1 to 2/1 for exterior slopes, depending on the results of the soil mechanics study
Impermeabilization of pond bottom	Clay liners should be used, preferably with $k < 10^{-9}$ m/s measured *in situ* and covered with a final layer of soil to protect the clay
Water balance	$Q_{med} \geq 0.001 \cdot A_T[(P - E) + I]$
Pretreatment	
Bar screens	Made of stainless or galvanized steel, or aluminum
Grit chamber	Two chambers in parallel, each with drainage and gates that seal well
Flow measurement	Prefabricated Parshall flume downstream of grit chamber; used to measure flowrates, and control horizontal velocity through bar screens and grit chamber channels
Pond hydraulics	
Inlets and outlets	*Facultative ponds*: multiple, water level adjustable, inlet and outlet devices. Exterior $L/W = 3/1$ *Maturation ponds*: baffled channels with a single, water level adjustable, inlet and outlet, with interior $L/W \geq 50/1$ to approximate plug flow
Hydraulic structures	
Distribution devices for parallel pond batteries	Open channels with adjustable weirs or sluice gates to control flowrates. Parshall flumes or weirs to measure each distributed flowrate
Distribution devices for multiple pond inlets	Divider boxes with adjustable gates
Inlets	Open channels of concrete
Outlets	Open concrete channels with adjustable bottom gate to control discharge depth, and adjustable rectangular weir to control surface water level
Final discharge pipe	Submerged pipe to avoid surface water foaming
Drainage structure for facultative ponds	Simple sluice gates or flash boards for gradual pond drainage for sludge removal
Bypass gate	Simple gate to open, close, and adjust
Bypass and stormwater runoff channels	Open channels. If possible, the same canal could serve for the diversion of high flows and also stormwater runoff
Embankment and slopes	
Interior slopes (revetment)	Concrete lining
Embankment crown	Well compacted with sufficient width for truck and machinery access
Interior access ramps for facultative ponds	Paved with concrete in all primary facultative ponds for heavy equipment access for sludge removal
Fences	Chain link topped with barbed wire
Operation building	Tool storage, potable water, bathroom and shower, first aid equipment, and laboratory facilities. Desirable to have electricity and telephone

Figure 9.1 Examples of well-constructed and maintained pond systems. The embankments are well maintained with the grass routinely cut. The interior slopes have a concrete revetment that serves to control growth of aquatic plants at the shoreline and prevent erosion by wave action. The surface of the water has no floating materials or scum, and the water level is maintained in the center of the revetment (top: Villanueva, Honduras; bottom: Danlí, Honduras).

Figure 9.2 A well-designed and constructed stormwater diversion channel. If the location of the pond system is downstream of stormwater runoff, channels must be used to divert the runoff to avoid erosion of the interior embankment slopes (see Figure 9.3), and the input of non-wastewater suspended solids is carried by the runoff (Santa Cruz de Yojoa, Honduras).

Figure 9.3 Stormwater runoff has to be controlled to protect the interior embankment slopes of ponds. Left: In this maturation pond, stormwater runoff was not controlled and caused serious erosion to the interior embankment, which did not have protection with a revetment (Zaragoza, El Salvador). Right: Wave action driven by strong winds erodes the unprotected interior embankment of this maturation pond (Santa Cruz, Bolivia).

9.3 GEOTECHNICAL INVESTIGATIONS

The main objectives of a geotechnical investigation are as follows:

(1) Ensure proper embankment design, including slope steepness.
(2) Determine the permeability of the soil to be able to calculate the water infiltration in the ponds.

In a geotechnical investigation, the depth to groundwater is measured first. Afterwards, soil samples are taken, at least four samples per hectare, down to a depth of 1 m greater than the bottom of the pond; these samples are used to represent the soil profile. The samples are then analyzed for the following soil parameters (Mara *et al.*, 1992):

(1) Classification by particle size.
(2) Modified Proctor test (maximum dry density and optimum moisture).
(3) Atterberg limits.
(4) Organic matter content.
(5) Coefficient of permeability.

Geotechnical investigation data are then used to design the embankment and interior/exterior slopes, and to determine if the soil permeability will enable an acceptable infiltration rate into the bottom and interior embankments of the pond.

The soil used for the construction of the embankment should be compacted in layers of 150–250 mm until it reaches 90% of the maximum dry density (determined by the Proctor test) (Mara *et al.*, 1992). After compaction, the soil should have a permeability coefficient determined *in situ* of less than 10^{-7} m/s (see discussion below). The interior slopes of the embankment are designed to have a 3/1 ratio (horizontal to vertical). The design of the exterior slopes is based on a soil mechanics analysis using the results of the soil tests and should vary from 1.5/1 to 2/1 (horizontal to vertical).

Determining the *in situ* permeability of the soil at the base of the pond is essential in calculating infiltration, performing a water balance of the pond system, and determining whether the system will need impermeabilization or not. Soil analysis methods are used to measure the permeability and calculate the infiltration (Cubillos, 1994).

9.4 WATER BALANCE

If a pond system is to maintain the optimal water level for proper operation, the following water balance must be met (Mara *et al.*, 1992):

$$Q_{\text{Mean}} \geq 0.001 \cdot A_T[(P-E)+I] \tag{9.1}$$

where Q_{Mean} is the mean influent flow to the pond system (m³/d), A_T is the total area of ponds (m²), P is the monthly mean rainfall converted to daily mean (mm/d), E is the monthly mean evaporation converted to daily average (mm/d), and I is the infiltration rate (mm/d).

If the water balance in Equation (9.1) is not satisfied, serious problems in operation and maintenance can arise as shown in Figure 9.4. Average monthly evaporation and precipitation data from major meteorological stations are very important to use in the water balance equation. Average daily precipitation and evaporation data are calculated for each month of the year to determine the critical month for using Equation (9.1).

Infiltration is calculated from the permeability measurement mentioned above. The maximum allowable permeability using Darcy's law is determined with Equation (9.2) (Mara *et al.*, 1992):

$$k = \frac{Q_I}{86\,400 \cdot A_b} \cdot \left[\frac{\Delta l}{\Delta h}\right] = \frac{(Q_{\text{med}} - 0.001 \cdot A_T[(P-E)+I])}{86\,400 \cdot A_b} \cdot \left[\frac{\Delta l}{\Delta h}\right] \tag{9.2}$$

where k is the maximum allowable permeability (m/s), Q_I is the maximum allowable infiltration ($= Q_{\text{med}} - 0.001 \cdot A_T[(P-E)+I]$) (m³/d), A_b is the area of the base of pond (m²), Δl is the depth of the layer below the pond to the most permeable stratum (m), and Δh is the hydraulic head (depth of water + Δl) (m).

As a general recommendation, the information in Table 9.2 can be consulted for interpretation of permeability values measured *in situ*.

If the measured permeability is greater than the maximum allowable, ponds will need impermeabilization. While the impermeabilization can be clay, soil, or synthetic membranes, the most recommended and most appropriate is clay as shown in Figure 9.5. The clay should be layered with a total thickness of at least 5–10 cm and covered with a layer of soil or sand to protect it; a mixture of clay with soil or sand can also be used instead of using pure clay (US EPA, 1983). The long-term clay waterproofing infiltration rate in a pond, after one year of operation, has been reported in the US to be 0.006 m³/m² d, which was approximately 13% of the hydraulic head (US EPA, 1983).

Geomembranes are often recommended and installed in pond systems, often with failed results after a short period of time. Frequent problems include

- Improper installation causing tears in fabric.
- Improper application, such as in tidal zones, where rising water can lift the membrane bottom, causing stresses and possible tears (Figure 9.6).
- Inadequate control of plant seedlings, which can tear the membrane fabric as they grow (Figure 9.6).
- Gradual deterioration of membrane above the waterline due to UV degradation, causing the material to crack and tear.
- Tearing of membrane during sludge removal when heavy equipment enters a drained pond.

Geomembranes should be used as a last resort in small cities where all of the above problems can easily occur.

Physical design and aspects of construction

Figure 9.4 Examples of the importance of the water balance and the effects of poor construction. In both ponds, the infiltration losses are excessive and water levels cannot be maintained. Both ponds lack impermeabilization on the bottom and interior embankment slopes (top: Zaragoza, El Salvador; bottom: La Ceiba, Honduras. Bottom photo courtesy of Ing. Iván Olivieri).

Table 9.2 General interpretations of permeability values, k, measured *in situ*.

k Value (m/s)	Significance
$>10^{-6}$	Soil is too permeable for a pond to fill and impermeabilization is needed
$<10^{-7}$	Infiltration occurs but not enough to prevent the pond from filling
$<10^{-8}$	Minimal infiltration occurs
$<10^{-9}$	There is low risk of contaminating groundwater
$>10^{-9}$	Hydrogeological studies are required if groundwater is used for drinking water

Source: Adapted from Mara and Pearson (1998).

Figure 9.5 Impermeabilization of ponds with clay which is essential to maintain the water balance and avoid excessive infiltration, with the possible contamination of the groundwater. The clay should be layered with a total thickness of at least 5–10 cm, and covered with a layer of soil to protect it. In this example, the contractor did not put clay on the interior slopes and the pond had excessive infiltration as seen in Figure 9.4 (La Ceiba, Honduras).

Figure 9.6 Problems encountered with geomembrane liners. In the top photos, emergent plants that were not controlled grew through the seams and tore the membrane liner at, and above, the waterline. In the bottom photo, the secondary facultative pond was built above a shallow water table influenced by tidal action. As the water table rose and fell, it formed bubbles in the membrane bottom that rose to the surface in a pond with a 1.5 m water depth. Eventually the membrane could tear from the continual stress (top photos: Roatán, Honduras; bottom photo: Puerto Cortés, Honduras).

Advantages of clay liners include:
- It is a common material used in construction projects in small cities worldwide.
- If a clay seal is damaged, it can easily be repaired with local knowledge.
- Heavy equipment can enter a drained pond without problems to remove sludge and repair interior embankment slopes.
- The cost of clay is very low relative to geomembrane liners.

9.5 PRELIMINARY TREATMENT AND FLOW MEASUREMENT

As discussed in detail in Chapter 3, each facility should have preliminary treatment with (i) bar screens made of stainless or galvanized steel; (ii) grit chambers with two channels in parallel, each with drainage and sluice gates that seal well; and (iii) a prefabricated Parshall flume following the grit chamber that is designed to control the horizontal velocity in the approach and grit chamber channels, and to measure flowrates.

Flow measurement is essential to determine the hydraulic and organic loadings to a pond system. A flow log allows the evaluation of the efficiency of treatment, the diagnosis of a pond with poor performance, the estimation of infiltration and illegal connections to the sewer system, and an approximate calculation of when the system will reach its peak loading. As discussed in Chapter 4, the most appropriate flow measurement device is a prefabricated Parshall flume. Other structures for measuring flowrates, such as weirs, are not as suitable for wastewater because solids can accumulate at their base, requiring more maintenance.

Problems frequently encountered in grit chambers include (i) construction of grit channels without drainage; (ii) use of sluice gates that do not seal and do not slide easily because of metal–concrete contact; and (iii) construction without a Parshall flume, without an adequate control of horizontal velocity, and without a working flow meter. In general, these problems are a result of poor design and lack of adequate supervision during construction. Figures 9.7 and 9.8 show examples of commonly encountered problems. Figure 9.9 shows an example of a properly installed prefabricated Parshall flume.

Figure 9.7 A grit chamber constructed without adequate channel drainage and well-functioning sluice gates. The channel taken out of service in the left photo was filled with stagnant water, with problems of floating scum, odors, and insects. Additionally, the sluice gate made of iron stuck tightly to the concrete track and it was supposed to slide in. The operator finally was able to pry open the sluice gate with a board, and operated both channels simultaneously – an incorrect method of operation. All grit chambers should have adequate drainage, and sluice gates that seal and slide easily (Santa Cruz de Yojoa, Honduras).

Physical design and aspects of construction

Figure 9.8 This grit chamber was built without drainage, without sluice gates for the downstream channels, and without a Parshall flume at the outlet to control horizontal velocity. The prefabricated Parshall flume was installed on the upstream instead of the downstream end of the grit chamber! This case is typical of problems encountered with inadequate design and construction supervision (Trinidad, Honduras).

Figure 9.9 Left: A properly installed, prefabricated Parshall flume used to measure influent flowrates to a facultative pond. Accurate flow measurement is essential for assessment of hydraulic and organic loadings to the pond system. Right: Well-designed and installed sluice gates in a dual channel grit chamber; the gates have metal-to-metal contact that are routinely greased for ease of operation (left photo: El Paraíso, Honduras; right photo: Suchitoto, El Salvador).

9.6 HYDRAULIC FLOW REGIME

The hydraulic flow regime is the key factor in the successful functioning of a waste stabilization pond system. The best treatment will always be with a hydraulic regime that approximates plug flow (Mangelson & Watters, 1972; Shilton & Harrison, 2003). If there are hydraulic dead zones with short-circuiting in the pond, however, the hydraulic retention time will be less than the calculated theoretical value, perhaps much less, seriously affecting the efficiency of the treatment process. The fundamental factor in the hydraulic design of facultative and maturation ponds is that the hydraulic regime approximates plug flow as much as possible.

Figure 9.10 shows examples of single inlet structures discharging above the water surface. This type of inlet structure causes the formation of dead zones in the corners of the ponds, and turbulence in

Figure 9.10 Single inlets with above waterline discharge cause turbulence from falling water, and large dead zones in the corners of ponds, resulting in hydraulic short-circuiting. As a result, the useful volume of the pond and the hydraulic retention time decrease, as does pathogen reduction efficiency. Multiple inlets and outlets at the water surface should be designed in facultative ponds, and a single inlet and outlet with baffles maturation ponds (top photo: Choluteca, Honduras; bottom photo: Catacamas, Honduras).

the center, promoting hydraulic short-circuiting. As a result, the ponds have mean hydraulic retention times much lower than theoretical values as a result of the hydraulic short-circuiting, significantly lowering the operating efficiency.

To avoid the problems of hydraulic short-circuiting and dead zones, the following designs should be used: (i) open inlet channels that discharge at water surface, (ii) multiple inlets and outlets in facultative ponds, and (iii) a single inlet and outlet with baffled channels in maturation ponds.

Figures 9.11 and 9.12 show examples of open inlet channels that discharge at the water level. This type of inlet device is preferable because it avoids turbulence caused by falling water if the inlet is above the water surface. Additionally, the incoming flow collides with the mass of water within the pond – promoting plug flow as seen in Figure 9.12.

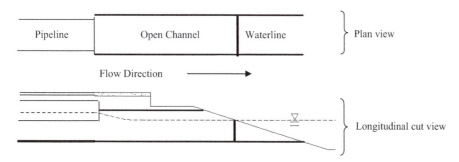

Figure 9.11 Inlets in ponds should be open channels that discharge at the water level, which promotes plug flow due to the collision between the incoming flow and the mass of water in the pond.

Figure 9.12 A good example of an inlet structure with open channel that enters this facultative pond at the water level. The raw wastewater plume entering the pond is clearly seen to be approximating plug flow (Granada, Nicaragua).

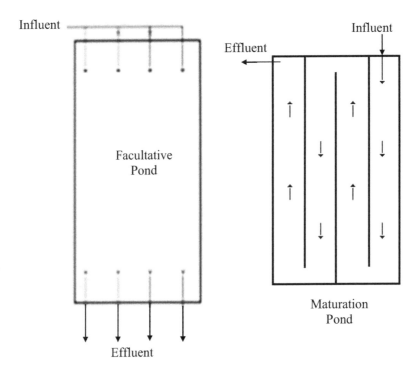

Figure 9.13 Facultative ponds should use multiple inlets and outlets to promote plug flow. Multiple inlets and outlets also promote uniform distribution of bottom sludge. In contrast, maturation ponds use baffles and a single inlet and outlet to promote plug flow.

Figure 9.13 presents examples of inlet and outlet configurations for facultative and maturation ponds. Baffles are not used in facultative ponds as the influent needs to be distributed throughout the entire surface area where daily production of oxygen through photosynthesis occurs. Multiple inlets and outlets also promote a more uniform distribution of sludge deposited in the pond (Franci, 1999; Nelson et al., 2004). Figure 9.14 shows an example of a well-designed facultative pond with multiple inlets and outlets. Maturation ponds, in contrast, use baffles with a single inlet and outlet to promote plug flow since the majority of BOD_L has been removed in the facultative pond.

Figure 9.15 shows examples of baffled maturation ponds with single inlets and outlets. Maturation ponds have low BOD_L loadings and the influent does not need to be distributed throughout the surface area of the pond. Research has shown that the use of baffles to approximate plug flow significantly improves treatment processes (Mangelson & Watters, 1972; Shilton & Harrison, 2003). Only a simple inlet and outlet structure is needed in the design. Baffles are not subject to large forces in the water column, and something as simple as a galvanized chain link fence covered with plastic sheeting can be used.

Figure 9.14 Multiple inlets (photo above) and outlets (photo below) are used in facultative ponds to approximate plug flow. Baffles cannot be used in facultative ponds because of the need to disperse the influent throughout the pond area where solar insolation promotes photosynthesis and oxygen production. Multiple inlets also promote the uniform distribution of deposited sludge in the first 30% or so of the pond's length (Chinendega, Nicaragua).

Figure 9.15 Examples of baffled maturation ponds using a single inlet and outlet to promote plug flow. Baffled maturation ponds should have a minimum length/width ratio of 50 to 1. In the top photo (Estelí, Nicaragua), the longitudinal channels have a length/width ratio ≈100/1. The bottom photo (Roatán, Honduras) shows transverse baffles, which better minimize the effects of wind at this site (top photo courtesy of Ing. Italo Gandini).

9.7 HYDRAULIC STRUCTURES

All hydraulic structures should be designed to be durable and easy to use, and should avoid valves and other mechanisms that deteriorate over time due to corrosion and lack of use (Yánez, 1992). Structures such as sluice gates, flashboards, and weirs are easily adjustable by the operator in order to control the few, but important, hydraulic processes in a pond system.

9.7.1 Flow distribution devices

Flow distribution is a key factor in the successful operation of ponds. The results of the Honduras Monitoring Project (Oakley, 2005) showed that the unequal division of flow among ponds in parallel can cause one or more to be overloaded. Also, the equal distribution of flow among multiple inlets and outlets within a single pond is a key factor in avoiding hydraulic short-circuiting and organic overloading in different sections of the pond.

All batteries of ponds in parallel should have adjustable flow distribution structures in open channels, preferably with prefabricated Parshall flumes after each structure to be able to measure the flowrate of the distribution. Also, all facultative ponds with multiple inlets and outlets should have flow distribution devices in open channels; it is preferable that the flow distribution device be adjustable as well so the operator can make fine adjustments to the distribution. Figures 9.16 and 9.17 show examples of distribution devices in operating ponds.

Figure 9.16 Examples of flow distribution devices between batteries of two facultative ponds in parallel. In the top photo (Masaya, Nicaragua), an adjustable partition is used (arrow). In the bottom photo (Granada, Nicaragua), a fixed partition between ponds in the center is combined with adjustable triangular weirs. The flows to each pond must be monitored to ensure equal distribution.

Figure 9.17 Examples of flow devices for adjusting the distribution of flows among multiple entrances in facultative ponds. In the top photo (Masaya, Nicaragua), an adjustable partition (arrow) that can easily be moved by hand is used. In the bottom photo (Chinendega, Nicaragua), a distribution chamber that cannot be adjusted is used; in this case the chamber has to be built and leveled with precision. In both cases, prefabricated Parshall flumes are used to measure the flow distribution accurately between ponds in parallel. Also note the use of open channels to facilitate flow distribution adjustment and channel maintenance.

9.7.2 Inlets and outlets

Each inlet and outlet must have open channels to facilitate adjustment and maintenance; the water level in the inlet channel should be at the same level as the water in the pond to prevent turbulence and promote plug flow. Each outlet should have an adjustable bottom sluice gate (water passes under rather than over) followed by an adjustable rectangular weir or flashboards. The bottom sluice gate is used to prevent floating scum in the effluent and to control the depth of discharge. The concentration of suspended solids in the form of algae can be lower below the algae band (Mara *et al.*, 1992). Because the band of algae can exist up to 60 cm deep, the best effluent quality can be obtained in being able to

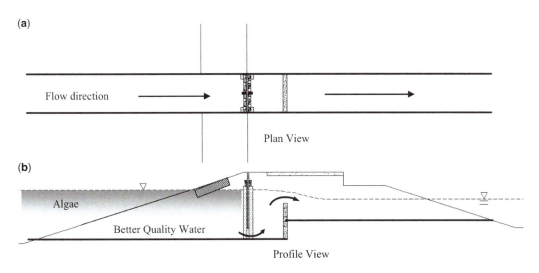

Figure 9.18 Example of an adjustable outlet structure. Each pond outlet should have an adjustable bottom sluice gate that serves to (i) prevent floating scum from discharging with the effluent and (ii) control the discharge depth – many times the concentration of suspended solids is lower below the algae band and better effluent quality can be obtained. Finally, each outlet should have an adjustable weir or flashboards to control the water level in the pond. See also Figure 9.19.

discharge below this level. Finally, each outlet must have an adjustable weir or flashboards to control the water level in the pond.

The rectangular outlet weir can be designed using the following equation (Mara et al., 1992):

$$q = 0.0567 h^{3/2} \tag{9.3}$$

where q is the flow per meter width of weir (L/s m) and h is the hydraulic head upstream of the weir (mm).

Figures 9.18 and 9.19 show examples of outlet designs with bottom sluice gates and adjustable effluent weirs. An example of the problems encountered in a pond without open channels and adjustable outlet devices is shown in Figure 9.20.

9.7.3 Discharge structures for final effluent
Some detergents and other surface-active agents, called surfactants, that are present in domestic wastewater are not biodegradable. As a result, if the final discharge is above the surface of the receiving canal, foam can be formed from the turbulence of the discharge as seen in Figures 9.21 and 9.22. While foam is not a serious problem from a contamination point of view, it is a visual and esthetic problem, and the public may think that the installation is not working well and contaminating the irrigation water. When the effluent is to be used for irrigation, the production of foam especially needs to be controlled.

The most appropriate way to control foam production in the final discharge is through subsurface discharge as presented in Figure 9.21.

9.7.4 Drainage structures for facultative ponds
An example of drainage structures installed in a battery of facultative ponds in parallel is shown in Figure 9.23. The structures are connected by drainage pipe to downstream maturation ponds,

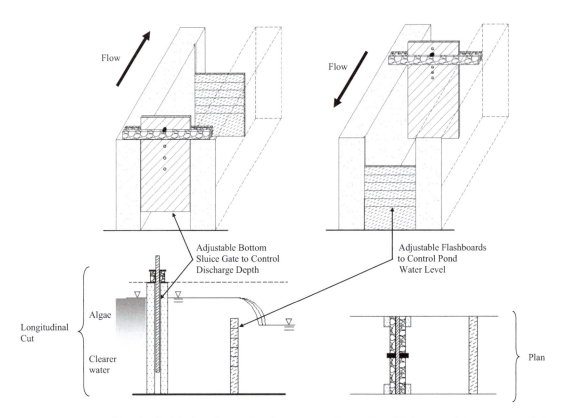

Figure 9.19 Details of the physical design of a pond outlet structure: (i) an adjustable bottom sluice gate to optimize effluent quality in terms of suspended solids caused by the algae band and (ii) an adjustable rectangular weir or flashboard to control the water level in the pond.

and use flashboards to slowly drain the water so turbulence does not cause settled sludge to rise. Drainage structures should be located on the discharge end of facultative ponds, with drainage into downstream ponds.

9.7.5 Overflow weirs and bypass channels

All pond installations should have overflow weirs and bypass channels to divert excessive flows, caused by infiltration and inflow during storm events, in order to protect the facility. Overflow weirs are structures located at the entrance of the pond system that divert hydraulic overloads of infiltrated stormwater to a bypass channel (Figure 9.24). If excessive flows enter the treatment system, biological processes can be washed out and hydraulic structures damaged or destroyed. Large quantities of inorganic solids can also be carried by stormwater, contributing to the premature filling of primary ponds with inorganic matter (sandy solids): a facultative pond in Nicaragua was extensively filled with

Figure 9.20 An example of poorly designed outlet devices, especially the outlet weir, in a maturation pond. The water level is above the concrete revetment, and the pond-length baffle is submerged. As a result, there is a hydraulic short circuit that allows the influent to pass directly to the outlet structure, negating the entire volume of the pond. The effluent weir is made of concrete and is not adjustable, and is located inside a concrete vault with a heavy concrete lid that could not be opened by the operator. The operator was unable to adjust the level of the pond to control the water level. Outlet structures should have adjustable sluice gates, adjustable flashboards, and open channels for operation and maintenance (Choloma, Honduras).

sandy solids in two years of operation due to a failure to bypass stormwater that had entered the sewer system (INAA, 1996).

The simplest overflow weir design uses an adjustable V-notch weir at the inlet structure of the pond system; the V-notch weir then discharges to a bypass channel when the flowrate increases with a rising head (Figure 9.24). If a stormwater diversion channel exists (see Figure 9.2), it can be combined with the bypass channel.

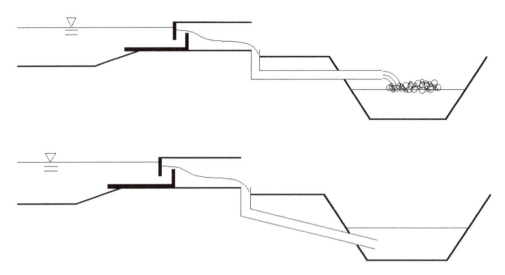

Figure 9.21 Discharge of the final effluent above the water level of the receiving canal produces foam, which arises from detergents and other non-biodegradable surfactants in the wastewater (see Figure 9.22). The most effective way to control foam is to use subsurface discharge in the receiving canal.

Figure 9.22 An example of the foam that can be formed by turbulence when the final discharge is above the water level in the receiving canal. Foam is caused by detergents and other non-biodegradable surfactants in wastewater. The way to control foam production is to use subsurface discharge devices as shown in Figure 9.21 (Villanueva, Honduras).

Figure 9.23 Drainage structures installed on the discharge end of a battery of two facultative ponds in parallel. The structures use flashboards that allow gradual lowering of the water level into a drainage pipe that discharges into the downstream maturation ponds. After draining, accumulated sludge can subsequently be removed with heavy equipment (Estanzuela, Guatemala).

Figure 9.24 Examples of overflow weirs and bypass channels to protect a pond system from hydraulic overloads during storms. The weir on the right (arrow) in the left photo can be adjusted by the operator when required; when the water level rises above normal during storm events, the excess flow is discharged into the bypass channel. (left photo: Granada, Nicaragua; right photo: Danlí, Honduras).

9.8 EMBANKMENTS AND SLOPES

9.8.1 Interior slopes

Figure 9.25 shows the recommended design for the interior slope of a pond. A 3/1 (horizontal/vertical) slope is normally used; this ratio can be changed if justified by geotechnical and soil mechanics investigations. The freeboard (vertical height) of the revetment must cover, with a safety factor, the water levels to be encountered in operation during the dry and rainy seasons; generally, the minimum free board varies between 0.5 and 1.0 m (Mendonça, 2000). The freeboard can be calculated using the following equation (Oswald, 1963):

$$F = (\log A_{med})^{1/2} - 1 \tag{9.4}$$

where F is the freeboard (m) and A_{med} is the mid-level area of pond (m²).

The revetment has two important purposes: (i) the protection of the slope from erosion caused by wave action in strong winds and (ii) to prevent the growth of aquatic plants at the shoreline. Figure 9.26 shows problems with aquatic plants when there is no adequate revetment. Figure 9.27 shows two examples of ponds with adequate revetment with sufficient freeboard.

9.8.2 Exterior slopes

Exterior slopes, as mentioned previously, must be designed according to the geotechnical and soil mechanics requirements of the site. The slope typically ranges from 1.5/1 to 2/1 (horizontal/vertical).

9.8.3 Embankment and access ramps

The crown of the embankment should be built wide enough to allow access for trucks and machinery for periodic maintenance. Each primary facultative pond in a pond system must include ramps for the access of machinery (excavators, front loaders, and dump trucks) for the removal of sludge as shown in Figures 9.28 and 9.29. The access ramps should be paved so that machinery has traction without damaging clay liners.

9.8.4 Fences

The area comprising the pond system should be fenced, preferably with barbed wire, to prevent the entry of animals and unauthorized persons. Figure 9.30 shows the problems that occur with entry of animals and unauthorized persons if the pond system does not have a security fence.

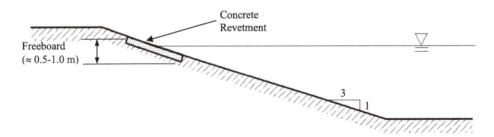

Figure 9.25 The interior slope design should include a concrete revetment to prevent aquatic plant growth and erosion from wave action. The freeboard must cover the water levels encountered in the operation during the dry and rainy seasons. Typically, the interior slope has a 3/1 (horizontal/vertical) in facultative ponds. It can be steeper in anaerobic and maturation ponds depending on soil characteristics and pond depth.

Figure 9.26 In the photo above (Tela, Honduras), the facultative pond has serious problems with the growth of aquatic plants at the shoreline due to lack of a revetment. In the photo below (Catacamas, Honduras), the poor construction of the revetment (note the contour lines at the water level) with insufficient freeboard does not prevent the growth of aquatic plants.

Figure 9.27 Examples of well-built pond revetments. The photo above (Catacamas, Honduras) shows the new revetment built for the pond shown in Figure 9.26 after rehabilitation. The photo below shows a well-built revetment for a maturation pond (Catacamas, Honduras).

Physical design and aspects of construction

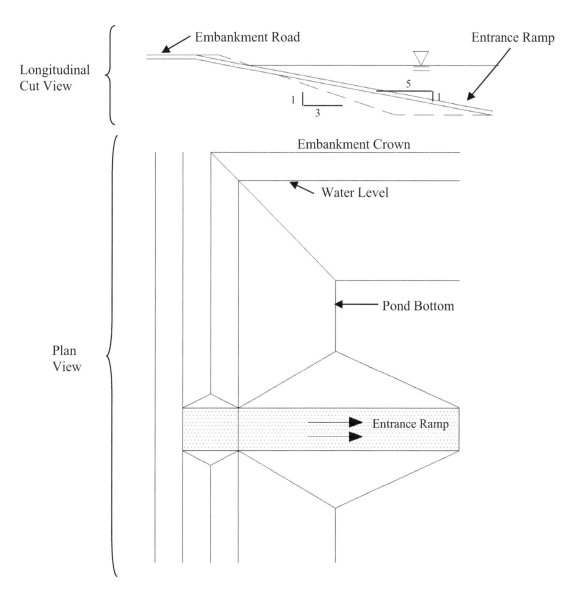

Figure 9.28 Each primary pond should include access ramps to the bottom for machinery such as excavators, front end loaders, and dump trucks, for sludge removal. The entrance ramps should be paved, so heavy machinery has traction without damaging the impermeabilization of the bottom and side slopes.

Figure 9.29 Examples of access ramps for sludge removal in primary ponds. Top photo: facultative pond, Santa Cruz de Yojoa, Honduras. Bottom photo: anaerobic pond, Danlí, Honduras.

Figure 9.30 The area encompassing the entire stabilization pond system should be fenced, preferably with barbed wire, to prevent the entry of animals, which can damage embankment slopes and serve as sources of infection, and that of unauthorized persons (photo above: Tela, Honduras; photo below: Choloma, Honduras).

9.8.5 Operation building

Every pond system, even the smallest, requires an operation building as seen in Figure 9.31. The purpose of the building is (1) the storage of tools, implements, and basic laboratory equipment for operation and maintenance; (2) provision of a toilet, shower, and dressing rooms; and (3) adequate supply of first aid equipment in case of an emergency. The building should have a source of drinking water, a telephone, and preferably a source of electricity for lighting and operation of laboratory equipment. The building can also be used by the security guard in charge of monitoring the facility.

Figure 9.31 Each pond system requires fencing, and an operation building for (1) the storage of tools and basic laboratory equipment; (2) provision of a toilet, shower, and dressing room; and (3) supply of first aid accessories for emergencies. In the top photo, the facility is well fenced with a gate, and the operation building has electricity (Choloma, Honduras). The bottom photo shows a small operation building without electricity (Danlí, Honduras).

doi: 10.2166/9781789061536_0317

Chapter 10
Operation and maintenance

10.1 INTRODUCTION

Routine operation and maintenance are critical to the proper functioning of waste stabilization pond systems. Although the main advantage of a waste stabilization pond system is its operational simplicity, this does not mean that operation and maintenance are not necessary. Indeed, a large number of pond installations worldwide have failed due to failures in basic operation and maintenance tasks (INAA, 1996; Yánez, 1992). This problem is not unique to ponds, and there are operation and maintenance problems in all types of wastewater treatment systems. Any technology, from the most complicated to the simplest, will fail, or not function correctly, without proper operation and maintenance. Since pond systems require less operational efforts than other technologies, the key task is to plan for the success of these efforts in the long term.

To avoid failure in the proper operation and maintenance of any pond system, it is required, at minimum, to have the following: (i) full-time personnel; (ii) personnel qualified in the basic facets of operation and maintenance; (iii) monitoring programs to evaluate efficiency; and (iv) an adequate plan for the removal, treatment, and final disposal of sludge every 10–15 years in facultative ponds and 1–3 years in anaerobic ponds. The key factor that can have a decided effect on successful operation and maintenance is the development and use of an operation and maintenance manual for each facility that should be included with the original design (and most often is not).

10.2 OPERATION AND MAINTENANCE MANUAL

An operation and maintenance manual should contain information that serves to fulfill the following objectives (INAA, 1996; Yánez, 1992):

(1) Standardization of operation and maintenance procedures.
(2) Procedures for rudimentary operation and maintenance, and the overall operation required to control the proper functioning of the entire installation.
(3) Operating procedures for initial start-up and for sludge removal and handling.
(4) Routine maintenance procedures.

© 2022 The Author. This is an Open Access book chapter distributed under the terms of the Creative Commons Attribution Licence (CC BY-NC-ND 4.0), which permits copying and redistribution for noncommercial purposes with no derivatives, provided the original work is properly cited (https://creativecommons.org/licenses/by-nc-nd/4.0/). This does not affect the rights licensed or assigned from any third party in this book. The chapter is from the book *Integrated Wastewater Management for Health and Valorization: A Design Manual for Resource Challenged Cities*, Stewart M. Oakley (Author).

(5) Hygienic measures for operators.
(6) The number and type of full-time and part-time personnel, including training requirements, required at the facility.
(7) Procedures to detect and analyze operational problems and how to remediate them.

The following sections discuss the most important aspects to include in the manual.

10.3 BASIC OPERATION
10.3.1 Initial start-up
The start-up of facultative and anaerobic ponds can present problems because the populations of microorganisms responsible for the treatment take time to develop. With this in mind, some very simple precautions can be taken to avoid complications during the start-up period.

(1) If the pond system has been designed for a population larger than the current one, only part of it should be put into operation, if possible. Generally, the project design should establish the initial pond configurations to be used in distinct phases of operation (MOPT, 1991). New ponds should not be left unfilled, however, and if the final size for the design population has been built, the entire system should be used at lower than design loading rates.
(2) If possible, facultative ponds should be initially filled with water from a nearby source, such as a lake or river, before raw wastewater is introduced into the system. Filling the ponds avoids the development of septic conditions if they were filled solely with raw wastewater and allows the gradual development of microorganism populations, particularly algal populations. In the event that a clean water source does not exist, facultative and maturation ponds should be filled with raw wastewater once, left unloaded for 20–30 d (maintaining water losses due to evaporation and infiltration with an addition of wastewater), and then filled and put into operation; this procedure gradually promotes the development of populations of microorganisms (Arthur, 1983; Mara *et al.*, 1992).
(3) Covered anaerobic ponds can be filled directly with raw wastewater.
(4) Ponds should be filled as soon as possible after they are built to avoid cracking from drying and from weed growth. All vegetation on the bottom and on interior embankment slopes should be removed before filling begins (MOPT, 1991).

10.3.2 Flow measurement
Flow measurement is of decisive importance in the evaluation of the pond system. A record of flow rates is essential in the determination of organic and hydraulic loading rates, hydraulic retention time, and as a result, the efficiency of the treatment system and its capacity. The operator should record the flow rates daily in order to have a flow history and to anticipate problems. As discussed in previous chapters, there has been significant failure in many overloaded facilities due to lack of flow measurement.

During rainy and dry seasons, a more intensive flow measurement regimen should be performed to obtain better data on flow behavior on treatment performance. Flow measurements should be made in 2-h periods for 3 consecutive days; the average flow for the sampling period can then be obtained. It is preferred that this activity includes Saturday and Sunday to know the behavior of weekend flows (INAA, 1996). It is very important to compare the differences between the wet and dry seasons to know the effect of infiltration on the physical (e.g., hydraulic retention time) and biological processes (e.g., pathogen reduction and BOD_5 removal).

The recommended type of flowmeter is the prefabricated Parshall flume. As discussed in Chapter 4, no *in situ* constructed Parshall flume is likely to meet the calibration standards required. The only option to solve this problem, and a less expensive solution too, is the use of prefabricated Parshall flumes. In Nicaragua, for example, prefabricated fiberglass Parshall flumes, which have a calibration chart, are used in all pond systems not only at the inlet but also in the partitions between parallel ponds and in the final discharge outlets (INAA, 1996).

10.3.3 Water level control

Each pond system is designed to have a fixed water level with a defined range of fluctuation. It is the operator's responsibility to maintain this level of the pond within this range or the pond will not function as it should (seen in Figure 9.4). If the operator cannot maintain the design water level with adjustable effluent weirs, which should have been incorporated in the design, the pond has to be evaluated to determine the cause of the problem.

10.3.4 Bypass channel operation

To protect the pond installation from excessive flows due to stormwater infiltration of the collection system, the operator must use the bypass channel when flows reach the previously defined critical level. This level is determined through investigations using results of measured flow data, and the results of laboratory analyses of suspended solids loads during rainfall events.

When influent flows approach the critical level, the operator must lower or remove the overflow weir to the bypass channel (Figure 9.24) to divert the flow to the bypass channel and raise the influent weirs to stop inflow to the pond system. Once the flow normalizes, the operator then reverses the operation, lowering the influent weirs and replacing the bypass weir at its normal level, which stops the emergency diversion. This operation requires a greater presence and vigilance of operators during the rainy season, frequently working in shifts over a 24 h period. In Nicaragua, for example, operators work 12 h shifts, 24 h/d, during the rainy season (INAA, 1996).

10.3.5 Adjusting the discharge level with bottom sluice gate

It is the operator's responsibility to adjust the discharge level of each discharge pond to obtain the best quality effluent if there are significant differences in algal concentrations with depth. The level of the algae band can change daily or weekly, depending on algal dynamics in each pond. The operator, or laboratory technician, should take samples of effluent with depth at the outlet structure and measure the concentration of suspended solids or algae. With these data, the optimal depth to adjust the bottom sluice gate can be determined.

10.3.6 Sensory detections: odors and colors

The sensory detection of odors and abnormal colors are very important in assessing the operating condition of a pond. The operator must be aware of odors and colors that are different from those that normally encountered in well-functioning ponds.

Facultative and maturation ponds should not have strong odors if they are functioning well. The color of wastewater at the entrance to a facultative pond should normally be gray; the color of the effluent at the outlet of facultative and maturation ponds should be a bright green due to the concentrations of algae produced in the pond.

10.3.7 Sludge depth measurement

The only way to verify the sludge accumulation calculations is to measure sludge depth in primary ponds (facultative or anaerobic) once a year. The accumulation of sludge is measured by submerging a pole long enough for the depth of the pond; it would need to be approximately 2.5 m for a facultative pond. The pole should have one end tied with absorbent white cloth. The pole is introduced into the pond, maintaining it in a vertical position, until it reaches the bottom; it is then slowly raised, again maintaining a vertical position, and the height colored with sludge is measured (sludge is easily retained on the fabric) (Mara *et al.*, 1992). Grids must be made with a boat on the surface of the pond in order to estimate the average depth and the volume of sludge. With the data obtained, the sludge accumulation rate and the volume of sludge in the pond can be estimated. Before the depth of the sludge reaches 0.5 m, and preferably 0.3 m, and before it occupies 25% of the volume of the pond, sludge removal should be planned during the next dry season.

10.4 ROUTINE MAINTENANCE

Routine maintenance of the pond facility should be the primary objective of the operator. If this maintenance is not carried out daily, the facility will deteriorate in short time, with dire consequences for effluent quality and reuse. The operator, therefore, must be aware that their work is very important for the proper functioning of the system.

10.4.1 Bar screens

Bar screen raking should be performed daily with the use of manual rakes (Figure 10.1). The removed material must be buried or covered daily onsite to avoid odors and the attraction of vectors (insects, rodents, and birds). The material should be covered with a layer of soil 0.1–0.3 m thick (INAA, 1996). It is advisable to have a designated excavation to bury the material little by little, covering it daily with lime or soil.

10.4.2 Grit chamber

Grit chamber maintenance consists of agitating the settled material in the operating channel twice a day, once in the morning and once in the afternoon; the purpose of agitation is to release the organic material trapped by the heavier inorganic solids (INAA, 1996). Once or twice a week or more frequently if the accumulated volume of grit solids demands it, the operating channel must be closed and drained, and the grit material removed and buried in a sanitary manner (Figure 10.1). The material can be buried in the same excavation used to bury the bar screenings material.

It is often noted that in most systems that have grit chambers, they are not operated correctly, as seen in Figure 10.2. Part of the problem is the poor design or construction of the entire grit chamber, and part is a problem of training of the operator in the correct operation of a grit chamber (see Figure 10.2).

10.4.3 Removal of scum and floating solids

The removal of scum and floating solids should be done routinely to avoid their buildup over the surface area of facultative ponds. Organic floating solids decompose and cause odor problems, and also insect problems by the formation of suitable sites for breeding (Figures 10.3 and 10.4).

Wind causes the buoyant scum and organic solids to eventually accumulate in the corners of a pond, where they should be removed with scum rake and hauled for burial in a wheelbarrow; these wastes should be buried in the same location where screening solids and grit are buried (Figure 10.5). Floating solids can also pass through outlet structures with the effluent and these structures must be cleaned routinely (Figure 10.6).

10.4.4 Grass, vegetation and weeds, and aquatic plants

Grass on embankments should not reach the water's edge in order to avoid problems with emergent plants (Figure 10.7), which promote insect growth. The operator must maintain a clean strip at least 20 cm above the water's edge. Weeds near and in the water must be removed and dried, and burned or buried. Special attention should be paid to the emergence of hyacinths and other aquatic plants, which should be removed, dried, and buried or burned as well.

A special problem that happens occasionally is the rapid growth of duckweed, which reaches a pond through wind, birds, or animals. The task of the operator is to remove the duckweed as quickly as possible before it covers the entire surface of the pond (Figure 10.8). Domestic ducks eat duckweed, and they can be brought to the installation to aid in the removal process.

Figure 10.1 In preliminary treatment, the operator's responsibility is to clean the bar screen daily, and the grit chamber when needed – typically once a week in small systems. It is essential to agitate the settled solids in operating grit chamber on a daily basis so that entrapped organic solids escape to the downstream pond, and inorganic solids remain in the channel. The solids from the bar screen and grit chamber should be buried and covered on a daily basis (Leon, Nicaragua).

Figure 10.2 In many facilities that have grit chambers, the operators do not operate them correctly. Part of the problem is poor design and poor construction, where the grit channels cannot be sealed or drained, and part is a problem of operator training in the correct operational procedure. In the photo to the left (Trinidad, Honduras), the operator left both channels in operation for lack of a way to close them with a sluice gate. In the photo on the right (Granada, Nicaragua), there were no sluice gates to close the channels, nor a way to drain a closed channel to remove the grit.

Water hyacinths have intentionally been planted in facultative and maturation ponds under the mistaken belief that they will aid the treatment process (Figure 10.9). Hyacinths are prime habitat for mosquitos and insects, grow rapidly, and will cover an entire pond quickly. When a pond is covered, most of the water column will be anaerobic, with no removal of bacterial and viral pathogens by UV radiation. The hyacinths soon block the inlet and outlet structures and must be harvested to maintain flows.

10.4.5 Mosquitoes, flies, rodents, and other animals

The proliferation of mosquitoes, flies, and rodents should be nil if pond scum is removed and buried, and emergent plants on the inside embankment and shoreline are controlled. If mosquitoes lay their eggs on the shoreline revetment, the water level can be lowered slightly to dry them out.

Amphibians and reptiles, birds, mammals, and fish all populate facultative and maturation ponds. Turtles usually do not cause problems (Figure 10.10), but they can burrow behind and under revetments (Figure 10.11). When significant turtle populations exist, the operator should routinely check the inside embankment and revetment and, when necessary, fill turtle excavations. Alligators and crocodiles are common in maturation ponds in tropical and subtropical climates (Figure 10.12), and can possibly help control turtle populations by eating them. Frogs and toads are common everywhere (Figure 10.13), as are numerous bird species. Mammals such as capybara are found in pond systems in Bolivia and Brazil (Figure 10.14). In short, waste stabilization ponds, in addition to being a combination of natural processes for treatment, are also a natural habitat for many varieties of wild species.

Figure 10.3 Scum, floating solids, and aquatic plants usually accumulate in the corners of ponds, driven by winds. An operator needs a scum rake and a wheelbarrow to remove these solids. If the solids are not removed frequently, they will emit odors due to decomposition and serve as suitable foci for the attraction and reproduction of insects (top photo: Lagartos, Brazil; bottom photo: Villanueva, Honduras).

10.4.6 Embankment slopes
The operator should inspect the condition of the embankment slopes once a week to verify if any settlement or erosion has occurred. Damages must be repaired with clay material and covered with protective grass on the outside slope and with revetment lining on the inside slope.

10.4.7 Fences and roads
As mentioned in Chapter 9, the entire stabilization pond system installation should be fenced, preferably chain-link fencing topped with barbed wire, to prevent the entry of domestic animals and unauthorized persons. Interior roads, especially those on top of embankments, must be well constructed and kept in good condition. When the conditions of fencing and roads deteriorate with age, the operator must notify the persons in charge of repairing these works as soon as possible.

Figure 10.4 Scum and floating solids, if not removed, foster the breeding of insects (top photo: Villanueva, Honduras; bottom photo: Guastatoya, Guatemala).

Operation and maintenance 325

Figure 10.5 Floating scum and solids accumulate in corners, where the operator can easily remove them with a scum rake and put them in a wheelbarrow. Afterwards, they must be buried, or covered with a layer of soil or lime. In the photo below the operators put the collected scum solids in a small excavation and cover them with lime to control odors. When the excavation is filled, it is covered with a layer of soil (Masaya, Nicaragua).

Figure 10.6 Left photo: the outlet of a facultative pond designed with a bottom sluice gate and a rectangular weir did not have the sluice gate in place; as a result, the floating solids were carried out with the effluent over the rectangular weir, and the removal efficiency of fecal coliforms was low (Catacamas, Honduras). Right photo: a circular baffle is used to control the discharge of scum in the effluent of a facultative pond (Chinendega, Nicaragua).

Figure 10.7 Top photo: the facultative pond has serious problems with overgrowth of weeds on the crown of the embankment down to the shoreline (note the man in the center of the photo) (Choluteca, Honduras). Bottom photo: a fundamental maintenance responsibility is the control of weed growth on pond embankments; in this photo a brigade of personnel from the National Water and Sewerage Administration (ANDA) of El Salvador cuts weeds on the embankment of a facultative pond on a supposed routine basis (Zaragoza, El Salvador).

Figure 10.8 A common example of duckweed, carried by the wind or by birds, that has partially covered the surface of a maturation pond. The operator should remove the duckweed as soon as possible before the pond is entirely covered. Domestic ducks rapidly eat duckweed and a few can be brought to a heavily covered pond to aid in its removal (Tela, Honduras).

Figure 10.9 Water hyacinths covering 50% of a maturation pond, with the baffle preventing further spread. The hyacinths were placed intentionally with the mistaken belief that they would improve treatment. Hyacinths are prime habitat for insects and should be removed immediately if they begin to grow in ponds. Domestic animals should not be permitted inside the treatment plant installation.

Figure 10.10 Significant turtle populations can be present in facultative and maturation ponds. Top photo: turtles on maturation pond revetment; lightened areas in pond are heads of turtles (Danlí, Honduras); bottom photo: turtle caught in inlet structure to facultative pond (León, Nicaragua).

Figure 10.11 Turtle excavations behind the revetment in a secondary facultative pond. Turtles can burrow behind and under geomembrane liners and the revetment to deposit their eggs. The operator should monitor the condition of a liner and revetment routinely when there are high turtle populations in the ponds (Danlí, Honduras).

Figure 10.12 A large crocodile (≈ 4 m) in a maturation pond. The presence of alligators and crocodiles is not uncommon in facultative and maturation ponds in tropical and subtropical climates. Both alligators and crocodiles have been known to eat turtles (Tela, Honduras).

Figure 10.13 Frog and toad populations are commonly found in facultative and maturation ponds (maturation pond, Trinidad, Honduras).

Figure 10.14 Capybaras in a maturation pond in Santa Cruz, Bolivia. Waste stabilization ponds, in addition to being a combination of natural processes, are also a natural habitat for wild species of birds, mammals, reptiles and amphibians, and fish.

Table 10.1 Equipment and tools required for a small waste stabilization pond facility.

Item	Quantity	Use
Rubber gloves	2 pair	Operator protection
High rubber boots	2 pair	Operator protection
Rubber capes	3	Operator protection
First aid kit	1	Operator protection
Life jacket	2	Operator protection
Field uniform	2	Operator protection
Crash helmet	2	Operator protection
Scum rake	2	Scum and floating solids collection
Shovel	2	Burial of grit, screenings, scum, and so on
Pick	2	Excavation for scum and grit burial
Wheelbarrow	1	Transport of scum and grit chamber solids
Lawn mower	1	Grounds maintenance
Hammer	1	General maintenance
Saw	1	General maintenance
Broom	1	General maintenance
Skimmer (3 m length)	2	Scum removal
Boat	1	Measurement of sludge, sampling, and so on
Life preservers	2	Boat use and shoreline work
Hose	1	General cleaning
Machete	2	Weed and brush control
Screwdriver	2	General maintenance
Buckets	2	Collection of scum and floating solids
12-in. Stilson wrench	2	General maintenance

Note: Adapted from INAA (1996).

10.4.8 Equipment and maintenance tools

Table 10.1 presents a list of basic equipment and tools that should be kept in the operation building (INAA, 1996).

10.5 FIELD RECORDS FOR BASIC OPERATION AND MAINTENANCE

Table 10.2 shows an example of the operational records and field reports of basic operation and routine maintenance that the operator should routinely record. Table 10.3 provides a general overview of the operation and maintenance activities and the frequency with which they should be carried out. Operators should keep monthly and annual digital photo files of each unit operation for documentation and trouble-shooting problems.

Table 10.2 Field observations for waste stabilization ponds.[a]

Pond installation: _____

Date: _____ Hour: _____ Name of the operator: _____

Air temperature: _____ Weather: _____

Flow (m³/d): _____ Bar rack status: _____

Grit chamber status: _____

Observation	Facultative	Maturation	Comments
Water color			
Odors			
Scum/floating solids			
Embankments			
Aquatic plants			
Slope erosion			
Insects			
Rodents			
Birds			
Reptiles			
Accumulated sludge			
Water levels			
Inlets			
Outlets			

Other observations (attach photos as needed):

[a] Operators should keep monthly and annual digital photo files of each unit operation for documentation and trouble-shooting operational problems.

Table 10.3 Frequency of routine operation and maintenance activities in waste stabilization pond installations.

Activity	Daily	Weekly	When Necessary	Observations
Basic operation				
Flow measurement	X			Recorded daily. Intensively measured during rainy season.
Water level control			X	Levels are recorded during rainy and dry seasons
Use of bypass channels			X	During hydraulic overloads
Adjustment of depth of discharge			X	Based on algae concentrations with depth
Sensory detections			X	You have to notice changes in smells and colors
Sludge depth measurement			X	Once per year in facultative ponds
Routine maintenance				
Bar screens	X			Bars cleaned and screenings buried on a daily basis or more frequently as needed
Grit chamber	X	X		Settled material should be agitated once a day and removed weekly, or as needed, and buried
Scum and floating solids in ponds	X		X	Removal with scum rake, transport with wheelbarrow and burial
Grass, vegetation, and weeds			X	Embankment slopes must be kept free of vegetation
Mosquitoes, flies, and rodents			X	Controlled by keeping embankments slopes clean without emergent vegetation at the waterline
Slopes, fences, and roads			X	Inspected at least monthly
Sludge depths			X	Should be measured in a grid network once a year in facultative ponds

10.6 OPERATION FOR PERFORMANCE CONTROL: ANALYTICAL MONITORING

The overall objective of treatment is to produce an effluent acceptable for reuse in agriculture and aquaculture. The objectives of the facultative pond process are: (1) removal of pathogens, especially helminths; (2) removal of raw wastewater total suspended solids (TSS); and (3) stabilization of raw wastewater organic matter through aerobic removal of soluble BOD_L (measured as BOD_5) in the aerobic zone, and anaerobic stabilization of settled TSS in the anaerobic zone. The main objective of the maturation pond process is the continued removal of pathogens. To achieve these objectives, it is necessary to carry out a series of measurements and analytical determinations:

(1) Concentration of BOD_L in the influent to the system and BOD_5 in the effluent of each pond in series.
(2) Concentration of TSS in the influent of each facultative pond to estimate sludge accumulation.
(3) The concentration of TSS in the effluent of each pond in series to determine algae concentrations.
(4) The concentration of helminth eggs and *E. coli* (or fecal coliforms) in the influent to the system and in the effluent of each pond in series to assess reuse categories.

(5) Concentrations of total nitrogen and total phosphorus in the final effluent to determine effluent reuse value.
(6) Flow measurement of raw wastewater influent, and the effluents of each pond in series.

With the results of this series of measurements, the following control parameters can be calculated:

(1) Hydraulic loading and the hydraulic retention time of each pond in the system.
(2) The surface organic loading to the facultative ponds.
(3) The final concentrations and removal efficiencies of helminth eggs and *E. coli* in relation to the WHO guidelines for restricted or unrestricted wastewater reuse in agriculture.
(4) The fertilizer value of the final effluent in kg total N/yr and kg total P/yr.
(5) The efficiency of BOD_5 and TSS removal in facultative ponds.
(6) The suspended solids load to the facultative pond and the sludge accumulation rate.

10.6.1 Laboratory sampling and testing program

Table 10.4 lists the process control parameters, sampling frequency, and sampling location. Table 10.5 presents the laboratory requirements for the analysis of each parameter. To implement a sampling and analysis program, the following aspects must be considered (Yánez, 1992):

(1) Type of measurement or analysis to be performed.
(2) Selection of analyses that can be performed at the facility.
(3) Preservation requirements of the samples.
(4) Waiting time until samples are taken to the laboratory.
(5) Variability of parameters and precision of analysis.
(6) Practical use of results for treatment and reuse objectives.

The application of correct sampling techniques is essential to obtain reliable data. A large number of studies of stabilization ponds have produced results that are unusable due to poor sampling techniques (MOPT, 1991). Therefore, it is essential that operators receive training in sampling techniques and have basic knowledge of the analyses to be performed in the laboratory (see Figures 10.15 and 10.16). The main role of the operator is to obtain representative samples and take the necessary precautions so that they reach the laboratory in the manner required for analysis (MOPT, 1991).

It is essential that the facility supervising engineer receives training in order to (i) select a qualified laboratory for sample analysis and (ii) be able to interpret the analytical results. The selected laboratory should train plant personnel in sampling protocols.

The parameters and their sampling frequency that are presented in Tables 10.4 and 10.5 are the minimum required to evaluate the performance of the treatment system and its ability to meet the requirements for reuse in agriculture or aquaculture.

10.6.2 Presentation and interpretation of the results of monitoring programs

It is fundamental that the supervising engineer and head operator develop the skills to analyze the monitoring results and to present them in a way that is easy to interpret. Table 10.6 presents an appropriate way to present the results to be able to interpret them clearly.

Most of the results that are obtained in Table 10.6 can be represented in the form of figures and graphs from which practical conclusions can be drawn about the operation of the system. Above all, it is important to have the results in a very clear and easy to interpret form for plant operating personnel to be able to interpret (MOPT, 1991).

Table 10.4 Suggested program for monitoring in facultative-maturation stabilization pond systems.

Parameters	Frequency			Sampling Location for Facultative-Maturation Ponds in Series			
	Daily	Monthly	Annual	Influent Facultative	Effluent Facultative	Effluent Maturation	Sludge Facultative
Hydraulic							
Mean flowrate (m³/d)	X			X	X	X	
Maximum flowrate (m³/d)	X			X	X	X	
Hydraulic retention time (d)	X			X	X	X	
Physical–chemical							
Odor	X			X	X	X	
Color	X			X	X	X	
Temperature (°C)	X			X	X	X	
Total suspended solids (mg/L)		X				X	
pH		X				X	
Total N (mg/L)		X				X	
Total P (mg/L)		X				X	
Total volatile solids in sludge (%)			X				X
Total fixed solids in sludge (%)			X				X
Biochemical							
Total BOD₅ (mg/L)		X		X		X	
Filtered BOD₅ (mg/L)		X				X	
Microbiological							
E. coli or fecal coliforms Colony Forming Units per 100 mL (CFU/100 mL)		X				X	
Helminth eggs (# eggs/L)		X				X	
Helminth eggs in sludge (# eggs/g dry weight)			X				X

Table 10.5 Sampling and preservation requirements for laboratory analysis.

Parameter	Container Type	Minimum Sample Volume	Sample Type	Preservation	Maximum Preservation Time
Temperature	None	–	*In situ*	Immediate analysis	Immediate analysis
pH	Plastic or glass	50 mL	Grab	Immediate analysis	Immediate analysis
BOD_5	Plastic or glass	1000 mL	Composite in 24 h	Cooling to 4°C	6 h
Suspended solids	Plastic or glass	200 mL	Composite in 24 h	Cooling to 4°C	7 d
Total, volatile and fixed solids in sludge	Plastic or glass	25 g (≈250 mL)	Grab	Cooling to 4°C	7 d
E. coli or fecal coliforms	Plastic or glass Sterilized	100 mL	Grab	Cooling to 4°C	6 h
Helminth eggs Water	Plastic or glass Sterilized	>5.0 L	Composite in 24 h	Cooling to 4°C	24 h
Sludge	Plastic or glass Sterilized	1.0 L	Grab	Cooling to 4°C	24 h

Source: APHA (1995).

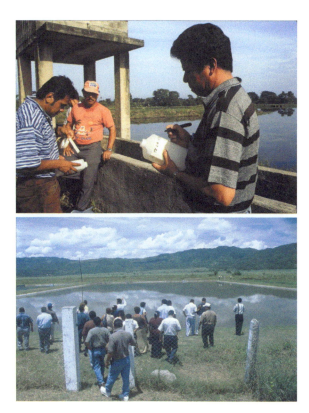

Figure 10.15 All personnel involved in the design, operation and maintenance, and monitoring will need training if the treatment system is to be successful. In these photos, groups of engineers and technicians receive training in the sampling and operation of stabilization ponds (top photo: Tela, Honduras; bottom photo: Villanueva, Honduras).

Figure 10.16 A series of intensive training courses, both in practice and theory, is recommended for all personnel involved in the operation of waste stabilization ponds. It would be very important to institutionalize the courses in an entity that could offer them annually, preferably in a place where there is a pond system in operation (left: Tela, Honduras; right: sampling in Trinidad, Honduras).

Operation and maintenance

Table 10.6 Table of monitoring results for facultative and maturation ponds in series.

Facility name: _____ Date: _____ Operator's name and signature: _____

Parameters	Unit	Influent	Sampling/Measurement Date	Facultative Pond	Effluent Facultative Pond	Maturation Pond	Effluent Maturation Pond
Area	m²						
Sludge depth	m						
Sludge volume	m³						
Water volume	m³						
Average flowrate	m³/d						
Hydraulic retention time	d						
Temperature	°C						
pH	Unitless						
Suspended solids	mg/L						
BOD$_5$ total	mg/L						
BOD$_5$ filtered	mg/L						
E. coli/fecal coliforms	CFU/100 mL						
Helminth eggs	Eggs/L						
Sludge:							
Total solids	%						
Volatile solids	%						
Fixed solids	%						
Helminth eggs	Eggs/g (dry weight)						

10.7 SLUDGE REMOVAL IN FACULTATIVE PONDS

The easiest and most economical way to remove sludge in facultative ponds is to drain the pond and dry the sludge *in situ*, and then remove it with heavy equipment. When the sludge dries to a total solids content from 18% to 20%, it changes from a liquid to a solid and a track loader or an excavator, with a dump truck, can be used to remove it (Oakley *et al.*, 2012).

During the period of pond draining and sludge drying, the influent must be diverted to another pond in parallel. After draining, the sludge is dried for a period of 1–3 months; when dried to a solid, sludge removal with machinery should take less than a week (Oakley *et al.*, 2012). The sludge must then be stored – in a place that does not present risks to the population and the environment – for a period of at least 5 years or longer to ensure destruction of viable helminth eggs. As soon as the sludge has been removed, the empty pond is filled to regain treatment capacity.

It is important to remove sludge from facultative ponds when the average accumulation is ≈ 0.5 m. If the sludge is much deeper, a hard crust will form on the surface and the sludge will not dry with depth, making it extremely difficult to remove with machinery.

The experiences of the Nicaraguan Institute of Aqueducts and Sewers with 25 waste stabilization pond systems show that normally loaded facultative ponds require sludge removal every 10 years, with shorter periods if ponds are overloaded (INAA, 1996). For this reason, it is essential to design facultative ponds in parallel to have one in operation while the other is desludged during several months. All stabilization pond installations need to have a program of annual sludge depth measurements, an operational plan drying and removal of sludge, long-term storage, and final disposal. If a sludge management program is not developed or implemented, ponds will begin to fail in less than 10–15 years of operation due to excessive accumulation of sludge.

10.8 REQUIRED PERSONNEL

In view of the significant investment in the construction of waste stabilization pond systems, there is an important need for training of personnel, engineers, and operators, in all aspects of design, monitoring, operation and maintenance, to develop the infrastructure to sustainably manage the system in the long term. All pond systems will fail if there are no trained personnel for operation and maintenance. Unfortunately, many already have failed in small to large cities around the world. Table 10.7 presents estimates of the personnel requirements for waste stabilization pond systems based on populations served. Table 10.8 lists the qualification requirements for personnel.

Note the personnel qualifications in Table 10.8 require training. A series of intensive courses is recommended to train personnel involved in pond design, operation and maintenance. It is very important to institutionalize the courses in an entity that can offer them on an annual basis. The formation of a training center where engineers, operators, and technicians can gain experience, both in practice and in theory, is highly recommended.

Table 10.7 Estimated personnel required for the operation and maintenance of stabilization ponds in municipalities up to 1 000 000 inhabitants.

Personnel	Population Served						
	10 000	25 000	50 000	100 000	200 000	500 000	1 000 000
Supervising engineer	0.25	0.5	0.5	1	1	1	2
Operators	1	1	1	2	3	3	3
Laborers	2	3	6	8	12	16	30
Security guard	1	1	1	2	2	2	2
Specialists as needed							
Total	4.25	5.5	8.5	13	18	22	37

Note: Adapted from Wagner (2010) and Yánez (1992).

Table 10.8 Qualifications for waste stabilization pond personnel.

Personnel	Qualifications[a]
Supervisory engineer	Sanitary or civil engineering degree with specialization in wastewater. Training in design, operation and maintenance of stabilization ponds. Training in first aid, occupational health and safety, wastewater monitoring, and interpretation of laboratory results. Experience in financing the operation of public works.
Operators	Approved secondary education. Skills for operation, maintenance, and basic monitoring of wastewater and stabilization ponds. Training in first aid, safety and occupational health, stabilization pond operation, flow monitoring, grit chamber operation, and basic wastewater sampling.
Laborers	Training in basic first aid
Security guard	Skills for security work. Training in basic first aid.
Specialists	Hired with necessary skills as needed for various activities such as multiparameter sampling, sludge sampling, sludge removal, and so on

[a]All personnel must receive regular training in first aid, safety and occupational health, must receive vaccination against tetanus, typhoid fever, and hepatitis A, and must be monitored once a year by a physician that includes analysis for intestinal infections with parasites.

10.9 HYGIENIC MEASURES FOR OPERATORS

It is essential to train operators in the health risks of their work, in the safety measures they should take to prevent accidents and infections, and first aid measures. Table 10.9 presents safety measures that have been recommended by the World Health Organization for stabilization pond operators (WHO, 1987).

Table 10.9 Recommended hygienic and safety measures for a waste stabilization pond facility.

(1) The facility must always have a source of clean water, soap, and chlorine. It is advisable to use disposable paper towels to avoid the need for transport for cleaning cloth towels, as they may remain unwashed for long periods and can serve as a source of infection.

(2) The operation building must have a first aid kit that includes, at a minimum, adhesive cloth, cotton, alcohol, disinfectants, a disinfectant detergent solution, scissors, tweezers, and repellent for mosquitoes and insects. The building should also have fire extinguishers and a cell phone for emergencies.

(3) All workers must have rubber gloves and boots, a work helmet, and at least two work overalls. All clothing used in the facility must remain in storage at the end of the working day.

(4) Whenever eating or drinking, workers should wash hands with soap and clean water. If any food is made on the premises of the facility, an area must be designated for that purpose. At all times workers should avoid eating at the same time that any work is being done that risks contact with contaminated waste. It is best not to eat near areas where liquid or solid waste is discharged or stored.

(5) All work tools should be washed with clean water before being put away after use

(6) Cuts, scratches, and bruises that the worker may suffer should be disinfected immediately after injury.

(7) If the site has electricity, and the workers must take care of the maintenance of electrical equipment, they should make sure that hands, clothes, and shoes are always dry.

(8) The entrance to the site must be kept closed when there are no authorized visits. Hygienic risks for visitors are high if they are not sufficiently informed.

(9) The facility must have a boat, rope, and at least two life jackets.

(10) Workers must be vaccinated against tetanus, typhoid fever, hepatitis A and B, and other possible diseases indicated by the health authorities of the area. Workers should also have a medical check-up at least once a year that includes tests for parasite infections.

(11) All workers should receive regular training in first aid, safety and occupational health.

Note: Modified from WHO (1987).

10.10 OPERATIONAL PROBLEMS AND THEIR SOLUTION

Stabilization ponds can occasionally have operational problems that the operator must be able to recognize in order to take the corresponding remedial measures.

10.10.1 Signs of well-functioning facultative and maturation ponds
The signs of good operation of both facultative and maturation ponds are the following:

(1) Pond and effluent water have a bright green color.
(2) The water surface is free of scum and floating solids.
(3) Absence of aquatic plants in the water and weeds on the interior embankment slopes.
(4) Absence of insects.
(5) Absence of strong odors.

10.10.2 Problems of operation in facultative and maturation ponds
The most frequent operating problems in facultative ponds are the accumulation of scum and floating solids; odors; development of brown, gray/black, dull yellow/green, pink, or red colorations, which are signs of pond overloading; weed growth; and the appearance of mosquitoes and other insects (MOPT, 1991; WEF, 1990).

10.10.3 Accumulation of scum and floating solids
The surfaces of ponds should be free of scum and floating organic solids. The presence of large areas covered with scum and floating material inhibits photosynthesis by restricting the passage of light, causes odors due to decomposition of floating organic solids, and attracts mosquitoes and other insects (Figures 10.3 and 10.4). The presence can be caused by the following factors:

(1) Insufficient maintenance to remove scum and solids on a routine, perhaps daily basis, if necessary.
(2) Flotation of anaerobic sludge lifted to the surface by methane and carbon dioxide bubbles. This could be a sign of excessive sludge accumulation and sludge depths should be checked.
(3) Excessive grease passing through preliminary treatment. If this is the case, a grease trap should be designed and installed in the preliminary treatment unit processes.

Scum and floating matter must be removed with a scum rake, and if the accumulation is widespread, a boat may be needed.

10.10.4 Odors
The most frequent reasons for the appearance of odors are as follows:

(1) BOD_L overloading causing anaerobic conditions. Overloading is caused by poor design, short hydraulic retention times due to hydraulic short-circuiting, and excessive accumulation of sludge causing anaerobic decomposition with releases of H_2S, NH_3, and CH_4 to the atmosphere.
(2) Presence of toxic substances from industrial wastewaters that reduce biological activity.
(3) Anaerobic decomposition of excessive scum and floating organic matter on the water surface (Figure 10.3).
(4) Shading of the pond surface by trees or structures that reduce photosynthesis and oxygen production (MOPT, 1991).

10.10.5 Abnormal colorations
Facultative and maturation ponds normally have a bright green coloration. The inlet to a facultative pond may have a gray/brown coloration from raw wastewater, but it should be bright green, a short distance into the pond. The following colorations are signs of operational problems:

Brown: reduction in photosynthesis activity.
Gray/black: Anaerobic conditions.
Yellow/dark green: Presence of blue-green algae; indicates low pH and oxygen levels.
Pink or red: Presence of photosynthetic sulfur bacteria; indicates anaerobic conditions.

10.10.6 Weed growth

Aquatic weed growth is caused by too shallow water depth; rooted aquatic plants do not grow in ponds deeper than 0.5 m. Floating aquatic plants such as duckweed or water hyacinths must be manually removed. Growth on the shoreline is a result of poor maintenance or a lack of an adequate revetment. Weed growth on slopes is caused by poor maintenance.

10.10.7 Mosquitoes and other insects

Facultative and maturation ponds should not have problems with mosquitoes or other insects as long as the inner embankments and the water surface are free of aquatic plants and floating matter, which serve as breeding foci for insects. The solution is to always keep pond surfaces and embankments free of emergent and aquatic plants, and floating material.

Table 10.10 presents a summary of potential pond operational problems and their solution.

Table 10.10 Operational problems in stabilization ponds and their solution.

Symptoms	Cause	Solution
Accumulation of scum and floating solids	Inadequate cleaning with scum rake	Proper maintenance with scum rake
	Flotation of anaerobic sludge by rising CH_4 and CO_2	Removal of accumulated sludge
Duckweed growth on pond surface	Contamination brought by wind, birds, or animals	Removal with scum rakes, and by introducing domestic ducks into ponds
Odors		
	Anaerobic conditions due to organic overload	Analysis of the cause of anaerobic conditions: • Excessive flow, industrial discharges, and decomposition of accumulated sludge
	Decomposition of scum and floating solids	Removal of scum and floating material
	Presence of toxic chemicals	Intensive monitoring to locate the cause
Abnormal colorations		
Green	Normal for facultative and maturation ponds	
Brown	Reduction in photosynthesis	Analyze for organic overload and toxic chemicals
Gray/black	Anaerobic conditions	Analyze for organic overload
Yellow/opaque green	Presence of blue-green algae	Low pH and dissolved oxygen from overload or toxic chemicals. Analyze for organic overload.
Pink/red coloration	Presence of photosynthetic sulfur bacteria due to anaerobic conditions	Analyze for organic overload
Weed growth in water	Water depth too shallow Lack of revetment Lack of maintenance	Water level control Revetment construction or maintenance Routine maintenance
Mosquitoes and insects	Breeding sites for their larvae	Removal of emergent aquatic plants and floating material Variation of water level to dry larvae on revetment

doi: 10.2166/9781789061536_0343

References

Acheson D. and Fiore A. E. (2004). Hepatitis A transmitted by food. *Clinical Infectious Diseases*, **38**(5), 705–715. https://doi.org/10.1086/381671
Ali S. (2013). Estimation of surface water temperature in small recharge pond from air temperature. *Indian Journal of Soil Science*, **41**(1), 1–7.
Allen R. G., Pereira L. S., Raes D. and Smith M. (1998). FAO Irrigation and Drainage Paper No. 56. Crop Evapotranspiration (Guidelines for Computing Crop Water Requirements), Food and Agriculture Organization of the United Nations, Rome.
Alvarado A., Sanchez E., Durazno G., Vesvikar M. and Nopens I. (2012). CFD analysis of sludge accumulation and hydraulic performance of a waste stabilization pond. *Water Science and Technology*, **66**(11), 2370–2377. https://doi.org/10.2166/wst.2012.450
Amahmid O., Asmama S. and Bouhoum K. (1999). The effect of waste water reuse in irrigation on the contamination level of food crops by Giardia cysts and Ascaris eggs. *International Journal of Food Microbiology*, **49**(1–2), 19–26. https://doi.org/10.1016/S0168-1605(99)00058-6
Anderson E. N. (1988). The Food of China, Yale University Press, New Haven, CT.
Angelakis A. N., Asano T., Bahri A., Jimenez B. E. and Tchobanoglous G. (2018). Water reuse: from ancient to modern times and the future. *Frontiers in Environmental Science*, **26**, 1–17.
APHA (American Public Health Association) (1995). Standard Methods for the Examination of Water and Wastewater, 19th Edition. American Public Health Association, Washington, D.C.
Arceivala S. J. and Asolekar S. R. (2007). Wastewater Treatment for Pollution Control and Reuse, 3rd edn, Tata McGraw Hill Education Private Limited, New Delhi.
Arceivala S. J., Lakshminarayana J. S. S., Alagarsamy S. R. and Sastry C. A. (1970). Waste Stabilization Ponds: Design, Construction & Operation in India, Central Public Health Engineering Research Institute, Nagpur.
Arthur J. P. (1983). Notes on the Design and Operation of Waste Stabilization Ponds in Warm Climates of Developing Countries, Technical Paper No. 7, The World Bank, Washington, DC.
ASCE (American Society of Civil Engineers). (1959). Sewage Treatment Plant Design, American Society of Civil Engineers and Federation of Sewage and Industrial Wastes Associations, New York.
ASCE/WPCF (American Society of Civil Engineers/Water Pollution Control Federation). (1977). Wastewater Treatment Plant Design, ASCE Manuals and Reports on Engineering Practice No. 36, New York.
Ashley K., Cordell D. and Mavinic D. (2011). A brief history of phosphorus: from the philosopher's stone to nutrient recovery and reuse. *Chemosphere*, **84**(6), 737–746. https://doi.org/10.1016/j.chemosphere.2011.03.001
Ayers R. S. and Westcot D. W. (1985). Water Quality for Agriculture, FAO Irrigation and Drainage Paper 29, Food and Agriculture Organization of the United Nations, Rome.
Ayres R. M., Alabaster G. P., Mara D. D. and Lee D. L. (1992). A design equation for human intestinal nematode egg removal in waste stabilization ponds. *Water Research*, **26**(6), 863–865.

Babbitt H. E. and Baumann E. R. (1958). Sewerage and Sewage Treatment, 8th edn, John Wiley & Sons, New York.

Barbeito Anzorena E. (2001). Campo Espejo del Aglomerado Gran Mendoza, República de Argentina, Estudio General de Caso, Proyecto Regional, Sistemas Integrados de Tratamiento y Uso de Aguas Residuales en América Latina: Realidad y Potencial, IDRC – OPS/HEP/CEPIS, República de Argentina.

Bartone C. R. (2012). Special Restricted Crop Area in Mendoza, Argentina, US Environmental Protection Agency, Washington, DC.

Ben Ayed L., Schijven J., Alouini Z. and Jemli M. (2009). Presence of parasitic protozoa and helminth in sewage and efficiency of sewage treatment in Tunisia. *Parasitology Research*, **105**(2), 393–406. https://doi.org/10.1007/s00436-009-1396-y

Bennett Engineering. (n.d.). Biggs Wastewater Treatment Plant Improvements. https://ben-en.com/biggs-wastewater-treatment-plant/ Accessed Auguste 31, 2021.

Bern C., Hernandez B., Lopez M. B., Arrowood M. J., de Mejia M. A., de Merida A. M., Hightower A. W., Venczel L., Herwaldt B. L. and Klein R. E. (1999). Epidemiologic studies of *Cyclospora cayetanensis* in Guatemala. *Emerging Infectious Diseases*, **5**(6), 766–774. https://doi.org/10.3201/eid0506.990604

Buetow D. E. (1962). Differential effects of temperature on the growth of *Euglena gracilis*. *Experimental Cell Research*, **27**(1), 137–142. https://doi.org/10.1016/0014-4827(62)90051-4

Buhr H. O. and Miller S. B. (1983). A dynamic model of the high-rate algal-bacterial wastewater treatment pond. *Water Research*, **17**(1), 29–37. https://doi.org/10.1016/0043-1354(83)90283-X

Castro de Esparza M. L., Aurazo M., Piscoya Z. and León G. (1992). Estudio preliminar de la remoción de Vibrio cholerae en aguas residuales tratadas mediante lagunas de estabilización, Informe Técnico 387, Centro Panamericano de Ingeniería Sanitaria (CEPIS), Lima.

CDC (Centers for Disease Control and Prevention). (2004). Outbreak of cyclosporiasis associated with snow peas – Pennsylvania, 2004. *MMWR Morbidity and Mortality Weekly Report*, **53**(37), 876–878.

CEPIS/OPS. (2000). Regional Project, Integrated Systems for the Treatment and Recycling of Waste Water in Latin America, Reality and Potential, Lima.

CNA (Comisión Nacional de Agua). (2007). Manual de Agua Potable, Alcantarillado y Saneamiento, Revisión 3a Edición, CNA, México, D.F.

Coronado O., Moscoso O. and Ruiz R. (2001). Estudio General del Caso: Ciudad de Cochabamba, Bolivia, Proyecto Regional de Sistemas Integrados de Tratamiento y Uso de Aguas Residuales en América Latina: Realidad y Potencia, Covenio IDRC – OPS/HEP/CEPIS.

Cubillos A. (1994). Lagunas de Estabilización, Centro Interamericano de Desarrollo e Investigación Ambiental y Territorial (CIDIAT), Mérida.

Davies-Colley R. (2005). Chapter 6: Pond disinfection, pp. 100–136. In: Pond Treatment Technology, A. Shilton (ed.), IWA Publishing, London, p. 479.

Departamento de Sanidad del Estado de Nueva York. (1993). Manual de Tratamiento de Aguas Negras, Editorial LUMUSA, México.

Dias D. F. C., Possmoser-Nascimento T. E., Rodrigues V. A. J. and von Sperling M. (2014). Overall performance evaluation of shallow maturation ponds in series treating UASB reactor effluent: ten years of intensive monitoring of a system in Brazil. *Ecological Engineering*, **71**, 206–214. https://doi.org/10.1016/j.ecoleng.2014.07.044

DIGESBA (Dirección General de Saneamiento Básico). (2001). Instalaciones Sanitarias–Alcantarillado Sanitario, Pluvial y Tratamiento de Aguas Residuales, Norma Boliviano NB 688, La Paz.

ECOMAC. (2004). Informes de Monitoreo: Lagunas de Estabilización en Honduras, (11 Volúmenes) Proyecto Monitoreo de Sistemas de Estabilización de Tratamiento de Aguas Negras, U.S. Army Corps of Engineers, Mobile District, Tegucigalpa, Honduras.

Egocheaga L. and Moscoso J. (2004). Una Estrategia para la Gestión de las Aguas Residuales Domésticos, CEPIS-/OPS, Lima.

Energy Star. (2015). Energy Star Portfolio Manager. www.energystar.gov/sites/default/files/tools/Wastewater_Trtmnt_Aug_2018_EN_508.pdf. Accessed on July 18, 2021.

Espinosa M. F., Von Sperling M. and Verbyla M. E. (2017). Performance evaluation of 388 full-scale waste stabilization pond systems with seven different configurations. *Water Science and Technology*, **75**(4), 916–927.

European Commission. (n.d.). Water Reuse. https://ec.europa.eu/environment/water/reuse.htm. Accessed on September 21, 2021.

Feachem R. G., Bradley D. J., Garelick H. and Mara D. D. (1983). Sanitation and Disease: Health Aspects of Excreta and Wastewater Management, John Wiley, London.

Foree E. G. and McCarty P. L. (1970). Anaerobic decomposition of algae. *Environmental Science &Technology*, **4**(10), 842–849. https://doi.org/10.1021/es60045a005

Foresi C. H. (2016). Wastewater reuse in Mendoza province. In: Safe Use of Wastewater in Agriculture: Good Practice Examples, H. Hettiarachchi and R. Ardakanian (eds.), United Nations University, Institute for Integrated Management of Material Fluxes and of Resources (UNU-FLORES), Argentina, pp. 251–265, 312 pp.

Franci R. (Coordenacão). (1999). Gerenciamento do Lodo de Lagoas de Estabilizacão Não Mecanizadas, Rede Cooperativa de Pesquisas, Rio de Janeiro.

Galloway J. N., Aber J. D., Erisman J. W., Seitzinger S. P., Howarth R. W., Cowling E. B. and Cosby B. J. (2003). The nitrogen cascade. *Bioscience*, **53**(4), 341–356. https://doi.org/10.1641/0006-3568(2003)053[0341:TNC]2.0.CO;2

Geradi M. H. (2015). The Biology and Troubleshooting of Facultative Lagoons, John Wiley & Sons, Hoboken, NJ.

GBD (Global Burden of Disease Collaborative Network) (2021). See https://ghdx.healthdata.org/organizations/global-burden-disease-collaborative-network. Accessed on November 21, 2021.

GFA (GFA Consulting Group GmbH). (2018). SFD Report Ouagadougou, Burkina Faso.

Gloyna E. F. (1971). Waste Stabilization Ponds, World Health Organization, Geneva.

Gonzalez A. and Thomas L. (2018). Taenia spp. In: Water and Sanitation for the 21st Century: Health and Microbiological Aspects of Excreta and Wastewater Management (Global Water Pathogen Project), J. B. Rose and B. Jiménez-Cisneros (eds.) (L. Robertson (eds), Part 3: Specific Excreted Pathogens: Environmental and Epidemiology Aspects – Section 4: Helminths), Michigan State University, E. Lansing, MI, UNESCO.

Grimason A. M., Smith H. V., Thitai W. N., Smith P. G., Jackson M. H. and Girdwood R. W. A. (1993). Occurrence and removal of Cryptosporidium spp. oocysts and Giardia spp. cysts in Kenyan waste stabilisation ponds. *Water Science and Technology*, **27**(3–4), 97–104. https://doi.org/10.2166/wst.1993.0329

Hanson R. S. and Hanson T. E. (1996). Methanotrophic bacteria. *Microbiological Reviews*, **60**(2), 439–471. https://doi.org/10.1128/mr.60.2.439-471.1996

Ho A. Y., Lopez A. S., Eberhart M. G., Levenson R., Finkel B. S., da Silva A. J., Roberts J. M., Orlandi P. A., Johnson C. C. and Herwaldt B. L. (2002). Outbreak of cyclosporiasis associated with imported raspberries, Philadelphia, Pennsylvania. *Emerging Infectious Diseases*, **8**(8), 783–788. https://doi.org/10.3201/eid0808.020012

Ho L., Jerves-Cobo R., Morales O., Larriva J., Arevalo-Durazno M., Barthel M., Six J., Bode S., Boeckx P. and Goethals P. (2021). Spatial and temporal variations of greenhouse gas emissions from a waste stabilization pond: effects of sludge distribution and accumulation. *Water Research*, **193**, 116858. https://doi.org/10.1016/j.watres.2021.116858

Hoefman S. (2013). From Nature to Nurture: Isolation, Physiology and Preservation of Methane-Oxidizing Bacteria. Doctoral dissertation, Ghent University, Ghent.

Howard G. and Bartram J. (2003). Domestic Water Quantity, Service Level and Health, World Health Organization, Geneva.

IEA. (2016). World Energy Outlook 2016, IEA, Paris. https://doi.org/10.1787/weo-2016-en

Ilic S., Drechsel P., Amoah P. and LeJeune J. T. (2009). Applying the multiple-barrier approach for microbial risk reduction in the post-harvest sector of wastewater-irrigated vegetables. In: Wastewater Irrigation and Health, A. Bahri, P. Drechel, L. Raschid-Sally and M. Redwood (eds.), Chapter 12, Routledge, London, pp. 265–286.

Imhoff K. and Fair G. M. (1956). Sewage Treatment, Wiley, New York.

INAA (Instituto Nicaragüense de Acueductos y Alcantarillados). (1996). Guía de Operación y Mantenimiento de Lagunas de Estabilización, Departamento de Calidad del Agua, Gerencia de Normación Técnica, Managua, Nicaragua.

Ingallinella A. M., Sanguinetti G., Fernández R. G., Strauss M. and Montangero A. (2002). Cotreatment of sewage and septage in waste stabilization ponds. *Water Science and Technology*, **45**(1), 9–15. https://doi.org/10.2166/wst.2002.0002

IPCC. (2019). 2019 Refinement to the IPCC Guidelines for National Greenhouse Gas Inventories. https://www.ipcc.ch/report/2019-refinement-to-the-2006-ipcc-guidelines-for-national-greenhouse-gas-inventories/. Accessed on November 16, 2021.

Jasper J. T., Nguyen M. T., Jones Z. L., Ismail N. S., Sedlak D. L., Sharp J. O., Luthy R. G., Horne A. J. and Nelson K. L. (2013). Unit process wetlands for removal of trace organic contaminants and pathogens from municipal wastewater effluents. *Environmental Engineering Science*, **30**(8), 421–436. https://doi.org/10.1089/ees.2012.0239

Jensen L. A., Marlin J. W., Dyck D. D. and Laubach H. E. (2009). Effect of tourism and trade on intestinal parasitic infections in Guatemala. *Journal of Community Health*, **34**(2), 98–101. https://doi.org/10.1007/s10900-008-9130-8

Jones E. R., van Vliet M. T., Qadir M. and Bierkens M. F. (2021). Country-level and gridded estimates of wastewater production, collection, treatment and reuse. *Earth System Science Data*, **13**(2), 237–254. https://doi.org/10.5194/essd-13-237-2021

Juanico M. (1996). The performance of batch stabilization reservoirs for wastewater treatment, storage and reuse in Israel. *Water Science and Technology*, **33**(10–11), 149–159. https://doi.org/10.2166/wst.1996.0671

Juanico M. and Shelef G. (1994). Design, operation and performance of stabilization reservoirs for wastewater irrigation in Israel. *Water Research*, **28**(1), 175–186. https://doi.org/10.1016/0043-1354(94)90132-5

Kawa N. C., Ding Y., Kingsbury J., Goldberg K., Lipschitz F., Scherer M. and Bonkiye F. (2019). Night soil: origins, discontinuities, and opportunities for bridging the metabolic rift. *Ethnobiology Letters*, **10**(1), 40–49. https://doi.org/10.14237/ebl.10.1.2019.1351

Kim M. J., Ki H. C., Kim S., Chai J. Y., Seo M., Oh C. S. and Shin D. H. (2014). Parasitic infection patterns correlated with urban–rural recycling of night soil in Korea and other East Asian countries: the archaeological and historical evidence. *Korean Studies*, **38**(1), 51–74.

Konaté Y., Maiga A. H., Basset D., Casellas C. and Picot B. (2013). Parasite removal by waste stabilisation pond in Burkina Faso, accumulation and inactivation in sludge. *Ecological Engineering*, **50**, 101–106. https://doi.org/10.1016/j.ecoleng.2012.03.021

Kumar D. and Asolekar S. R. (2016). Significance of natural treatment systems to enhance reuse of treated effluent: a critical assessment. *Ecological Engineering*, **94**, 225–237. https://doi.org/10.1016/j.ecoleng.2016.05.067

Kumwenda S., Msefula C., Kadewa W., Ngwira B. and Morse T. (2017) Estimating the health risk associated with the use of ecological sanitation toilets in Malawi. *Journal of Environmental and Public Health*, **13**.

Lai F. (2008). Review of Sewer Design Criteria and RDII Prediction Methods, Office of Research and Development, U.S. Environmental Protection Agency, Cincinnati, OH.

Laubach H. E., Bentley C. Z., Ginter E. L., Spalter J. S. and Jensen L. A. (2004). A study of risk factors associated with the prevalence of Cryptosporidium in villages around Lake Atitlan, Guatemala. *Brazilian Journal of Infectious Diseases*, **8**(4), 319–323.

Li W. W., Yu H. Q. and Rittmann B. E. (2015). Chemistry: reuse water pollutants. *Nature*, **528**(7580), 29–31. https://doi.org/10.1038/528029a

Libhaber M. and Orozco-Jaramillo Á. (2012). Sustainable Treatment and Reuse of Municipal Wastewater, IWA Publishing, London.

Ling B. (1994). Safe use of treated night-soil. *ILEIA Newsletter*, **10**(3), 10–11.

Ling B., Den T. X., Lu Z. P., Min L. W., Wang Z. X. and Yuan A. X. (1993). Use of night soil in agriculture and fish farming. *World Health Forum*, **14**(1), 67–70.

Liran A., Juanico M. and Shelef G. (1994). Coliform removal in a stabilization reservoir for wastewater irrigation in Israel. *Water Research*, **28**(6), 1305–1314. https://doi.org/10.1016/0043-1354(94)90295-X

Lloyd B. J., Leitner A. R., Vorkas C. A. and Guganesharajah R. K. (2003a). Under-performance evaluation and rehabilitation strategy for waste stabilization ponds in Mexico. *Water Science and Technology*, **48**(2), 35–43. https://doi.org/10.2166/wst.2003.0080

Lloyd B. J., Vorkas C. A. and Guganesharajah R. K. (2003b). Reducing hydraulic short-circuiting in maturation ponds to maximize pathogen removal using channels and wind breaks. *Water Science and Technology*, **48**(2), 153–162. https://doi.org/10.2166/wst.2003.0109

Lofrano G. and Brown J. (2010). Wastewater management through the ages: a history of mankind. *Science of the Total Environment*, **408**(22), 5254–5264. https://doi.org/10.1016/j.scitotenv.2010.07.062

Maiga Y. (2006). Study of the Physico-Chemical and Microbiological Quality of Wastewater at the City of Ouagadougou, Burkina Faso Wastewater Treatment Plant. Masters thesis, University of Ouagadougou, Ouagadougou (In French).

Maiga Y., Wethe J., Denyigba K. and Ouattara A. S. (2009). The impact of pond depth and environmental conditions on sunlight inactivation of *Escherichia coli* and enterococci in wastewater in a warm climate. *Canadian Journal of Microbiology*, **55**(12), 1364–1374. https://doi.org/10.1139/W09-104

Mangelson K. and Watters G. (1972) Treatment efficiency of waste stabilization ponds. *Journal of the Sanitary Engineering Division*, **98**(SA2), 407–425. https://doi.org/10.1061/JSEDAI.0001401

Mara D. (1976). Sewage Treatment in Hot Climates, John Wiley & Sons, New York.

Mara D. (2003). Domestic Wastewater Treatment in Developing Countries, Earthscan, London.

Mara D. and Pearson H. (1998). Design Manual for Waste Stabilization Ponds in Mediterranean Countries, Lagoon Technology International, Leeds.

Mara D., Pearson H. W., Alabaster G. P. and Mills S. W. (1992). Waste Stabilization Ponds: A Design Manual for Eastern Africa, Lagoon Technology International, Leeds.

Marais G. V. R. and van Haandel A. C. (1996). Design of grit channels controlled by Parshall flumes. *Water Science and Technology*, **33**(3), 195–210. https://doi.org/10.2166/wst.1996.0071

Marka S L. G. (2012). Producción de Aguas Servidas, Tratamiento y Uso en Bolivia, Proyecto de Desarrollo de Capacidades para el Uso Seguro de Aguas Servidas en Agricultura, FAO, WHO, UNEP, UNU-INWEH, UNW-DPC, IWMI e ICID, Dirección General de Riego, Viceministerio de Recursos Hídricos y Riego, La Paz.

McCarty P. L., Bae J. and Kim J. (2011). Domestic wastewater treatment as a net energy producer – can this be achieved? *Environmental Science & Technology*, **45**(17), 7100–7106. https://doi.org/10.1021/es2014264

McNeill J. R. and Winiwarter V. (2004). Breaking the sod: humankind, history, and soil. *Science*, **304**(5677), 1627–1629. https://doi.org/10.1126/science.1099893

Mendonça S. (2000). Sistemas de Lagunas de Estabilización, McGraw-Hill Interamericana, Santa Fé de Bogotá.

Mercado A., Coronado O. and Iriarte M. (2013). Evaluación de la eficiencia la planta de tratamiento de aguas residuales de Punata, Cochabamba, Bolivia. Importancia de la operación mantenimiento. XV Congreso Bolivariano de Ingenieria Sanitaria y Medio Ambiente, 20–22 November 2013, AIDIS, Cochabamba.

Metcalf and Eddy. (1981). Wastewater Engineering: Collection and Pumping of Wastewater, McGraw-Hill, New York.

Metcalf and Eddy. (1991). Wastewater Engineering: Treatment, Disposal, and Reuse, 3rd edn, McGraw-Hill, New York.

Metcalf and Eddy/AECOM. (2006). Water Reuse, McGraw-Hill, New York.

Metcalf and Eddy/AECOM. (2014). Wastewater Engineering, 5th edn, McGraw-Hill, New York.

Middlebrooks E. J. (1988). Review of rock filters for the upgrade of lagoon effluents. *Journal Water Pollution Control Federation*, **60**(9), 1657–1662.

Middlebrooks E. J., Middlebrooks C. E., Reynolds T. H., Watters G. Z., Reed S. C. and George D. B. (1982). Wastewater Stabilization Lagoon Design, Performance and Upgrading, Macmillan, New York.

Middlebrooks E. J., Adams V. D., Bilby S. and Shilton A. (2005). Chapter 11: Solids removal and other upgrading techniques, pp. 218–249. In: Pond Treatment Technology, A. Shilton (ed.), IWA Publishing, London, p. 479.

Miguez C. B., Shen C. F., Shen Bourque D., Guiot S. R. and Groleau D. (1999). Monitoring methanotrophic bacteria in hybrid anaerobic-aerobic reactors with PCR and a catabolic gene probe. *Applied and Environmental Microbiology*, **65**(2), 381–388. https://doi.org/10.1128/AEM.65.2.381-388.1999

Ministerio de Vivienda, Construcción y Saneamiento. (n.d.). Plantas de Tratamiento de Aguas Residuales, NORMA OS.090, Lima, Perú.

MOPT (Ministerio de Obras Públicas y Transportes). (1991). Depuración por Lagunaje de Aguas Residuales: Manual de Operadores, Monografías de la Secretaría de Estado para las Políticas del Agua y el Medio Ambiente, Ministerio de Obras Públicas y Transportes, Madrid.

Moscoso J. (2016). Manual de Buenas Prácticas para el Uso Seguro y Productivo de las Aguas Residuales Domésticas, Autoridad Nacional de Agua/Ministerio de Agricultura y Riego, Lima, p. 220.

MWH. (2005). Water Treatment: Principles and Design, 2nd edn, John Wiley & Sons, Hoboken, NJ.

Nalley J. O., O'Donnell D. R. and Litchman E. (2018). Temperature effects on growth rates and fatty acid content in freshwater algae and cyanobacteria. *Algal Research*, **35**, 500–507. https://doi.org/10.1016/j.algal.2018.09.018

Nelson K., Jiménez Cisneros B., Tchobanoglous G. and Darby J. (2004). Sludge accumulation, characteristics, and pathogen inactivation in four primary waste stabilization ponds in central Mexico. *Water Research*, **38**(1), 111–127. https://doi.org/10.1016/j.watres.2003.09.013

Nemerow N. L. and Dasgupta A. (1991). Industrial and Hazardous Waste Treatment, Van Nostrand Reinhold, New York.

Nitiema L. W., Boubacar S., Dramane Z., Kabore A. and No P. J. (2013). Microbial quality of wastewater used in urban truck farming and health risks issues in developing countries: case study of Ouagadougou in Burkina Faso. *Journal of Environmental Protection*, **4**(6), 575–584. https://doi.org/10.4236/jep.2013.46067

Oakley S. (2018). Preliminary treatment and primary sedimentation. In: Water and Sanitation for the 21st Century: Health and Microbiological Aspects of Excreta and Wastewater Management (Global Water Pathogen Project), J. B. Rose and B. Jiménez-Cisneros (eds.) (J. R. Mihelcic and M. E. Verbyla (eds), Part 4: Management of Risk from Excreta and Wastewater – Section: Sanitation System Technologies, Pathogen Reduction in Sewered System Technologies), Michigan State University, E. Lansing, MI, UNESCO, 21 pp.

Oakley S. M. (2004). Monitoring of Wastewater Stabilization Ponds in Honduras. Proceedings of the XXIX Inter--American Congress of Sanitary and Environmental Engineering, Asociación Interamericana de Ingeniería Sanitaria (AIDIS), San Juan, Puerto Rico, August 2004.

Oakley S. M. (2005). Manual de Lagunas de Estabilización en Honduras, USAID, RRAS-CA, FHIS, Tegucigalpa, Honduras.

Oakley S. and Mihelcic J. R. (2019). Pathogen reduction and survival in complete treatment works. In: Water and Sanitation for the 21st Century: Health and Microbiological Aspects of Excreta and Wastewater Management (Global Water Pathogen Project), J. B. Rose and B. Jiménez-Cisneros (eds.) (J. R. Mihelcic and M. E. Verbyla (eds), Part 4: Management of Risk from Excreta and Wastewater – Section: Sanitation System Technologies, Pathogen Reduction in Sewered System Technologies), Michigan State University, E. Lansing, MI, UNESCO, 60 pp.

Oakley S. M. and Saravia P. (2021). Manejo integrado de aguas residuales dentro y fuera la cuenca del Lago Atitlán. *Agua, Saneamiento & Ambiente*, **16**(1), 46–63.

Oakley S. and von Sperling M. (2017). Media filters: trickling filters and anaerobic filters. In: Water and Sanitation for the 21st Century: Health and Microbiological Aspects of Excreta and Wastewater Management (Global Water Pathogen Project), J. B. Rose and B. Jiménez-Cisneros (eds.) (J. R. Mihelcic and M. E. Verbyla (eds), Part 4: Management of Risk from Excreta and Wastewater – Section: Sanitation System Technologies, Pathogen Reduction in Sewered System Technologies), Michigan State University, E. Lansing, MI, UNESCO, 11 pp.

Oakley S. M., Mendonça L. C. and Mendonça S. R. (2012). Sludge removal from primary wastewater stabilization ponds with excessive accumulation: a sustainable method for developing regions. *Journal of Water, Sanitation and Hygiene for Development*, **2**(2), 68–78. https://doi.org/10.2166/washdev.2012.093

Oakley S., von Sperling M. and Verbyla M. (2017). Anaerobic sludge blanket reactors. In: Water and Sanitation for the 21st Century: Health and Microbiological Aspects of Excreta and Wastewater Management (Global Water Pathogen Project), J. B. Rose and B. Jiménez-Cisneros (eds.) (J. R. Mihelcic and M. E. Verbyla (eds), Part 4: Management of Risk from Excreta and Wastewater – Section: Sanitation System Technologies, Pathogen Reduction in Sewered System Technologies), Michigan State University, E. Lansing, MI, UNESCO, 16 pp.

Öberg G., Merlinsky M. G., LaValle A., Morales M. and Tobias M. M. (2014). The notion of sewage as waste: a study of infrastructure change and institutional inertia in Buenos Aires, Argentina and Vancouver, Canada. *Ecology and Society*, **19**(2), 8 pp. https://doi.org/10.5751/ES-06531-190219

OECD (2016), "Water-energy nexus", in World Energy Outlook 2016, OECD Publishing, Paris, Retrieved at https://doi.org/10.1787/weo-2016-11-en.

Oliveira S. C. and Von Sperling M. (2008). Reliability analysis of wastewater treatment plants. *Water Research*, **42**(4–5), 1182–1194. https://doi.org/10.1016/j.watres.2007.09.001

Oswald W. J. (1963). Fundamental factors in stabilization pond design. In: Advances in Biological Waste Treatment, W. W. Eckenfelder and B. J. McCabe (eds.), Macmillan, New York, pp. 357–393.

Oswald W. J. and Gotaas M. (1957). Photosynthesis in sewage treatment. *Transactions of the American Society of Civil Engineers*, **122**(1), 73–105. https://doi.org/10.1061/TACEAT.0007483

Paredes M. G., Güereca L. P., Molina L. T. and Noyola A. (2015). Methane emissions from stabilization ponds for municipal wastewater treatment in Mexico. *Journal of Integrative Environmental Sciences*, **12**(Suppl. 1), 139–153. https://doi.org/10.1080/1943815X.2015.1110185

Pepper I. L., Gerba C. P. and Gentry T. J. (2015). Environmental Microbiology, 3rd edn, Academic Press, San Diego, CA.

Pham-Duc P., Nguyen-Viet H., Hattendorf J., Zinsstag J., Phung-Dac C., Zurbrügg C. and Odermatt P. (2013). *Ascaris lumbricoides* and *Trichuris trichiura* infections associated with wastewater and human excreta use in agriculture in Vietnam. *Parasitology International*, **62**(2), 172–180. https://doi.org/10.1016/j.parint.2012.12.007

Picot B., Sambuco J. P., Brouillet J. L. and Riviere Y. (2005). Wastewater stabilisation ponds: sludge accumulation, technical and financial study on desludging and sludge disposal case studies in France. *Water Science and Technology*, **51**(12), 227–234. https://doi.org/10.2166/wst.2005.0469

Pullan R. L., Smith J. L., Jasrasaria R. and Brooker S. J. (2014). Global numbers of infection and disease burden of soil transmitted helminth infections in 2010. *Parasites & Vectors*, **7** (37), 19 pp. https://doi.org/10.1186/1756-3305-7-37

Qadir M., Drechsel P., Jiménez Cisneros B., Kim Y., Pramanik A., Mehta P. and Olaniyan O. (2020). Global and regional potential of wastewater as a water, nutrient and energy source. *Natural Resources Forum*, **44**(1), 40–51. https://doi.org/10.1111/1477-8947.12187

Qu J., Wang H., Wang K., Yu G., Ke B., Yu H. Q., Ren H., Zheng X., Li J., Li W.W. and Gao S. (2019). Municipal wastewater treatment in China: Development history and future perspectives. *Frontiers of Environmental Science & Engineering*, **13**(6), 1–7.

Ramo A., Del Cacho E., Sánchez-Acedo C. and Quílez J. (2017). Occurrence and genetic diversity of *Cryptosporidum* and *Giardia* in urban wastewater treatment plants in north-eastern Spain. *Science of the Total Environment*, **598**, 628–638. https://doi.org/10.1016/j.scitotenv.2017.04.097

Rauek T. F. (2020). Gestión del Uso de Aguas Residuales Urbanas Tratadas en Agricultura, Mendoza, Argentina. Presentation. https://cdn.www.gob.pe/uploads/document/file/1483420/Gesti%C3%B3n%20del%20Uso%20de%20Aguas%20Residuales%20Urbanas%20Tratadas%20en%20Agricultura%20en%20Mendoza%2C%20Argentina.pdf

Reynolds T. D. and Richards P. A. (1996). Unit Operations and Processes in Environmental Engineering, 2nd edn, PWS Publishing Company, Boston, MA.

Reynolds J. H., Swiss R. E., Macko C. A. and Middlebrooks E. J. (1977). Performance Evaluation of an Existing Seven Cell Lagoon System, EPA-600/2-77-086, Office of Research and Development, US Environmental Protection Agency, Cincinnati, OH.

Ritchie H. and Roser M. (2020). Energy. Published online at OurWorldInData.org. https://ourworldindata.org/energy [Online Resource]. Accessed on January 11, 2022.

Rittmann B. E. and McCarty P. L. (2001). Environmental Biotechnology: Principles and Applications, McGraw-Hill, New York.

Ruano R. (2005). Determinación del Período Apropiado de Corte del Pasto Napier (Penisetum purpureum) Después del Riego por Aspersión con Agua Residual Sedimentada en la Planta de Tratamiento Aurora II para Reducir la Contaminación Bacteriológica Foliar y Propiciar Buenas Características Agronómicas, Maestro en Recursos Hidráulicos, Escuela Regional de Ingeniería Sanitaria y Recursos Hidráulicos, Universidad de San Carlos de Guatemala, Guatemala.

Sánchez de León E. M. (2001). Proyecto regional. Sistemas integrados de tratamiento y uso de aguas residuales en América Latina: realidad y potencial, Estudio General del caso Sololá, Guatemala, IDRC-OPS/HEP/CEPIS.

Sawyer C., McCarty P. and Parkin G. (2002). Chemistry for Environmental Engineering and Science, 5th edn, McGraw-Hill Education, New York.

Schulte P. (2014). Defining Water Scarcity, Water Stress, and Water Risk, Pacific Institute. https://pacinst.org/water-definitions/. Accessed on March 17, 2022.

Shaban A. and Sharma R. N. (2007). Water consumption patterns in domestic households in major cities. *Economic and Political Weekly*, 2190–2197.

Shilton A. and Harrison J. (2003). Guidelines for the Hydraulic Design of Waste Stabilization Ponds, Institute of Technology and Engineering, Massey University, Palmerston North.

Shilton A. and Sweeney D. (2005). Chapter 12: Hydraulic design, pp. 188–217. In: Pond Treatment Technology, A. Shilton (ed.), IWA Publishing, London, p. 479.

Somé Y. S. C., Kpoda W. N., Gampini E. E. and Kabré G. B. (2014). Microbiology and parasitology assessement of irrigation waterused for production of vegetables in Ouagadougou. Burkina Faso. *Res. Rev. BioSci*, **9**(3), 7.

Stott R. (2003). Chapter 31: Fate and behavior of parasites in wastewater treatment systems, pp. 491–521. In: The Handbook of Water and Wastewater Microbiology, D. Mara and N. Horan (eds.), Academic Press, London, p. 819.

Stratton H., Lemckert C., Roiko A., Zhang H., Wilson S., Gibb K., van der Akker B., Macdonald J., Melvin S., Sheludchenko M., Li M., Xie J., Padovan A. and Lehmann R. (2015). Validation of Maturation Ponds in Order to Enhance Safe and Economical Water Recycling, Australian Water Recycling Centre of Excellence, Brisbane.

Swanson G. R. and Williamson K. J. (1980). Upgrading lagoon effluents with rock filters. *Journal of the Environmental Engineering Division*, **106**(6), 1111–1129.

Sweeney D. G., Cromar N. J., Nixon J. B., Ta C. T. and Fallowfield H. J. (2003). The spatial significance of water quality indicators in waste stabilization ponds-limitations of residence time distribution analysis in predicting treatment efficiency. *Water Science and Technology*, **48**(2), 211–218. https://doi.org/10.2166/wst.2003.0123

Tchobanoglous G. and Schroeder E. D. (1985). Water Quality, Addison-Wesley Publishing Company, Reading, MA.

Thebo A. L., Drechsel P., Lambin E. F. and Nelson K. L. (2017). A global, spatially-explicit assessment of irrigated croplands influenced by urban wastewater flows. *Environmental Research Letters*, **12**(7), 074008. https://doi.org/10.1088/1748-9326/aa75d1

Thirumurthi D. (1969). Design principles of waste stabilization ponds. *Journal of the Sanitary Engineering Division*, **95**(2), 311–332. https://doi.org/10.1061/JSEDAI.0000952

Torihara K. and Kishimoto N. (2015). Evaluation of growth characteristics of *Euglena gracilis* for microalgal biomass production using wastewater. *Journal of Water and Environment Technology*, **13**(3), 195–205. https://doi.org/10.2965/jwet.2015.195

UN (2015), Sustainable Development Goals, Retrieved on April 23, 2022 from https://sdgs.un.org/goals

UN (United Nations). (2019). World Urbanization Prospects: The 2018 Revision, Department of Economic and Social Affairs, Population Division, New York.

UNEP (United Nations Environment Programme). (2016). A Snapshot of the World's Water Quality: Towards a Global Assessment, United Nations Environment Programme, Nairobi.

UNEP (United Nations Environment Programme). (2017). The United Nations World Water Development Report 2017, Wastewater: The Untapped Resource. https://wedocs.unep.org/20.500.11822/20448. Accessed on January 11, 2022.

UNICEF (United Nations Children's Fund) and WHO (World Health Organization). (2019). Progress on Household Drinking Water, Sanitation and Hygiene 2000–2017, Special Focus on Inequalities, New York.

USEPA. (1981). Process Design Manual: Land Treatment of Municipal Wastewater, Center for Environmental Research Information, US Environmental Protection Agency, Cincinnati, OH.

USEPA. (1983). Process Design Manual: Municipal Wastewater Stabilization Ponds, US Environmental Protection Agency, Center for Environmental Research Information, Cincinnati, OH.

USEPA. (2021). Inventory of US Greenhouse Gas Emissions and Sinks, EPA 430-R-19-001. https://www.epa.gov/ghgemissions/inventory-us-greenhouse-gas-emissions-and-sinks. Accessed on November 11, 2021.

Van der Linde E. R. C. (2009). Nitrogen Removal Mechanisms and Pathways in Facultative Waste Stabilization Ponds in the United Kingdom. Doctoral dissertation, School of Civil Engineering, University of Leeds, Leeds.

Vélez O. R., Fasciolo G. E. and Bertranou A. V. (2002). Domestic wastewater treatment in waste stabilization ponds for irrigation in Mendoza, Argentina: policies and challenges. *Water Science and Technology*, **45**(1), 127–132. https://doi.org/10.2166/wst.2002.0017

Verbyla M. E., Iriarte M. M., Guzmán A. M., Coronado O., Almanza M. and Mihelcic J. R. (2016). Pathogens and fecal indicators in waste stabilization pond systems with direct reuse for irrigation: fate and transport in water, soil and crops. *Science of the Total Environment*, **551**, 429–437. https://doi.org/10.1016/j.scitotenv.2016.01.159

Verbyla M., von Sperling M. and Maiga Y. (2017). Waste stabilization ponds. In: Water and Sanitation for the 21st Century: Health and Microbiological Aspects of Excreta and Wastewater Management (Global Water Pathogen Project), J. B. Rose and B. Jiménez-Cisneros (eds.) (J. R. Mihelcic and M. E. Verbyla (eds), Part 4: Management of Risk from Excreta and Wastewater – Section: Sanitation System Technologies, Pathogen Reduction in Sewered System Technologies), Michigan State University, E. Lansing, MI, UNESCO, 19 pp.

Viceministerio de Vivienda y Construcción. (1997). Plantas de Tratamiento de Aguas Residuales, Reglamento Nacional de Construcciones, Norma de Saneamiento S.090, Lima.

Viessman W. and Hammer M. J. (2005). Water Supply and Pollution Control, 7th edn, Pearson/Prentice Hall, Upper Saddle River, NJ.

Von Sperling M. (2005). Modelling of coliform removal in 186 facultative and maturation ponds around the world. *Water Research*, **39**(20), 5261–5273. https://doi.org/10.1016/j.watres.2005.10.016

Von Sperling M. (2007). Waste Stabilization Ponds, Volume 3, Biological Wastewater Treatment Series, IWA Publishing, London.

Wagner W. (2010). Recomendaciones para la elección de plantas de tratamiento de aguas residuales aptas para Bolivia, ANESAPA, GTZ/PROAPAC, La Paz.

WEF (Water Environment Federation). (1990). Operation of Municipal Wastewater Treatment Plants, 2nd edn, Manual of Practice No. 11, WEF, Alexandria, VA, Vol. II.

Wehner J. F. and Wilhelm R. H. (1956). Boundary conditions of flow reactor. *Chemical Engineering Science*, **6**(2), 89–93. https://doi.org/10.1016/0009-2509(56)80014-6

WHO (World Health Organization). (1973). Reuse of Effluents: Methods of Wastewater Treatment and Health Safeguards, Report of a WHO Meeting of Experts, Technical Report Series No. 517, World Health Organization, Geneva.

WHO (World Health Organization). (1987). Wastewater Stabilization Ponds: Principles of Planning and Practice, WHO EMRO Technical Publication No. 10, World Health Organization, Regional Office for the Eastern Mediterranean, Alexandria.

WHO (World Health Organization). (1989). Health Guidelines for the Use of Wastewater in Agriculture and Aquaculture, Report of a WHO Scientific Group, Technical Report Series, No. 778, World Health Organization, Geneva.

WHO (World Health Organization). (2006). Guidelines for the Safe Use of Wastewater, Excreta, and Wastewater, Wastewater Use in Agriculture, Geneva, Vol. 2.

Wood S. W. and Cowie A. (2004). A Review of Greenhouse Gas Emission Factors for Fertiliser Production. For IEA Bioenergy Task 38. http://large.stanford.edu/courses/2014/ph240/yuan2/docs/wood.pdf. Accessed on February 11, 2021.

WPCF (Water Pollution Control Federation). (1969). Design and Construction of Sanitary and Storm Sewers, Joint Committee of the Water Pollution Control Federation and the American Society of Civil Engineers, Washington, DC.

Yihdego Y. and Webb J. A. (2018). Comparison of evaporation rate on open water bodies: energy balance estimate versus measured pan. *Journal of Water and Climate Change*, **9**(1), 101–111. https://doi.org/10.2166/wcc.2017.139

Yu C., Huang X., Chen H., Godfray H. C. J., Wright J. S., Hall J. W., Gong P., Ni S., Qiao S., Huang G. and Xiao Y. (2019). Managing nitrogen to restore water quality in China. *Nature*, **567**(7749), 516–520. https://doi.org/10.1038/s41586-019-1001-1

ND - #0060 - 110123 - C370 - 246/189/21 - PB - 9781789061529 - Gloss Lamination